Electrochemical Material Science and Electrode Processes

Electrochemical Material Science and Electrode Processes

Guest Editors

Natalia Tsyntsaru
Henrikas Cesiulis

MDPI

Basel • Beijing • Wuhan • Barcelona • Belgrade • Novi Sad • Cluj • Manchester

Guest Editors
Natalia Tsyntsaru
Institute of Applied Physics
Moldova State University
Chisinau
Moldova

Henrikas Cesiulis
Faculty of Chemistry and
Geosciences
Vilnius University
Vilnius
Lithuania

Editorial Office
MDPI AG
Grosspeteranlage 5
4052 Basel, Switzerland

This is a reprint of the Special Issue, published open access by the journal *Materials* (ISSN 1996-1944), freely accessible at: https://www.mdpi.com/journal/materials/special_issues/095Y987GPI.

For citation purposes, cite each article independently as indicated on the article page online and as indicated below:

Lastname, A.A.; Lastname, B.B. Article Title. *Journal Name* **Year**, *Volume Number*, Page Range.

ISBN 978-3-7258-5725-8 (Hbk)
ISBN 978-3-7258-5726-5 (PDF)
https://doi.org/10.3390/books978-3-7258-5726-5

Cover image courtesy of Natalia Tsyntsaru

Contents

About the Editors

Natalia Tsyntsaru

Natalia Tsyntsaru, Dr. is the associate professor and leading researcher at the Institute of Applied Physics, Moldova State University, and Vilnius University, Lithuania. Dr. Tsyntsaru's areas of expertise are electrodeposition, anodization, nanostructurization, tribology, materials science, alloys, mechanical, and corrosion properties. She completed her PhD in electrochemistry at the Institute of Applied Physics, Moldova, in 2007. After researching thermokinetics phenomena during the electrodeposition of alloys as a Marie Curie Fellow, she joined KU Leuven, Belgium, in 2010, where she carried out research on the tribological, mechanical, and corrosion properties of Co-W alloys, including the anodization of aluminum. She has collaborated with many research centers, including those from Lithuania, Belgium, Spain, USA. This led to the elaboration of multifunctional coatings/materials that could be proposed for versatile applications. She is a member of the scientific committee of four international conferences (International Conference on Nanomaterials Science and Mechanical Engineering, Portugal; International Conference "Balttrib", Lithuania; International Conference "MSCMP", Moldova, and International Conference on Nanotechnology Research and Innovation, Portugal). She is the author or co-author of more than 70 papers in peer-reviewed journals. She has been involved in 17 projects, including being a principal investigator in 5 of them.

Henrikas Cesiulis

Henrikas Cesiulis, Dr. is a Professor in the Department of Physical Chemistry at Vilnius University (Lithuania). He is also the Director of JSC Elektronikos Perdirbimo Technologijos (Electronics Treatment Technologies) in Vilnius, Lithuania. His area of expertise is in applied electrochemistry and thermodynamics. He works in the fields of metals and alloys electrodeposition, enforced corrosion, metals leaching and recovery from WEEE from environmentally friendly solutions. He developed a concept for the induced co-deposition of tungsten and molybdenum alloys with iron-group metals. He proposed the use of metal foams for electrowinning processes. He is the co-author of 81 scientific articles that are cited >1900 times. He gave ~ 20 keynotes and invited lectures at international conferences and universities abroad. He has been the supervisor of 18 national and international research projects. He is a member of the editorial board of the journal *Surface Engineering and Applied Electrochemistry* and a committee member of the International Conferences on Nanomaterials Science and Mechanical Engineering, University of Aveiro, Portugal.

Preface

Numerous articles and scholarly books have been published concerning electrochemical materials and electrode processes, covering a wide range of topics in depth. However, the field is continually evolving, and new data and innovative methodologies that warrant further exploration and application are presented for researchers in this Special Issue Reprint.

This Reprint is meticulously organized to showcase recent advancements in electrochemical materials and electrode technologies that hold promising potential for applications in several emerging areas. Key areas of focus include the following:

1. Water splitting: research is ongoing into efficient catalysts that can drive the electrolysis of water, leading to sustainable hydrogen production. The innovations described can contribute significantly to renewable energy solutions.

2. Hydrogen production: as a clean energy carrier, hydrogen is crucial, and advancements in electrode materials enable more efficient and cost-effective production methods, which could enhance energy transition efforts.

3. Supercapacitive devices: the development of high-performance supercapacitors requires advanced materials that exhibit excellent charge storage capabilities. Research in this area aims to improve energy density and cycling stability.

4. Recovery of metals: efficient extraction and recovery methods for metals from industrial waste streams and water are becoming increasingly important. Electrochemical techniques are being refined to promote sustainability and reduce environmental impacts.

5. Sensing targeted ions: the development of selective ion sensors can revolutionize environmental monitoring and diagnostics. Recent breakthroughs in electrode materials and design are enhancing sensitivity and specificity.

By integrating these cutting-edge developments, this Reprint aims to serve as a valuable resource for researchers seeking to push the boundaries of electrochemical science and explore their practical applications for contemporary challenges.

Natalia Tsyntsaru and Henrikas Cesiulis
Guest Editors

materials

MDPI

Review

Design of Iron-Based Multifunctional Alloys Electrodeposited from Complexing Electrolytes

Natalia Tsyntsaru [1,2,*], Henrikas Cesiulis [1,*] and Oksana Bersirova [1]

1 Faculty of Chemistry and Geosciences, Vilnius University, 03225 Vilnius, Lithuania; oksana.bersirova@chgf.vu.lt
2 Institute of Applied Physics, Moldova State University, 2028 Chisinau, Moldova
* Correspondence: ashra_nt@yahoo.com (N.T.); henrikas.cesiulis@chgf.vu.lt (H.C.)

Abstract: There is a growing focus on sustainability, characterized by making changes that anticipate future needs and adapting them to present requirements. Sustainability is reflected in various areas of materials science as well. Thus, more research is focused on the fabrication of advanced materials based on earth-abundant metals. The role of iron and its alloys is particularly significant as iron is the second most abundant metal on our planet. Additionally, the electrochemical method offers an environmentally friendly approach for synthesizing multifunctional alloys. Thus, iron can be successfully codeposited with a targeted metal from complexing electrolytes, opening a large horizon for a smart tuning of properties and enabling various applications. In this review, we discuss the practical aspects of the electrodeposition of iron-based alloys from complexing electrolytes, with a focus on refractory metals as multifunctional materials having magnetic, catalytic, mechanical, and antimicrobial/antibacterial properties with advanced thermal, wear, and corrosion resistance. Peculiarities of electrodeposition from complexing electrolytes are practically significant as they can greatly influence the final structure, composition, and designed properties by adjusting the electroactive complexes in the solution. Moreover, these alloys can be further upgraded into composites, multi-layered, hybrid/recovered materials, or high-entropy alloys.

Keywords: iron alloys; electrodeposition; complexing electrolytes; refractory metals alloys; advanced properties

1. Introduction: Iron-Based Alloys and Materials for Unlocking Sustainable Versatile Applications

Nowadays, there has been growing consideration on sustainability, characterized by viable changes that enable foreseeing future needs and adjusting them to the present necessities [1]. Sustainability aspects are rising in different branches of materials science as well. Thus, more research is focused on the fabrication of advanced materials based on earth-abundant metals (iron, nickel, and cobalt), especially for catalytic applications [2,3]. Moreover, the role of iron and its compounds is significant because iron is the second most earth-abundant metal. They can be used in various applications, which will be discussed below, but certainly, the magnetic [4] and catalytic [5,6] properties of iron-based advanced materials are acknowledged foremost.

Furthermore, iron-based materials with high theoretical capacity and good mechanical and thermal stability have attracted research interest as electrode materials for alkali metal-ion batteries (AMIBs). Advances in the chemical synthesis and structural design of iron oxide-containing materials for these purposes are discussed elsewhere [7]. In addition, other materials like the FeCo alloy have found applications in lithium–sulfur batteries (LiSBs). Thus, it was confirmed that the FeCo alloy produced by the melt-diffusion method enhances the redox kinetics of polysulfide conversion. In addition to its strong adsorption capabilities for lithium polysulfides (LiPSs), this alloy effectively restricts the migration of LiPSs and significantly reduces their accumulation [8].

Another pertinent field of iron-based materials applicability is linked to their proven biocompatibility and mechanical properties. Thus, iron and iron-based alloys are excellent source materials for clinical cardiac and vascular applications as biodegradable stents [9,10]. However, its relatively low degradation rate limits its use for the healing and remodeling of diseased blood vessels. To address these issues, a multi-purpose fabrication process was employed to develop a bilayer material composed of electroformed iron and iron–phosphorus alloy [11]. Fe-P has good biocompatibility and a favorable effect on the iron degradation rate [12]. Fe–Mn–Pd alloys also reveal a degradation resistance that is one order of magnitude lower than that observed for pure iron [13]. In addition, alloys possessing antimicrobial/antibacterial properties such as Fe-Ag and Fe-Au have been a point of interest for the research community [14–16].

The synthesis of iron-based materials and alloys can be realized by various solid-state techniques, among which electrodeposition is often considered a smart alternative. Indeed, electrodeposition is a well-accepted surface modification method that improves the decorative and functional characteristics of a wide variety of materials. Thus, the alternation of variables such as electrolyte composition, pH, temperature, agitation, current density, and potential [17–23], as well as the application of a magnetic field [24], can tune the composition, microstructure, and interrelated properties of the deposited materials in a wide range.

Furthermore, electrodeposition provides a significant advantage over other fabrication methods for synthesizing micro- and nanomaterials, and various aspects of this technique can be explored further in other studies [25]. It enables the production of materials and small devices with complex geometries that do not require additional processing, resulting in an effective device that is nearly "ready to use". Another peculiarity discussed in detail in the next section is linked to the electrodeposition of materials from complexing electrolytes. Complexing (chelating) is a powerful tool that can alter the electrode potentials of the metals in the electrolyte by converting the simple ions of more noble metals into complex ions with lower equilibrium potentials. Therefore, it is possible to codeposit metals possessing different standard reduction potentials because the negative shift of equilibrium potential due to the formation of complexes would decrease the exchange current density and improve the deposited layer's uniformity [26].

The electrodeposition of several types of iron alloys has been under significant attention in recent decades for magnetic applications. Alloys of iron with nickel and cobalt are most frequently used for these purposes [27]. In addition, electroplating methods for depositing industrially valuable magnetic alloys such as Fe-Ni, Fe-Zn, Fe-Co [27,28], and Fe-Pt were investigated [29]. Moreover, Fe-Pd alloys possessing high perpendicular magnetic anisotropy and ferromagnetic shape memory [30,31], Fe-Ga magnetostrictive alloys [32,33], and soft magnetic alloys like Fe-W, Fe-Mo, Fe-Co-W (Mo), and Ni-Fe-W (see for ex.: Refs. [34–40]) were also studied.

Nonetheless, according to the European Chemicals Agency, nickel and Co(II) salts are substances of high concern [41]. In this view, electrodeposited Fe-Sn alloys were proposed as an alternative to conventional nickel- or cobalt-based magnetic materials [42,43]. In this case, the intermetallic compounds are formed, and ferromagnetic iron-rich alloys of the compositions Fe_3Sn_2, Fe_5Sn_3, and Fe_3Sn are of interest to industrial applications due to their affordable cost and sustainability aspects [44]. Other magnetic iron alloys, such as Fe-P alloys [45–48] and less common Fe-Re alloys [49], were also investigated.

The advancements in green chemistry and engineering are driving toward the implementation of new processes or redesigning existing ones. The development of 3d transition metal-based catalysts, particularly based on earth-abundant metals like iron, has been scientifically proven over decades and is even more important now [50,51]. Due to their adjustable chemical reactivity, catalytic efficiency, and corrosion resistance [52], iron-based catalysts have been widely studied for various applications, including water purification (e.g., Fe-Mn alloys and electrodeposited zero-valent iron for Fenton reaction [53–55]), CO_2 removal (e.g., Fe-Ni nanoalloys [56]), and oxygen reduction (e.g., Fe-Co-Pt alloys [57]).

When considering corrosion properties, fabricating composite materials is one of the common approaches. This approach allows for the combination of the advantageous properties of the second-phase particles within the metallic matrix, resulting in novel materials possessing tailored characteristics. Composite coatings with alumina particles have been the most extensively studied among particle-reinforced composites. Thus, electrodeposited composite iron-based coatings such as Fe-W/Al_2O_3 [58,59] or Ni-Fe/Al_2O_3 [60] have been studied and evaluated as an effective sustainable alternative to electrodeposited hard chromium coatings. Also, Ni-Fe/WC composites have improved resistance to corrosion in comparison with their Ni-Fe counterparts [61].

One of the main peculiarities of alloys obtained by electrodeposition is the formation of solid solution phases of metals, which often cannot be predicted from a phase diagram, i.e., the formation of oversaturated solutions. Thus, during electrodeposition, the thermodynamically unstable structures of alloys are forming. The structural relaxation and recrystallization processes after annealing lead to changes in the grain size, microstructure, and properties of electrodeposited alloys. In the case of iron-based advanced materials, the recrystallization temperature is rather high, typically ranging from 400 to 600 °C [62–65]. The accurate control of heating temperature allows for tuning of the alloy's properties, making these materials suitable for thermal-resistant applications.

Though several review articles have been published on different aspects of iron-group materials, the present review aims to discuss the peculiarities and possibilities of the electrochemical design of iron-containing alloys from complexing electrolytes, the advanced properties associated with these alloys, and the perspectives on the future development of the electrodeposited materials.

2. Codeposition of Metals and Peculiarities of Iron-Based Alloys Electrodeposition from Complexing Electrolytes

The thermodynamic conditions for the alloy electrodeposition are quite well understood from the macroscopic standpoint. This approach is comprehensively discussed in [22,26,66–68]. The key point in the macroscopic thermodynamic description of alloy electrodeposition is an analysis of the Nernst equation written for the equilibria between metal ions in the solution and metal (origin or foreign). Due to the activities of elements in the alloy differing from the unit, the Nernst equations for the two equilibria alloy/solution of the electroreduction of a binary (For simplicity, the binary alloys are described. However, the same approach can be applied to the multicomponent alloys (ternary, quarterly, etc.)) alloy for metals A and B can be presented as:

$$\begin{array}{c} A^{z_A+} + z_A e^- \rightarrow A_{alloy} \\ B^{z_B+} + z_B e^- \rightarrow B_{alloy} \end{array} \quad (1)$$

where the equilibrium potentials $E^{(i)}_{eq,alloy}$ for each alloying component can be written as follows:

$$E^{(A)}_{eq,alloy} = E^{0'}_{A} + \frac{RT}{z_A F} \ln \frac{a_{A^{z_A+}}}{a_{A,alloy}}$$
$$E^{(B)}_{eq,alloy} = E^{0'}_{B} + \frac{RT}{z_B F} \ln \frac{a_{B^{z_B+}}}{a_{B,alloy}} \quad (2)$$

where $E^{0'}_{A}$ and $E^{0'}_{B}$ are formal potentials of the components mentioned above, *A* and *B*, respectively; and

$a_{A^{z_A+}}$, $a_{B^{z_B+}}$, $a_{A,alloy}$, $a_{B,alloy}$ is the activity of ions in the solution and components *A* and *B* in the alloy, respectively. Other marks are standard.

The most favorable conditions for the electrodeposition of an alloy consisting of two metals A and B are usually achieved when the equilibrium potentials of those metals are close. In this case, the equilibrium composition of the alloys can be expressed by equilibration of Equations (1) and (2), and it might be represented as follows:

$$\frac{a^{z_A}{}_{A,alloy}}{a^{z_B}{}_{B,alloy}} = \frac{a^{z_A}{}_{A^{z_A+}}}{a^{z_B}{}_{B^{z_B+}}} \exp\left[\frac{F}{RT}\left(E^{0'}_{B} - E^{0'}_{A}\right)\right] \quad (3)$$

As seen from Equation (3) (please note that the equation is expressed differently in [26]), the composition of alloys depends on the activities (or concentrations) of metal ions in the solution. Notably, the activity coefficients can be neglected for moderately dilute solutions (up to 1–2 M), and concentrations of components are used instead of ones as they appear in the Nernst equation in the logarithmic form. Thus, for example, if the activity coefficient is only 0.8, instead of an assumed value of 1.0, the resulting error in the value of E_{eq} is only (5.7/z) mV (where "z" is the number of electrons transferred in the reaction per molecule). Therefore, in many cases, the concentrations are used instead of activities in the thermodynamic equations [66].

Due to various complexing agents in the solutions, the concentrations of solvated "simple" metal ions decrease, leading to a negative shift in the equilibrium potentials. Here, we should dive into how this negative shift is formed in the complexing electrolytes of a metal (M).

The equilibrium in the solution containing complexes is settled between the free metal ions M^{z+}, free ligand L, and various complexes $ML_j{}^{z+}$:

$$M^{z+} + jL^{m} \Leftrightarrow ML_j^{z+jm} \quad (4)$$

where j = 1 ... n; m is a charge of a ligand.

In the simplest case, when a single complex can be formed in the solution, the equilibria may be represented by the equilibrium constant called cumulative stability constant (β_n), which can be presented as:

$$\beta_n = \frac{\left[ML_n^{z+nm}\right]}{\left[M^{z+}\right]\left[L^{m}\right]^{j}} \quad (5)$$

If stable complexes are formed, the value $\beta_n >> 1$. Usually, the concentration of ligands in the solution is higher than the total metal concentration. Therefore, it essentially reduces the concentration of free metal ions in the solution

$$\left[M^{z+}\right] = \frac{1}{\beta_n} \cdot \frac{\left[ML_n^{z+nm}\right]}{\left[L^{m}\right]^{j}} \quad (6)$$

and causes an essential negative shift in equilibrium (thermodynamic) potential

$$E_{eq}^{(M)} = E_{M}^{0'} + \frac{RT}{zF} \ln\left(\frac{[\mathrm{ML_n^{z+nm}}]}{[\mathrm{L^m}]^j}\right) - \frac{RT}{zF} \ln \beta_n \tag{7}$$

Consequently, it also causes the negative shift of the metal deposition potential, E_{dep}, to more negative values:

$$E_{dep}^{(M)} = E_{M}^{0'} + \frac{RT}{zF} \ln\left(\frac{[\mathrm{ML_n^{z+nm}}]}{[\mathrm{L^m}]^j}\right) - \frac{RT}{zF} \ln \beta_n + \eta \tag{8}$$

The term η in Equation (8) represents the difference between deposition and equilibrium potentials:

$$\eta = E_{dep}^{(M)} - E_{eq}^{(M)} \tag{9}$$

Understanding the possibilities of alloy electrodeposition is crucial. However, predicting the composition of electrodeposited alloys based solely on thermodynamic data is quite complicated due to the kinetic peculiarities of metal electroreduction. The complexing electrolytes for metals and alloys electrodeposition are essential because the formation of complexes decreases the exchange current density (η increases) and could lead to the improved uniformity, smoothness, and brightness of the deposits [66].

On the other hand, to successfully investigate the electrochemical processes at the metal/solution interphase in the presence of complexing agents, attention should also be paid to the solution chemistry. In other words, evaluating the distribution of complexes and ligands in the solution volume is necessary. For this purpose, it is possible to extract concentrations of all species formed in the solution after the dissolution of corresponding salts by solving the system of equations using commercially available software, e.g., Maple6, MathCad, MINDTEQ, Meduse, or others. The set of the system of equations has to include the following relations and quantities:

(i) The equilibrium constants for all compounds added to or formed in the solutions: acids deprotonation, hydrolysis, polymerization, stability constants of metal complexes with ligands, etc.;
(ii) The mass balance equations $[J]_{tot} = \sum\left[J_i^{n+/-}\right]$ for all forms in the equilibrium mixture, and
(iii) The charge balance "Cat" and "An" denote cation and anion, respectively.

Certainly, the calculated distributions of species in the electrolyte can vary depending on the concentration of precursors and pH, leading to changes in the deposit composition and properties, respectively. For example, when assessing the species distribution in citrate- and tungstate-based electrolytes, 20 to 32 different species can be identified and included in the calculations.

This approach has proven effective in:

(i) The formulation of a thermodynamically stable solution for the electrodeposition of Fe-rich or W-rich alloys from the Fe(III)-based glycolate-citrate electrolytes (Figure 1). Additionally, the distribution of electrochemically active complexes varies with pH, which in turn affects current efficiency (CE) and the properties of the resulting materials [69,70];
(ii) Understanding the correlation between W content in the Co-W alloys as a function of pH and various citrate complexes of WO_4^{2-} [34];
(iii) Determination of conditions that allowed for the increase in the deposition rate of Ni-W alloys [71];
(iv) Impeding the deposition of Ni during Ni-W alloy electrodeposition processes [66];

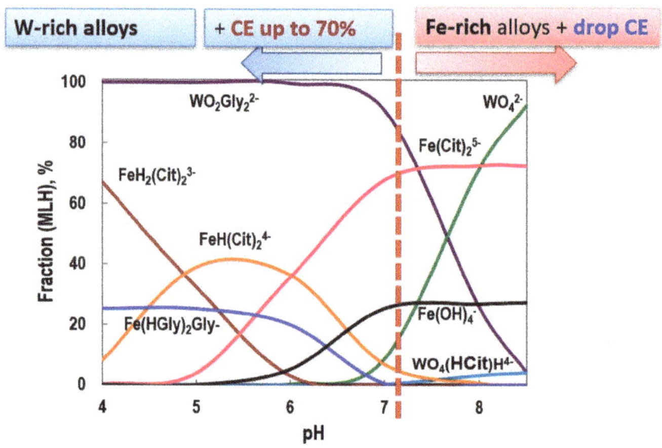

Figure 1. Calculated distribution of various complexes in the glycolate-citrate electrolyte vs. pH (based on the data reported in [69]) and variation of tungsten/iron content and current efficiency of Fe-W electrodeposits. WO_4^{2-}, $Fe(Cit)_2^{5-}$, $Fe(OH)_4^{2-}$, $WO_4(HCit)H^{4-}$, $WO_2Gly_2^{2-}$, $FeH_2(Cit)_2^{3-}$, $FeH(Cit)_2^{4-}$, $Fe(HGly)_2Gly^{-}$ are the species formed in the glycolate-citrate solutions at Fe-W alloy electrodeposition. The distribution of species is represented by the colorful lines with the tag of the corresponding species next to it. The red dashed line corresponds to the transition from W-rich to Fe-rich alloys.

Furthermore, the evaluation of species distribution in the complexing electrolytes allowed authors to clarify the following:

(v) The pH range for stable ferric-stannous tartrate-chloride-sulfate electrolytes for Fe-Sn electrodeposition [42];

(vi) The role of glycine in the Fe-P alloy electrodeposition [72];

(vii) Codeposition conditions, e.g., for CoFeNiCu alloys electrodeposition from citrate solutions [73];

(viii) Pourbaix diagrams for systems containing few metals and complexing agents, e.g., for Cu-Co-Fe in acetate solutions [74] and for Fe-V-O-H metastable materials [75].

It is important to acknowledge that complexes significantly influence the structure, composition, and properties of materials. This is due to their direct involvement in forming electroactive species either within the bulk material or on the surface of the electrodes during the deposition process. Thus, the well-known peculiarity associated with the electrodeposition of materials containing iron group metals with refractory metals mentioned above is linked to the mechanism of codeposition of such materials, which is classified as induced codeposition [76,77]. This is unlike anomalous deposition [78], which is a common case for the codeposition of iron-group metals (e.g., FeNi alloys) or the codeposition of Fe with Zn or Cd.

The understanding of the induced codeposition mechanism and the role of complexes in the case of refractory metals codeposition was overviewed elsewhere [34,66]. We will not enter into a deep discussion on this matter here, but it can only be specified that either "bulk" and/or "adsorbed" complexes formed during induced codeposition could be correlated to the experimentally observed results.

However, while developing iron-based materials with rhenium under an induced codeposition mechanism, some particularities should be considered: namely, perrhenate ions (the source of rhenium) do not form polynuclear complexes in acidic solutions in contrast to WO_4^{2-}. Therefore, rhenium codeposition with iron group metals is possible from weakly acidic or neutral solutions, preferably using citrate as a complexing agent. The compositions of electrolytes, electrodeposition conditions, compositions of obtained alloys, and their properties are reviewed in [66,79,80].

Also, it should be considered that the Nernst potential (E_o^{red}) of Re in the perrhenate ions solutions depends on the pH. It is +0.363 V at pH 0, −0.110 V at pH 7, and –0.584 V at pH 14.0. These values are more positive than those for the iron group metals, e.g., E_o^{red} for Fe^{2+}/Fe and Fe^{3+}/Fe are −0.447 V and −0.037 V, respectively [81].

However, the electroreduction of ReO_4^- is highly irreversible. The ReO_4^- reduction wave occurs just before the onset of electrolyte decomposition, but the electroreduction in weakly acidic citrate or oxalate solutions is facile, resulting in a polarographic wave at much more positive potentials [82]. It is explained by the formation of electrochemically active $(ReO_4{\cdot}H_2Cit)^{2-}$ complexes at pH above 3. The role of this complex in the codeposition mechanism of Re with Fe has been discussed in [66,80]. However, based on the UV–vis absorption spectroscopy data obtained in the citrate-perrhenate system at pH 5, it is indicated that the ReO^{4-} ions are unlikely to chemically bind to the dissociated citric acid. Therefore, it is assumed that even in the presence of citrate ions, the ReO_4^- ion remains hydrated [83]. In that respect, it is an interesting consideration that the overall reaction leading to the formation of rhenium is influenced by the presence of H^+ ions, which depend on the adsorption process and therefore on the electrode surface nature. This means that the surfaces with high hydrogen adsorption could facilitate the reduction of perrhenate ions to Re [84].

Moreover, to explain some experimental results of induced codeposition, a new theoretical approach was developed that assumes the formation of an N-dimensional fractal cluster in the solution [85]. This theory enables us to avoid the difficulties of the classical theory of nucleation operating with the concepts of surface energy, which is intrinsic to a geometrical surface. In this case, the electrochemical kinetics of nucleation rather than electrochemical reaction is considered. In this sense, the nucleation stage should be considered as the step following the formation of an ad-particle on the surface.

By applying a non-linear approximation of nanonucleation from fractal solutions for the interpretation of induced codeposition of iron group alloys with refractory metals, it was possible to explain that a high rate of propagation of crystallization front had influence on:

(i) Nanocrystallinity of the electrodeposited new phases;
(ii) Formation of composites containing oxide-hydroxide inclusions;
(iii) Absorbed hydrogen at high current density.

Another peculiarity that should be taken into account in the developed electrolytes is the stability of Fe(II) species [86] because it can often hinder the usage of iron-based materials for scale-up applications. Electrolytes based on Fe(II) salts are unstable thermodynamically, and the solution content is governed by the Fe(II) oxidation kinetics [27].

It was shown that citrates and ammonia [27] inhibit the kinetics of Fe(II) oxidation to Fe(III) but cannot prevent it. Some attempts were made to prolong bath life by adding reducing agents, such as L-ascorbic acid [86,87], but those compounds can affect the bath maintenance and coatings properties. In the case of Fe-Ag alloy electrodeposition, an instability caused by the reduction of Ag(I) by Fe(II) was partially resolved by adding 1% Fe^{3+} [16]. The solution containing Fe(II) will oxidize gradually unless used regularly. The time and efforts required to restore the electrolyte to an operable condition may outweigh the economic benefit of depositing a lower-cost material; e.g., it is necessary to reduce the Fe(III) ion before using a freshly prepared bath, and this process might require 24–48 h [27]. Therefore, the electrodeposition of Fe alloys from the solutions based on Fe(III) compounds was also investigated for the fabrication of Fe-W, Fe-Ni, Fe-Pt, and ternary alloys [69,88–91].

3. Versatile Materials Based on Iron-Based Alloys

In this section, the main focus will be on the practical aspects of electrodeposited iron-based alloys containing refractory metals, primarily tungsten-based alloys, as these alloys can be adapted for various applications.

From an application point of view, iron-based alloys with refractory metals are an important class of functional materials that are characterized by a distinctive set of properties linked to the presence of high melting refractory metal, such as increased heat resistance and stability at high temperatures, high corrosion resistance, and excellent mechanical properties [34,92–94]. These alloys are advanced materials for strategic components in many industries (hard coatings, catalysts, aerospace, chemical, nuclear, defense, biomedical, etc.), including as a component of superalloys [95]. The codeposition of refractory metals with iron group metals leads to the obtainment of high-quality electrodeposits with higher current efficiencies [63,69,96], and in many cases, they can replace high-cost materials. However, the research on iron alloys is still growing compared with other metals due to the sustainability aspects and advanced properties of iron-based materials.

3.1. Structural Peculiarities, Mechanical, Thermal Resistance, and Magnetic Properties

3.1.1. Structural Peculiarities and Mechanical Properties

The key factors affecting the properties of the electrodeposited material are directly linked to its structure and chemical composition. The electrochemical approach induces the precise scenario for an alloy synthesis with targeted amorphous and/or nano-structure, composition, internal ordering, and texture. Thus, the main characteristic of electrodeposited iron-based materials with refractory metals is their nanocrystalline structure, which primarily impacts those alloys' hardness and is correlated to alloy composition in the electrodeposit and reduced crystallite size [36,38,58,63]. According to the Hall–Petch relation, the mechanical properties typically improve as the grain size decreases due to the increase in grain boundaries, which hinders dislocation motion. Moreover, effects such as nanostructuring and solid solution strengthening will also be involved.

However, there is a critical value of crystallite size at which the Hall–Petch relation will be distorted. In the case of Co-W [97] and Ni-W [98] alloys, the breakdown occurs at crystallite sizes of 5 nm and 10 nm, respectively. Notably, the same tendency was also observed in the case of more complex systems such as, e.g., high-entropy NiFeCo-W alloys [99]. Remarkably, it often corresponds to the transition to the amorphous-like or ultra-nanocrystalline structure of electrodeposited materials [100], which promotes the diverging of the shear bands that lead to the nanohardness decrease. However, the ultra-nanocrystalline structure brings other benefits, e.g., it can increase catalytic properties, which will be discussed later below.

The hardness of Fe-W alloys increases from 4.1 GPa for the Fe-6 at.% W alloy to 10.4 GPa for the Fe-25 at.% W alloy. Along with this, the elastic modulus also rises significantly, increasing from 83 GPa to 216 GPa. This trend is strongly associated with the development of a stable intermetallic Fe_2W phase in alloys that contain more than 16 at.% of tungsten. [36]. As a result, smart tuning of mechanical properties can be accomplished by varying the alloy composition and applied experimental conditions (Figure 2). Concerning alloy composition, it can be underlined that under similar testing conditions and refractory metal content, generally, iron-based alloys have higher hardness than cobalt- or nickel-based electrodeposited materials [63], which encourages the use of iron-based materials for hard coating applications.

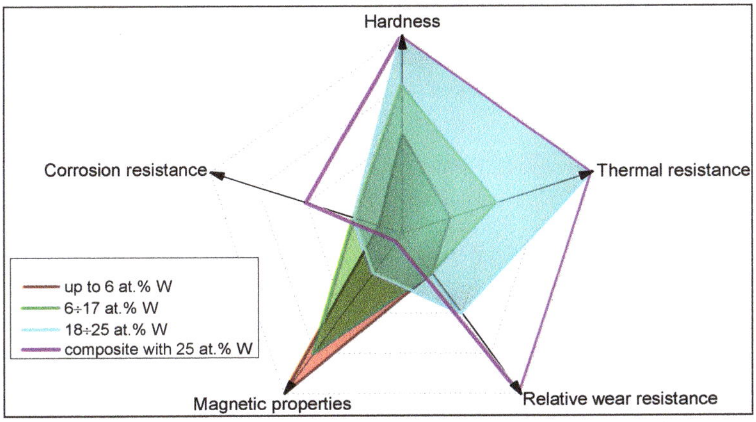

Figure 2. Properties mapping of electrodeposited Fe-W alloys as a function of tungsten content in the coating. In addition, properties of Fe-W/Al_2O_3 composited coating are presented for comparison. The black arrows show the increase in the corresponding characteristics (a.u.). The Fe-W alloys were electrodeposited from environmentally sustainable glycolate-citrate electrolytes [69].

On the other hand, experimental conditions such as volume current density (VCD) can also affect drastically the hardness and even corrosion, as demonstrated in [101–104]. The VCD describes the ratio of current to the volume of an electrolyte (I/V). It is an important characteristic, especially taking into account the upscaling of lab processes, because usually only common characteristics such as pH, composition, temperature, current density, or potential are considered. The dependence of alloy composition on VCD is linked to a different rate of the metal complex (gluconate, citrate, etc.) concentration change during induced codeposition. VCD is mainly specific to the induced codeposition of tungsten alloys, and it does not influence the electrodeposition of industrially important metals such as nickel or chromium from non-complexing electrolytes [101].

During the codeposition of refractory metals, significant differences in composition, morphology, and microhardness may occur depending on the electrolyte volume at either a constant electrode potential or current density [102]. Thus, depending on the VCD, electrodeposited nanocrystalline alloys contain not only iron group and refractory metals but also different amounts of intercalated oxides and hydroxides, and even alloyed hydrogen (e.g., Fe-W-H) [102,105], that will ultimately have an impact on the properties. Another important experimental parameter is the nature of the counter electrode (material from which the anode is created from and whether it is insoluble or soluble). It affects the intrinsic and extrinsic characteristics of refractory metal alloys during induced codeposition from complexing electrolytes, including the rate of deposition and hardness [103,106].

3.1.2. Thermal Resistance

The presence of inclusions and abundant evolution of hydrogen during electrodeposition can promote the brittleness of the obtained iron-based materials [93]. In this view, annealing can decrease the internal stress of the deposits and advance the mechanical properties. During annealing, electrodeposited refractory alloys will recrystallize, forming coarser grains and new phases (oxides, intermetallic phases, carbides, etc.) [62–64,70,107,108]. Typically, refractory metals-rich alloys, in comparison with alloys having higher iron content (Figure 3a–c), show improved thermal resistance linked to the formation of thermodynamically stable nanostructures with decreased energies of segregation and grain boundaries.

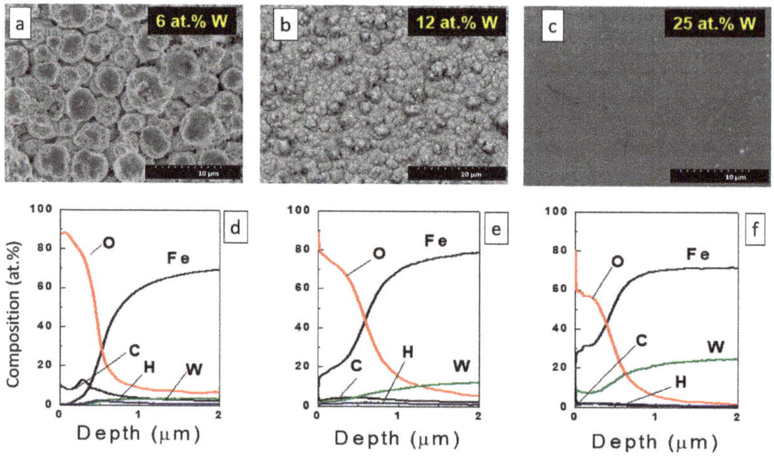

Figure 3. The representative SEM images of electrodeposited Fe-W alloys and the distribution of elements in the top layers of the 10 μm thick coatings. SEM images and the corresponding distribution of elements: (**a**,**d**) 6 at.% of W, (**b**,**e**) 12 at.% of W, (**c**,**f**)—25 6 at.% of W. Inspired from [109].

For instance, the XRD amorphous-like structure of Fe-W alloys can withstand temperatures of up to 600 °C, which is higher than that for Ni-W or Co-W alloys with the same tungsten content [63,70]. Above this temperature, a partial recrystallization occurs in the Fe-W alloys, but some amorphous peaks are still presented in XRD patterns even at 800 °C [109]. The nanocrystalline peaks of annealed alloys usually are attributed to α-Fe, Fe_2W, $FeWO_4$, and Fe_6W_6C phases [109,110], although the Fe_7W_6 phase [111] can also be present in the as-deposited coating, having a more coarse-grained structure and lower tungsten content. However, tungsten-rich alloys, electrodeposited from citrate-based electrolytes and annealed at 700 °C [110] or 800 °C [109], contain uncommon phases such as Fe_6W_6C and Fe_3W_3C.

The appearance of hydrogen-, carbon-, and oxygen-containing phases is practically independent of W content (Figure 3d–f), and it is attributed mostly to the deposition from complexing electrolytes having organic ligands, which afterwards affect the microstructure and properties of as-deposited and annealed alloys [109,110,112,113]. Moreover, the light elements are typically present in the top layers of the deposits, and their contents essentially decrease when moving deeper from the surface up to 2 μm (see Figure 3). This effect does not depend on the content of W in the alloys or the morphology of the surface. The ratio of content W to Fe remains almost constant along the entire thickness of electrodeposits.

Moreover, the dependence of annealed alloy properties (hardness, wear, and corrosion resistance) on temperature usually will have an optimum. This optimum for heat treatment will depend on the initial tungsten content in the coating and the "amorphous state" of as-deposited Fe-W coatings, which is connected to a shift in the annealing temperature from 500 °C [111] to 600 °C [109] and 700 °C [110]. Also, it is connected to the precipitation and dispersion of fine α-Fe crystallite and Fe_3W_3C particles in the Fe-W alloys [114]. Thus, for the as-deposited coatings having 25 ÷ 36 at.% of W and annealed ones, this leads to an increase in hardness ranging from ~8 ÷ 10 GPa to ~16 ÷ 20 GPa. At higher temperatures than 800 °C, coarse grains are formed, which negatively impacts the hardness of the material [109,113]. Overall, we should emphasize that the increased content of refractory metal in iron alloys will facilitate a transition to a more thermal-resistant state of deposited materials (Figure 2).

3.1.3. Magnetic Properties

Electrodeposited Fe-based alloys are extensively studied for applications in magnetic devices because it is possible to tune both the composition and structure of obtained materials. To achieve the optimal soft magnetic properties, grain size control is critical. According to a model proposed by Herzer [115], coercivity H_c is proportional to the sixth power of average grain size, i.e., the so-called "*D6*" law. As was shown in [116], for Co-Fe alloys electrodeposited from citrate-based baths, the average grain sizes in deposits are around 10–20 nm, which is quite fine and beneficial for achieving optimal soft magnetic properties. So, the CoFe films electrodeposited from citrate-added baths exhibited better soft magnetic properties than those deposited from citrate-free baths under the applied experimental conditions. A similar behavior was obtained for $Ni_{86.0}Fe_{9.8}W_{1.3}Cu_{2.9}$ alloy electrodeposited from citrate baths: during annealing, the mean crystallite size increased from 12 to 29 nm, resulting in reduced magnetization and broadened magnetic hysteresis and thus increased loss in the active power and decreased strength of the coercive field [117]. CoFeNi thick films electrodeposited from citrate baths also obey a "*D6*" law for the correlation between crystallite size and coercivity, but the saturation magnetization is not influenced by the crystallite size and depends on the chemical composition [118]. However, in the case of electrodeposited Fe-Pd alloys from citrate-ammonia baths, deposited alloys are almost disordered, whereas annealing at 400–600 °C promotes crystallinity, and the magnetic coercivity increases up to 10 times [119].

Exploration of soft magnetic iron-based alloys such as NiFeCo can be further extended by the incorporation of rare earth elements (e.g., Tb and Dy), which contribute to the giant magnetostriction of electrodeposited materials that can be engaged in resonant frequency sensing devices [120]. In addition, a magnetic field applied during the electrodeposition can provide a feasible way to produce oriented magnetostrictive CoFe materials for dual-mode flexible sensors [121].

In the case of iron-based materials with tungsten, the magnetic properties are strongly dependent on the electrochemical design. Thus, the experimental conditions (pH, temperature, current/potential) will affect the structure, composition and, in the end, the transition to a semi-hard or soft magnetic nature of deposited materials. Thus, hard and soft magnetic properties of refractory binary [97,122,123] and ternary [40,124–126] alloys and more complex systems [127] with iron have been reported. The hard and soft magnetic properties are more likely to be found in cobalt-based alloys [97], while iron-based alloys with refractory metals are generally characterized by a low coercive field ($Hc < 200$ Oe). Therefore, soft or semi-soft magnetic properties could be expected, which can be applied to sensors, read/write heads in hard discs, and microelectromechanical systems.

Here, we should underline that, commonly with an increase in the non-magnetic phase (refractory metals in our case), the saturation magnetization will decrease. Thus, Fe-W alloys electrodeposited at room temperature display a much sharper decrease in saturation magnetization for deposits having 12 at.% of W, but deposition at 65 °C (16 at.% of W) leads to the retention of rather high magnetization [36]. This behavior is attributed to the thermodynamic constraints of nucleation and crystallite phase formation from complexing electrolytes at room and elevated bath temperatures. An amorphous-like structure can be formed at room temperature at a lower content of refractory metals (12–14 at.% of W). However, at higher temperatures, the XRD amorphous peaks start to appear only at 16–18 at.% of W.

Furthermore, the nanocrystalline deposits have a higher value of the saturation magnetization caused by the presence of the α-Fe phase in the Fe-W alloy, while XRD amorphous phases such as solid solution of tungsten in iron, W(Fe), and intermetallic Fe_2W contribute to a decrease in saturation [36]. The coercivity evaluation of such materials illustrates an overall semi-soft magnetic character with values in the range of 10 ÷ 190 Oe depending on tungsten content (Figure 2).

3.2. Wear and Corrosion Resistance

One of the important aspects of research on the iron group metals refractory alloys is connected to finding materials that can replace electrodeposited chromium, whose synthesis involves restricted compounds of Cr(VI) [128]. Hence, iron-base alloys are an important pillar that addresses sustainable issues linked to Cr(VI) and the subsequent use of chromium as a hard and corrosion-resistant coating. Additionally, our group's research indicates that the wear rate of tungsten-based alloys under identical testing conditions is comparable with that of electrodeposited chromium [34]. Other researchers obtained similar data using different test setups [129]. In general, comparing the tribological behavior of different materials is quite complicated due to the involvement of different tribo-systems and test conditions. Usually, the evaluation of tribological properties is mostly linked to material characteristics, namely the ratio between hardness (or plasticity) and wear [98,130]. The higher ratio between the hardness and elastic modulus (elastic strain to failure) provides a material with enhanced resistance to permanent plastic deformation, which could diminish wear caused by this deformation [130]. Indeed, the elastic-plastic behavior was the prevailing cause of wear under dry friction conditions of electrodeposited Ni-W alloys [98,131]. Moreover, the formation of a stable hexagonal close-packed (hcp) structure that impedes the plastic flow in Co-W electrodeposits led to an improved wear resistance at high applied loads [34,132].

Nevertheless, the tribological behavior of Fe-W alloys under dry friction conditions is primarily influenced not only by their mechanical properties but also by the chemical stability of iron during fretting, which involves the tribo-oxidation of the alloys [133]. Iron oxide particles that form in the wear track act as an abrasive third body, which increases the asperity contact between the material and counter-body, thus resulting in a high coefficient of friction and larger wear volume than in the case of Ni- or Co-based alloys. It should be taken into account that increased refractory metal content (low iron content) in the deposits and heat treatment of as-deposited alloys will increase the relative wear and corrosion resistance of iron-based materials [66,98,133].

Another possibility to improve the wear and corrosion resistances of iron-based materials like Fe-W is to introduce a new alloying element, e.g., Zn, P, and La, into binary alloys. Although the tribological and corrosion properties usually improve with the increasing content of newly alloyed components, this approach will provide only a temporary remedy until iron particles start to oxidize [134–136].

However, using an environmentally sustainable thin layer (1 μm) of rapeseed oil significantly reduced the tribo-oxidation and wear of Fe-W alloys under lubricating conditions [137]. Thus, the particles of iron oxide were absent in the wear track and at the surface of the counter-body, and the roughness inside the wear track was very close to the initial surface roughness. Hence, the lubricating conditions can considerably improve the tribological characteristics of Fe-W coatings. Nevertheless, the liquid lubricant in this case must be injected regularly to extend the lifetime of the coating, which can only apply to some specific devices.

In this view, composites containing conductive or non-conducting compounds provide a larger stage for innovative solutions [60]. Moreover, composites combine the properties of an alloy with the chemical and physical properties of second-phase particles, thus producing a material with advanced characteristics compared with its analog having the same composition (Figure 2) [58,59]. Usually, it is possible to increase the hardness of the composites, which is positively affected by grain refining, dispersion strengthening, and the texturing effect induced by particles. In addition, due to the self-lubricating effect, the wear resistance can also be improved by choosing the appropriate composites based on a metal matrix and particles such as ZrO_2, SiC, or Al_2O_3 [138].

It is important to note that introducing particles does not guarantee that the resulting composite will have improved characteristics. Frequently, it can lead to increased porosity, low adhesion to the metallic matrix, and peculiarities of codeposition, which could increase, for example, the corrosion rate. Thus, the Al_2O_3 particles in the Ni-Fe/Al_2O_3 composite had an overall unfavorable influence on the corrosion resistance due to the increased porosity, roughness, and inhomogeneity of the deposits [60]. Therefore, smart design of composites is needed.

Another approach involving a trilaminar structured composite has been applied to protect the Fe-W alloy. Namely, a silane cross-linked graphene oxide layer and hydrophobic organosilane layer were formed on top of the Fe-W amorphous alloy layer to protect it against corrosion [139]. This approach of adjusting the morphology of the electrodeposited Fe-W alloy layer provided the possibility to decrease the corrosion rate by three orders of magnitude and increase overall hydrophobic performance (the contact angle was 141.7 degrees).

3.3. Catalysts Based on Electrodeposited Iron-Based Materials and Upgraded Systems

Currently, electrodeposited alloys are extensively studied as electrocatalysts for water electrolysis and other chemical reactions. A crucial factor in electrocatalysis is the choice of electrode materials that exhibit high electrocatalytic activity. This quality is essential for reducing the costs of the process. The efficiency depends on two main factors: the composition and structure of electrodeposited material and the electrochemically active area of the catalyst. It was confirmed that a cauliflower-like (Figure 4a), ultra-nanocrystalline structure (crystal grain size below 5 nm) (Figure 4b) of the electrode material leads to an enhancement in the electrocatalytic performance. The fine crystallites of the material increase its surface-over-volume ratio and, thus, a larger surface area is accessible for reactants in the electrochemical reactions [100].

We should highlight an important factor that influences the catalytic activity of electrodeposited coatings, namely the electrochemically active surface area (EASA). By knowing the EASA, it is possible to evaluate the intrinsic catalytic activity of the material and distinguish which of the parameters contributed to the catalytic performance of investigated electrodes, especially in the case of 3D electrodes (electrodeposition on porous electrode, metallic foam).

Electrochemical impedance spectroscopy (EIS) is an effective method for evaluating EASA. Various parameters of EIS data can be used for the EASA estimation depending on the reactions of the catalytic process: (a) capacitance of the double electric layer (reaction involves adsorbed intermediates); (b) Warburg impedance (reaction with limiting mass transfer rate); and (c) charge transfer resistance (reactions with slow charge transfer rate, e.g., HER, OER) [140–142]. For instance, by using a 3D electrode (Ni-foam) and co-doping FeW-based catalysts with Ni_3S_2/NiS, it was possible to improve the catalytic performance of elaborated materials toward OER and urea oxidation [143] or towards both HER and OER [144].

As discussed earlier, electrodeposited nanocrystalline iron-based coatings with tungsten and other refractory metals are versatile materials that can also be applied in catalysis. The analysis of the electronic configuration of d-metals shows that the codeposition of two or more such metals from two branches of Balandin's volcano curve could improve and even exceed their individual electrocatalytic properties [145]. Thus, iron-based materials have been employed as catalysts for: methanol and ethanol oxidation [146,147]; hydrogen

(HER) and oxygen (OER) evolution reactions [100,148,149]; and photoelectrochemical water splitting by dealloying of the electrodeposited Fe-W alloy [150].

The catalytic activity of iron-based materials with tungsten is closely linked to the composition, structure/phases of the alloys, and pH of the working medium. Moreover, a significant improvement in the catalytic activity leading to a visible reduction in the hydrogen overpotential and an increase in the apparent exchange current density was noticed with the increase in the temperature of the working medium [100]. Generally, the high content of the refractory metal and ultra-nanocrystalline structure (crystallite size below 10 nm) determines the enhanced catalytic properties of such alloys coupled with reasonable corrosion resistance in acidic solutions, which for example are needed for methanol oxidation [100,146,151].

The high catalytic activity of the ultra-nanocrystalline tungsten alloys (containing 30 at.% of W) for HER application can also be ascribed to the formation of stable intermetallic phases [152], ensuring optimal metal distribution over the surface and producing larger catalytically active sites. However, the catalytic activity of CoW-based alloys is usually higher than that of iron-containing analog [100]; see Figure 4c.

A considerable part of research involving iron-based materials is focused on the design of effective catalysts based on Fe binary and ternary alloys with Mo [153,154] because they have the potential to be applied as an efficient HER electrocatalyst in alkaline media. Among effective Mo-rich (52–54 at.%) alloys, the apparent exchange current density of hydrogen evolution for Co-Mo electrodeposits (Figure 4c) was considerably higher than that for Ni–Mo and Fe-Mo coatings [155].

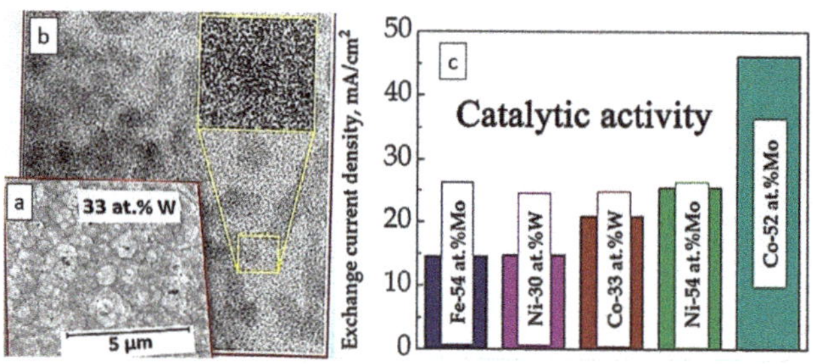

Figure 4. Representative SEM (**a**) and TEM (**b**) images and exchange current densities (c) of ultra-nanocrystalline tungsten and molybdenum-rich iron group metal alloys, based on the data [100,155].

The overvoltage of a material during the hydrogen evolution reaction (HER) at a specific current density is often used as a criterion to assess its catalytic activity. This allows for a comparison of the catalytic performance of different materials for HER and OER. Thus, at room temperature, the overvoltage of HER at 100 mA cm^{-2} (η_{100}) in alkaline media for Fe-Mo alloys is -0.22 V [155]; for Fe-W, it is -0.3 V [100]. These values are higher than those obtained for alloys containing only iron-group metals FeCoNi or FeCoNiP, where η_{100} varies from -0.196 V to -0.135 V, respectively [156].

Furthermore, the molybdenum-rich coatings also show high catalytic activity for the oxygen evolution reaction [157]. There are limited data on the catalytic activity of Fe-Re alloys for HER and OER or other electrochemical reactions, but these alloys show superior catalytic performance compared with monometallic Fe and Re catalysts in hydrogenation reactions [158].

Iron-based high-entropy alloys (HEAs) unlock new possibilities to control the functional properties of electrodeposits. For example, HEAs typically exhibit higher catalytic activity for HER compared with electrodeposited binary alloys. This is due to a larger number of active sites having different electronic structures and adsorption energies [159].

Commonly, the catalytic behavior of HEAs is associated with four core effects: a high entropy effect in thermodynamics [160], lattice distortion effect in structure [161], hysteresis diffusion effect in dynamics, and cocktail effect in properties [162].

Thus, due to the interaction of different elements and the high mixing entropy, the electrodeposited Ni-Fe-Cu-Co-W high-entropy alloys demonstrate a cauliflower-like structure and high catalytic activity. The excellent performance for OER can be attributed to several factors: (1) the synergistic effect among various metals enhances their intrinsic catalytic activity; (2) the binder-free self-supporting structure creates a shorter diffusion pathway for electrolyte penetration and facilitates faster electron transfer; and (3) the cauliflower structure increases the contact area between the electrode and the electrolyte. As a result, Ni-Fe-Cu-Co-W alloys, when electrodeposited at an optimal current density, achieve a lower overpotential value of 100 mV at 0.247 V, which is even smaller than that of many active OER catalysts such as IrO_2 [163].

The synergetic effects observed in the electrodeposited FeCoNiMoW high-entropy alloy also have a significant impact on the magnetic and corrosion properties of these materials. Due to the unique microstructure of the deposited coatings and the relatively high content of non-magnetic elements (over 20 at.% of W and Mo), the coercivity (Hc) value is low, fluctuating between 11 and 18 Oe. Additionally, these alloys demonstrate excellent corrosion resistance, exhibiting a noble corrosion potential of $E_{corr} = -0.55$ V and a low corrosion current density of $I_{corr} = 3.2\ \mu A/cm^2$ [164].

Another possibility and advantage of exploring ferromagnetic alloys as catalysts is linked to the magnetic field application during the catalytic reaction. Thus, due to the synergistic effect of catalytic and magnetic action, it was possible to reduce by 10% the electricity needed for hydrogen generation compared with the base case scenario [165]. Moreover, the perspectives of iron-based magnetic materials as catalysts are not limited only to HER or OER reactions. For instance, the Fe-Ce-W magnetic catalyst can be smartly tuned to increase selectivity towards the reduction of various nitrogen oxides [166]. Additionally, assessing the stability of Fe-based catalysts is vital for their long-term application in practical engineering, particularly in oxidation processes [167], aggressive media [168–170], and fuel cells [171].

4. Summary and Future Perspectives

We explored the potential of utilizing earth-abundant materials, specifically focusing on iron-based and refractory metal alloy electrodeposits. The properties of these materials can be easily adjusted through careful electrochemical design. For example, by simply shifting the pH of the solution, we can affect the distribution of complex species and obtain alloys that are rich in iron or tungsten with varying current efficiency (CE), as shown in Figure 5. This process leads to different compositions, morphologies, and structures, which can also be further fine-tuned by precisely controlling the temperature of the electrolytic baths. This control allows for transitions between nanocrystalline, ultra-nanocrystalline, or amorphous-like structures.

One of the main factors impacting the mechanical, magnetic, catalytic, and tribological properties of Fe-alloys is linked to the nanocrystalline structure of electrodeposited materials obtained from complexing solutions. Moreover, thermodynamic approaches can facilitate and optimize the electrosynthesis of nanocrystalline alloys. Namely, the knowledge of the complex speciation (Figure 5) in the solutions can help to: (a) define the compositional stability ranges of electrodepositing baths; (b) correct Pourbaix diagrams for particular conditions; (c) find the codeposition conditions and ways for process intensification; and (d) understand the influence of the composition of the bath on the composition and structure of electrodeposited alloys.

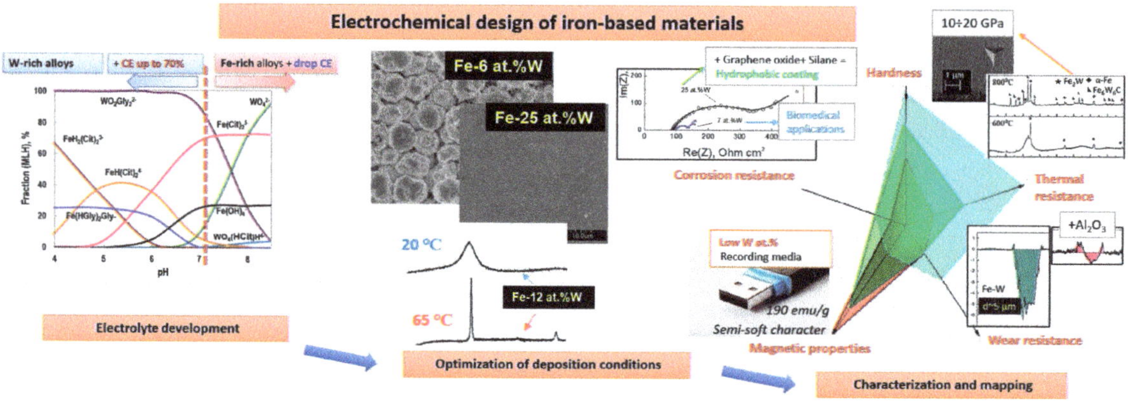

Figure 5. The outline of the electrochemical design of Fe-W alloys: from bath elaboration towards characterization and application.

This interplay between parameters of electrochemical design is an opportunity to tailor such materials for versatile applications, which can be further enhanced by exploring the potential of micro- and nano-electrodeposits [172]. Namely, the excellent mechanical properties and thermal resistance of rich-W iron alloys can be explored for fusion reactors or other applications requiring thermal resistance. The soft magnetic Fe-rich alloys, along with their electrical properties, magnetostriction, and biocompatibility, can be explored for applications in micro-transformers, sensors, and NEMS fabrication. The mechanical properties and advanced biodegradability of iron-based scaffolds show potential for bone tissue engineering.

To adopt these alloys for tribological applications, the lubricated conditions should be applied to prevent the tribo-oxidation of iron-based alloys or to produce composites with added inert hard particles such as Al_2O_3, TiO_2, or others. The alternative way to increase the mechanical robustness and wear resistance of electrodeposited materials is carburization, where the thermal decomposition of ethanol in an inert gas atmosphere at high temperatures is used to create carbide phases such as, e.g., Fe_3W_3C [173]. As we discussed earlier, these phases essentially contribute to increased hardness. Additionally, for targeted applications, the corrosion resistance of Fe-W alloys can be enhanced by increasing their hydrophobicity through the formation of silane cross-linked graphene oxide and hydrophobic organosilane layers on the base material (Figure 5). The catalytic properties can be tuned by the formation of nanostructured electrodeposits (e.g., Ni-Fe nanosheets) that can be rationalized afterwards by W doping [174]. Furthermore, electrodeposited NiFeW hydroxide electrocatalysts for efficient water oxidation can outperform conventional 20 wt.% Ir/C electrocatalysts [175].

Moreover, taking into account the growing focus on sustainability worldwide, the synthesis and discovery of new applications for earth-abundant materials will increase. However, iron-based alloys containing rare metals like Pt and palladium (Pd) demonstrate superior catalytic activity and stability compared with other alternatives. Their chemo-physical properties can be easily adjusted by altering the synthetic parameters. It is essential to choose the appropriate system, as the selection depends on the active phase needed for a specific reaction [176].

A sustainable method for obtaining the mentioned materials could involve conducting galvanic replacement reactions on iron or iron alloys using solutions that contain platinum or palladium derived from their waste-leaching processes. Thus, research has demonstrated the effectiveness of the galvanic replacement method for producing catalysts based on a

Fe-Zn substrate [177]. However, substantial effort is still needed to fully understand and implement the eco-friendly electrochemical recovery of such materials.

Author Contributions: Conceptualization, N.T. and H.C.; writing—original draft preparation, N.T., H.C. and O.B.; writing—review and editing, H.C. and N.T.; illustration—N.T. All authors have read and agreed to the published version of the manuscript.

Funding: This study was partially funded by the Research Council of Lithuania, grant no. S-MIP-24-11; the Moldavian National project ANCD project 011204; and the European Union MSCA4Ukraine project, grant no. 1233494.

Acknowledgments: The authors are thankful to all partners of the H2020 "SELECTA" and "SMART-ELECTRODES" projects for the fruitful collaboration on the accomplished research.

Conflicts of Interest: The authors declare no conflicts of interest.

References

1. Our Common Future, Chapter 2: Towards Sustainable Development. UN Documents Gathering a Body of Global Agreements. Available online: http://www.un-documents.net/ocf-02.htm (accessed on 11 July 2024).
2. Fajardo, A.S.; Westerhoff, P.; Sanchez-Sanchez, C.M.; Garcia-Segura, S. Earth-Abundant Elements: A Sustainable Solution for Electrocatalytic Reduction of Nitrate. *Appl. Catal. B Environ.* **2021**, *281*, 119465. [CrossRef]
3. Chirik, P.J.; Engle, K.M.; Simmons, E.M.; Wisniewski, S.R. Collaboration as a Key to Advance Capabilities for Earth-Abundant Metal Catalysis. *Org. Process Res. Dev.* **2023**, *27*, 1160–1184. [CrossRef]
4. Bhatt, G.; Ghatak, A.; Murugavel, R. Futuristic Storage Devices: Single Molecular Magnets of Rare Earths versus Spin Crossover Systems of Earth-Abundant Metals. *J. Chem. Sci.* **2023**, *135*, 40. [CrossRef]
5. Cardoso, R.M.F.; Esteves Da Silva, J.C.G.; Pinto Da Silva, L. Application of Engineered Nanomaterials as Nanocatalysts in Catalytic Ozonation: A Review. *Materials* **2024**, *17*, 3185. [CrossRef] [PubMed]
6. Docherty, J.H.; Peng, J.; Dominey, A.P.; Thomas, S.P. Activation and Discovery of Earth-Abundant Metal Catalysts Using Sodium Tert-Butoxide. *Nature Chem.* **2017**, *9*, 595–600. [CrossRef] [PubMed]
7. Li, J.; Wang, C.; Wang, R.; Zhang, C.; Li, G.; Davey, K.; Zhang, S.; Guo, Z. Progress and Perspectives on Iron-Based Electrode Materials for Alkali Metal-Ion Batteries: A Critical Review. *Chem. Soc. Rev.* **2024**, *53*, 4154–4229. [CrossRef] [PubMed]
8. Li, H.; Fei, L.; Zhang, R.; Yu, S.; Zhang, Y.; Shu, L.; Li, Y.; Wang, Y. FeCo Alloy Catalysts Promoting Polysulfide Conversion for Advanced Lithium–Sulfur Batteries. *J. Energy Chem.* **2020**, *49*, 339–347. [CrossRef]
9. Gąsior, G.; Szczepański, J.; Radtke, A. Biodegradable Iron-Based Materials—What Was Done and What More Can Be Done? *Materials* **2021**, *14*, 3381. [CrossRef]
10. Moravej, M.; Mantovani, D. Biodegradable Metals for Cardiovascular Stent Application: Interests and New Opportunities. *Int. J. Mol. Sci.* **2011**, *12*, 4250–4270. [CrossRef] [PubMed]
11. Mostavan, A.; Paternoster, C.; Tolouei, R.; Ghali, E.; Dubé, D.; Mantovani, D. Effect of Electrolyte Composition and Deposition Current for Fe/Fe-P Electroformed Bilayers for Biodegradable Metallic Medical Applications. *Mater. Sci. Eng. C* **2017**, *70*, 195–206. [CrossRef] [PubMed]
12. Wegener, B.; Sievers, B.; Utzschneider, S.; Müller, P.; Jansson, V.; Rößler, S.; Nies, B.; Stephani, G.; Kieback, B.; Quadbeck, P. Microstructure, Cytotoxicity, and Corrosion of Powder-Metallurgical Iron Alloys for Biodegradable Bone Replacement Materials. *Mater. Sci. Eng. B* **2011**, *176*, 1789–1796. [CrossRef]
13. Schinhammer, M.; Hänzi, A.C.; Löffler, J.F.; Uggowitzer, P.J. Design Strategy for Biodegradable Fe-Based Alloys for Medical Applications. *Acta Biomater.* **2010**, *6*, 1705–1713. [CrossRef]
14. Huang, T.; Cheng, J.; Bian, D.; Zheng, Y. Fe–Au and Fe–Ag Composites as Candidates for Biodegradable Stent Materials. *J. Biomed. Mater. Res.* **2016**, *104*, 225–240. [CrossRef] [PubMed]
15. Akbarzadeh, R.; Ghaedi, M.; Nasiri Kokhdan, S.; Jannesar, R.; Sadeghfar, F.; Sadri, F.; Tayebi, L. Electrochemical Hydrogen Storage, Photocatalytical, and Antibacterial Activity of Fe Ag Bimetallic Nanoparticles Supported on TiO_2 Nanowires. *Int. J. Hydrogen Energy* **2018**, *43*, 18316–18329. [CrossRef]
16. Sun, Y.; Zangari, G. Rational Compositional Control of Electrodeposited Ag–Fe Films. *Inorg. Chem.* **2020**, *59*, 5405–5417. [CrossRef] [PubMed]
17. Mažeika, K.; Reklaitis, J.; Nicolenco, A.; Vainoris, M.; Tsyntsaru, N.; Cesiulis, H. Magnetic State Instability of Disordered Electrodeposited Nanogranular Fe Films. *J. Magn. Magn. Mater.* **2021**, *540*, 168433. [CrossRef]
18. Khazi, I.; Mescheder, U. Micromechanical Properties of Anomalously Electrodeposited Nanocrystalline Nickel-Cobalt Alloys: A Review. *Mater. Res. Express* **2019**, *6*, 082001. [CrossRef]

19. Torabinejad, V.; Aliofkhazraei, M.; Sabour Rouhaghdam, A.; Allahyarzadeh, M.H. Tribological Behavior of Electrodeposited Ni–Fe Multilayer Coating. *Tribol. Trans.* **2017**, *60*, 923–931. [CrossRef]
20. Gamburg, Y.D.; Zangari, G. *Theory and Practice of Metal Electrodeposition*; Springer: New York, NY, USA, 2011; ISBN 978-1-4419-9668-8.
21. Shojaei, Z.; Khayati, G.R.; Darezereshki, E. Review of Electrodeposition Methods for the Preparation of High-Entropy Alloys. *Int. J. Miner. Metall. Mater.* **2022**, *29*, 1683–1696. [CrossRef]
22. Zangari, G. Electrodeposition of Alloys and Compounds in the Era of Microelectronics and Energy Conversion Technology. *Coatings* **2015**, *5*, 195–218. [CrossRef]
23. Bertero, E.; Manzano, C.V.; Pellicer, E.; Sort, J.; Ulfig, R.M.; Mischler, S.; Michler, J.; Philippe, L. 'Green' Cr(III)–Glycine Electrolyte for the Production of FeCrNi Coatings: Electrodeposition Mechanisms and Role of By-Products in Terms of Coating Composition and Microstructure. *RSC Adv.* **2019**, *9*, 25762–25775. [CrossRef] [PubMed]
24. Białostocka, A.M.; Klekotka, M.; Klekotka, U.; Kalska-Szostko, B. Tribological Properties of the FeNi Alloys Electrodeposited with and Without External Magnetic Field Assistance. *Eksploat. Niezawodn. Maintenance Reliab.* **2022**, *24*, 687–694. [CrossRef]
25. Tovar-Oliva, M.S.; Tudela, I. Electrodeposition of Nano- and Micro-Materials: Advancements in Electrocatalysts for Electrochemical Applications. *Results Eng.* **2024**, *24*, 103285. [CrossRef]
26. Li, X. Electrodeposition of Multi-Component Alloys: Thermodynamics, Kinetics and Mechanism. *Curr. Opin. Electrochem.* **2023**, *39*, 101289. [CrossRef]
27. Schlesinger, M. *Modern Electroplating*, 5th ed.; The ECS Series of Texts and Monographs; Wiley: Hoboken, NJ, USA, 2011; ISBN 978-0-470-16778-6.
28. Gonçalves, S.; Andrade, V.; Sousa, C.T.; Araújo, J.P.; Belo, J.H.; Apolinário, A. Tunable Iron–Cobalt Thin Films Grown by Electrodeposition. *Magnetochemistry* **2023**, *9*, 161. [CrossRef]
29. Ying, Y.; Wang, H.; Zheng, J.; Yu, J.; Li, W.; Qiao, L.; Cai, W.; Che, S. Preparation, Microstructure, and Magnetic Properties of Electrodeposited Nanocrystalline L10 FePt Films. *J. Supercond. Nov. Magn.* **2020**, *33*, 3563–3570. [CrossRef]
30. Annamalai, S.; Chelvane, A.J.; Mohanty, J. Enhancement of Magnetic and Surface Properties in Magneto-Pulse Electrodeposited Fe-Pd Alloy Thin Films at Various Deposition Potentials. *Mater. Res. Express* **2019**, *6*, 066110. [CrossRef]
31. Rezaei, M.; Haghshenas, D.F.; Ghorbani, M.; Dolati, A. Electrochemical Behavior of Nanostructured Fe-Pd Alloy During Electrodeposition on Different Substrates. *J. Electrochem. Sci. Technol.* **2018**, *9*, 202–211. [CrossRef]
32. Nicolenco, A.; Gómez, A.; Chen, X.-Z.; Menéndez, E.; Fornell, J.; Pané, S.; Pellicer, E.; Sort, J. Strain Gradient Mediated Magnetoelectricity in Fe-Ga/P(VDF-TrFE) Multiferroic Bilayers Integrated on Silicon. *Appl. Mater. Today* **2020**, *19*, 100579. [CrossRef]
33. Nicolenco, A.; Chen, Y.; Tsyntsaru, N.; Cesiulis, H.; Pellicer, E.; Sort, J. Mechanical, Magnetic and Magnetostrictive Properties of Porous Fe-Ga Films Prepared by Electrodeposition. *Mater. Des.* **2021**, *208*, 109915. [CrossRef]
34. Tsyntsaru, N.; Cesiulis, H.; Donten, M.; Sort, J.; Pellicer, E.; Podlaha-Murphy, E.J. Modern Trends in Tungsten Alloys Electrodeposition with Iron Group Metals. *Surf. Eng. Appl. Electrochem.* **2012**, *48*, 491–520. [CrossRef]
35. Mukhtar, A.; Wu, K. Coupled Electrodeposition and Magnetic Properties of Ternary CoFeW Alloys. *Mater. Charact.* **2022**, *192*, 112246. [CrossRef]
36. Nicolenco, A.; Tsyntsaru, N.; Fornell, J.; Pellicer, E.; Reklaitis, J.; Baltrunas, D.; Cesiulis, H.; Sort, J. Mapping of Magnetic and Mechanical Properties of Fe-W Alloys Electrodeposited from Fe(III)-Based Glycolate-Citrate Bath. *Mater. Des.* **2018**, *139*, 429–438. [CrossRef]
37. Ghaferi, Z.; Sharafi, S.; Bahrololoom, M.E. The Role of Electrolyte pH on Phase Evolution and Magnetic Properties of CoFeW Codeposited Films. *Appl. Surf. Sci.* **2016**, *375*, 35–41. [CrossRef]
38. Yermolenko, I.Y.; Ved', M.V.; Sakhnenko, N.D.; Shipkova, I.G.; Zyubanova, S.I. Nanostructured Magnetic Films Based on Iron with Refractory Metals. *J. Magn. Magn. Mater.* **2019**, *475*, 115–120. [CrossRef]
39. Cirovic, N.; Spasojevic, P.; Ribic-Zelenovic, L.; Maskovic, P.; Spasojevic, M. Synthesis, Structure and Properties of Nickel-Iron-Tungsten Alloy Electrodeposits—Part I: Effect of Synthesis Parameters on Chemical Composition, Microstructure and Morphology. *Sci. Sinter.* **2015**, *47*, 347–365. [CrossRef]
40. Cirovic, N.; Spasojevic, P.; Ribic-Zelenovic, L.; Maskovic, P.; Maricic, A.; Spasojevic, M. Synthesis, Structure and Properties of Nickel-Iron-Tungsten Alloy Electrodeposits—Part II: Effect of Microstructure on Hardness, Electrical and Magnetic Properties. *Sci. Sinter.* **2016**, *48*, 1–16. [CrossRef]
41. Candidate List of Substances of Very High Concern for Authorisation. Available online: https://echa.europa.eu/candidate-list-table?p_p_id=disslists_WAR_disslistsportlet&p_p_lifecycle=1&p_p_state=normal&p_p_mode=view&p_p_col_id=column-1&p_p_col_pos=2&p_p_col_count=3&_disslists_WAR_disslistsportlet_javax.Portlet.action=searchDissListsNo (accessed on 11 May 2024).
42. Mrkonjić Zajkoska, S.; Dobročka, E.; Hansal, S.; Mann, R.; Hansal, W.E.G.; Kautek, W. Tartrate-Based Electrolyte for Electrodeposition of Fe–Sn Alloys. *Coatings* **2019**, *9*, 313. [CrossRef]

43. Zajkoska, S.M.; Mann, R.; Hansal, W.; Roy, S.; Kautek, W. Electrodeposition of Fe-Sn from the Chloride-Based Electrolyte. *Trans. IMF* **2019**, *97*, 247–253. [CrossRef]
44. Giefers, H.; Nicol, M. High Pressure X-Ray Diffraction Study of All Fe–Sn Intermetallic Compounds and One Fe–Sn Solid Solution. *J. Alloys Compd.* **2006**, *422*, 132–144. [CrossRef]
45. García-Arribas, A.; Fdez-Gubieda, M.L.; Barandiarán, J.M. Comparative Study of the Structure and Magnetic Properties of Co-P and Fe-P Amorphous Alloys. *Phys. Rev. B* **2000**, *61*, 6238–6245. [CrossRef]
46. Liu, T.; Ji, B.; Wu, Y.; Liu, Z.; Wang, W. Effects of the pH Value on the Electrodeposition of Fe–P Alloy as a Magnetic Film Material. *J. Phys. Chem. C* **2022**, *126*, 15472–15484. [CrossRef]
47. Kovalska, N.; Tsyntsaru, N.; Cesiulis, H.; Gebert, A.; Fornell, J.; Pellicer, E.; Sort, J.; Hansal, W.; Kautek, W. Electrodeposition of Nanocrystalline Fe-P Coatings: Influence of Bath Temperature and Glycine Concentration on Structure, Mechanical and Corrosion Behavior. *Coatings* **2019**, *9*, 189. [CrossRef]
48. Kovalska, N.; Hansal, W.E.G.; Tsyntsaru, N.; Cesiulis, H.; Gebert, A.; Kautek, W. Electrodeposition and Corrosion Behaviour of Nanocrystalline Fe–P Coatings. *Trans. IMF* **2019**, *97*, 89–94. [CrossRef]
49. Naor, A.; Eliaz, N.; Gileadi, E. Electrodeposition of Alloys of Rhenium with Iron-Group Metals from Aqueous Solutions. *ECS Trans.* **2010**, *25*, 137–149. [CrossRef]
50. Li, X.; Chen, Y.; Zheng, X.; Zhu, Y.; Wang, Z.; Wang, Y. One-Step Electrodeposition of Composition-Controllable Dendritic NiFe Alloy Electrocatalysts for Oxygen Evolution Reaction. *J. Alloys Compd.* **2023**, *968*, 172313. [CrossRef]
51. Xiong, Y.; He, P. A Review on Electrocatalysis for Alkaline Oxygen Evolution Reaction (OER) by Fe-Based Catalysts. *J. Mater. Sci.* **2023**, *58*, 2041–2067. [CrossRef]
52. Kutyła, D.; Skibińska, K.; Wojtysiak, M.; Salci, A.; Kołczyk-Siedlecka, K.; Wojtaszek, K.; Wojnicki, M.; Żabiński, P.; Solmaz, R. Electrochemical Preparation and Alkaline Water Splitting Activity of Ternary Co–Fe–Mo Non-Crystalline Coatings. *Int. J. Hydrogen Energy* **2024**, *51*, 1450–1459. [CrossRef]
53. Ni, R.; Xu, W.; Wang, C.; Man, Z.; Cheng, X. The Application of Iron-Manganese Compounds towards Micropollutants Purification in Water: A Critical Review. *Chem. Eng. J.* **2024**, *493*, 152553. [CrossRef]
54. Xia, Q.; Jiang, Z.; Wang, J.; Yao, Z. A Facile Preparation of Hierarchical Dendritic Zero-Valent Iron for Fenton-like Degradation of Phenol. *Catal. Commun.* **2017**, *100*, 57–61. [CrossRef]
55. Vainoris, M.; Nicolenco, A.; Tsyntsaru, N.; Podlaha-Murphy, E.; Alcaide, F.; Cesiulis, H. Electrodeposited Fe on Cu Foam as Advanced Fenton Reagent for Catalytic Mineralization of Methyl Orange. *Front. Chem.* **2022**, *10*, 977980. [CrossRef] [PubMed]
56. Theofanidis, S.A.; Galvita, V.V.; Konstantopoulos, C.; Poelman, H.; Marin, G.B. Fe-Based Nano-Materials in Catalysis. *Materials* **2018**, *11*, 831. [CrossRef] [PubMed]
57. Protsenko, V.S.; Danilov, F.I. Current Trends in Electrodeposition of Electrocatalytic Coatings. In *Methods for Electrocatalysis*; Inamuddin, Boddula, R., Asiri, A.M., Eds.; Springer International Publishing: Cham, Switzerland, 2020; pp. 263–299, ISBN 978-3-030-27160-2.
58. Mulone, A.; Nicolenco, A.; Imaz, N.; Fornell, J.; Sort, J.; Klement, U. Effect of Heat Treatments on the Mechanical and Tribological Properties of Electrodeposited Fe–W/Al_2O_3 Composites. *Wear* **2020**, *448–449*, 203232. [CrossRef]
59. Nicolenco, A.; Mulone, A.; Imaz, N.; Tsyntsaru, N.; Sort, J.; Pellicer, E.; Klement, U.; Cesiulis, H.; García-Lecina, E. Nanocrystalline Electrodeposited Fe-W/Al_2O_3 Composites: Effect of Alumina Sub-Microparticles on the Mechanical, Tribological, and Corrosion Properties. *Front. Chem.* **2019**, *7*, 241. [CrossRef] [PubMed]
60. Starosta, R.; Zielinski, A. Effect of Chemical Composition on Corrosion and Wear Behaviour of the Composite Ni–Fe–Al_2O_3 Coatings. *J. Mater. Process. Technol.* **2004**, *157–158*, 434–441. [CrossRef]
61. Jin, H.; Ji, R.; Dong, T.; Liu, S.; Zhang, F.; Zhao, L.; Ma, C.; Cai, B.; Liu, Y. Efficient Fabrication and Characterization of Ni-Fe-WC Composite Coatings with High Corrosion Resistance. *J. Mater. Res. Technol.* **2022**, *16*, 152–167. [CrossRef]
62. Tsyntsaru, N.; Cesiulis, H.; Budreika, A.; Ye, X.; Juskenas, R.; Celis, J.-P. The Effect of Electrodeposition Conditions and Post-Annealing on Nanostructure of Co–W Coatings. *Surf. Coat. Technol.* **2012**, *206*, 4262–4269. [CrossRef]
63. Tsyntsaru, N.; Bobanova, J.; Ye, X.; Cesiulis, H.; Dikusar, A.; Prosycevas, I.; Celis, J.-P. Iron–Tungsten Alloys Electrodeposited under Direct Current from Citrate–Ammonia Plating Baths. *Surf. Coat. Technol.* **2009**, *203*, 3136–3141. [CrossRef]
64. Wang, S.; Zeng, C.; Ling, Y.; Wang, J.; Xu, G. Phase Transformations and Electrochemical Characterizations of Electrodeposited Amorphous Fe–W Coatings. *Surf. Coat. Technol.* **2016**, *286*, 36–41. [CrossRef]
65. Wodarz, S.; Hashimoto, S.; Kambe, M.; Zangari, G.; Homma, T. Templated Electrochemical Synthesis of Fe–Pt Nanopatterns for High-Density Memory Applications. *ACS Appl. Nano Mater.* **2018**, *1*, 2317–2323. [CrossRef]
66. Eliaz, N.; Gileadi, E. Induced Codeposition of Alloys of Tungsten, Molybdenum and Rhenium with Transition Metals. In *Modern Aspects of Electrochemistry*; Vayenas, C.G., White, R.E., Gamboa-Aldeco, M.E., Eds.; Springer: New York, NY, USA, 2008; Volume 42, pp. 191–301, ISBN 978-0-387-49488-3.
67. Survila, A. *Electrochemistry of Metal Complexes: Applications from Electroplating to Oxide Layer Formation*; Wiley: Weinheim, Germany, 2015; ISBN 978-3-527-69125-8.

68. Heusler, K.E. Multicomponent Electrodes. *Electrochim. Acta* **1996**, *41*, 411–418. [CrossRef]
69. Nicolenco, A.; Tsyntsaru, N.; Cesiulis, H. Fe (III)-Based Ammonia-Free Bath for Electrodeposition of Fe-W Alloys. *J. Electrochem. Soc.* **2017**, *164*, D590–D596. [CrossRef]
70. Mulone, A.; Nicolenco, A.; Imaz, N.; Martinez-Nogues, V.; Tsyntsaru, N.; Cesiulis, H.; Klement, U. Improvement in the Wear Resistance under Dry Friction of Electrodeposited Fe-W Coatings through Heat Treatments. *Coatings* **2019**, *9*, 66. [CrossRef]
71. Cesiulis, H.; Baltutiene, A.; Donten, M.; Donten, M.; Stojek, Z. Increase in Rate of Electrodeposition and in Ni(II) Concentration in the Bath as a Way to Control Grain Size of Amorphous/Nanocrystalline Ni-W Alloys. *J. Solid State Electrochem.* **2002**, *6*, 237–244. [CrossRef]
72. Kovalska, N.; Pfaffeneder-Kmen, M.; Tsyntsaru, N.; Mann, R.; Cesiulis, H.; Hansal, W.; Kautek, W. The Role of Glycine in the Iron-Phosphorous Alloy Electrodeposition. *Electrochim. Acta* **2019**, *309*, 450–459. [CrossRef]
73. Mehrizi, S.; Sohi, M.H.; Saremi, M. Effect of Sodium Citrate as Complexing on Electrochemical Behavior and Speciation Diagrams of CoFeNiCu Baths. *Ionics* **2013**, *19*, 911–918. [CrossRef]
74. Fortas, G.; Ouir, S.; Manseri, A.; Djerir, W.; Smaili, F.; Gabouze, N.; Haine, N. Influence of Deposition Parameters on the Electrodeposited Ternary CoFeCu Coatings. *J. Mater. Sci. Mater. Electron.* **2024**, *35*, 1462. [CrossRef]
75. Singh, A.K.; Zhou, L.; Shinde, A.; Suram, S.K.; Montoya, J.H.; Winston, D.; Gregoire, J.M.; Persson, K.A. Electrochemical Stability of Metastable Materials. *Chem. Mater.* **2017**, *29*, 10159–10167. [CrossRef]
76. Brenner, A. *Electrodeposition of Alloys: Principles and Practice*; Academic Press: New York, NY, USA, 1963; ISBN 978-1-4832-2311-7.
77. Podlaha, E.J.; Landolt, D. Induced Codeposition: I. An Experimental Investigation of Ni-Mo Alloys. *J. Electrochem. Soc.* **1996**, *143*, 885–892. [CrossRef]
78. Zech, N.; Podlaha, E.J.; Landolt, D. Anomalous Codeposition of Iron Group Metals: I. Experimental Results. *J. Electrochem. Soc.* **1999**, *146*, 2886–2891. [CrossRef]
79. Greco, V.P. *Rhenium Alloys—Iron Group Metals (Electrodeposition and Properties)*; Defense Technical Information Center: Fort Belvoir, VA, USA, 1971.
80. Zhulikov, V.V.; Gamburg, Y.D. Electrodeposition of Rhenium and Its Alloys. *Russ. J. Electrochem.* **2016**, *52*, 847–857. [CrossRef]
81. Electrochemical Series. In *Corrosion: Materials*; Cramer, S.D., Covino, B.S., Eds.; ASM International: Almere, The Netherlands, 2005; pp. 665–671, ISBN 978-1-62708-183-2.
82. Vajo, J.J.; Aikens, D.A.; Ashley, L.; Poeltl, D.E.; Bailey, R.A.; Clark, H.M.; Bunce, S.C. Facile Electroreduction of Perrhenate in Weakly Acidic Citrate and Oxalate Media. *Inorg. Chem.* **1981**, *20*, 3328–3333. [CrossRef]
83. Contu, F.; Taylor, S.R. Further Insight into the Mechanism of Re–Ni Electrodeposition from Concentrated Aqueous Citrate Baths. *Electrochim. Acta* **2012**, *70*, 34–41. [CrossRef]
84. Vargas-Uscategui, A.; Mosquera, E.; Cifuentes, L. Analysis of the Electrodeposition Process of Rhenium and Rhenium Oxides in Alkaline Aqueous Electrolyte. *Electrochim. Acta* **2013**, *109*, 283–290. [CrossRef]
85. Baranov, S.A.; Dikusar, A.I. Kinetics of Electrochemical Nanonucleation during Induced Codeposition of Iron-Group Metals with Refractory Metals (W, Mo, Re). *Surf. Eng. Appl. Electrochem.* **2022**, *58*, 429–439. [CrossRef]
86. Torabinejad, V.; Aliofkhazraei, M.; Assareh, S.; Allahyarzadeh, M.H.; Rouhaghdam, A.S. Electrodeposition of Ni-Fe Alloys, Composites, and Nano Coatings—A Review. *J. Alloys Compd.* **2017**, *691*, 841–859. [CrossRef]
87. Park, D.-Y.; Yoo, B.Y.; Kelcher, S.; Myung, N.V. Electrodeposition of Low-Stress High Magnetic Moment Fe-Rich FeCoNi Thin Films. *Electrochim. Acta* **2006**, *51*, 2523–2530. [CrossRef]
88. Arai, S.; Tomiita, K.; Shimizu, M.; Narita, H. Electrodeposition of Fe-Ni Alloy Films Having Invar Compositions with Fe^{3+} as the Sole Iron Source. *J. Electrochem. Soc.* **2023**, *170*, 112506. [CrossRef]
89. Liang, D.; Mallett, J.J.; Zangari, G. Electrodeposition of Fe−Pt Films with Low Oxide Content Using an Alkaline Complexing Electrolyte. *ACS Appl. Mater. Interfaces* **2010**, *2*, 961–964. [CrossRef]
90. Liang, D.; Zangari, G. Fe–Pt Magnetic Multilayers by Electrochemical Deposition. *Electrochim. Acta* **2011**, *56*, 10567–10574. [CrossRef]
91. Yar-Mukhamedova, G.; Ved, M.; Sakhnenko, N.; Karakurkchi, A.; Yermolenko, I. Iron Binary and Ternary Coatings with Molybdenum and Tungsten. *Appl. Surf. Sci.* **2016**, *383*, 346–352. [CrossRef]
92. Ramaprakash, M.; Jerom Samraj, A.; Neelavannan, M.G.; Rajasekaran, N. The Induced Co-Deposition of Ni–Mo–W Ternary Alloy; Coatings for Hardness and Corrosion Resistance Applications. *Results Surf. Interfaces* **2024**, *15*, 100235. [CrossRef]
93. He, F.; Yang, J.; Lei, T.; Gu, C. Structure and Properties of Electrodeposited Fe–Ni–W Alloys with Different Levels of Tungsten Content: A Comparative Study. *Appl. Surf. Sci.* **2007**, *253*, 7591–7598. [CrossRef]
94. Wang, S.; Ling, Y.; Zhang, J.; Wang, J.; Xu, G. Microstructure and Properties of Hydrophobic Films Derived from Fe-W Amorphous Alloy. *Int. J. Miner. Metall. Mater.* **2014**, *21*, 395–400. [CrossRef]
95. Dobrzańska-Danikiewicz, A.D.; Wolany, W. A Rhenium Review—From Discovery to Novel Applications. *Arch. Mater. Sci. Eng.* **2016**, *82*, 70–78. [CrossRef]
96. Jones, T. Rhenium Plating. *Met. Finishing* **2003**, *101*, 86–96. [CrossRef]

97. Tsyntsaru, N.; Cesiulis, H.; Pellicer, E.; Celis, J.-P.; Sort, J. Structural, Magnetic, and Mechanical Properties of Electrodeposited Cobalt–Tungsten Alloys: Intrinsic and Extrinsic Interdependencies. *Electrochim. Acta* **2013**, *104*, 94–103. [CrossRef]
98. Sriraman, K.R.; Ganesh Sundara Raman, S.; Seshadri, S.K. Synthesis and Evaluation of Hardness and Sliding Wear Resistance of Electrodeposited Nanocrystalline Ni–W Alloys. *Mater. Sci. Eng. A* **2006**, *418*, 303–311. [CrossRef]
99. Haché, M.J.R.; Tam, J.; Erb, U.; Zou, Y. Electrodeposited NiFeCo-(Mo,W) High-Entropy Alloys with Nanocrystalline and Amorphous Structures. *J. Alloys Compd.* **2023**, *952*, 170026. [CrossRef]
100. Vernickaite, E.; Tsyntsaru, N.; Sobczak, K.; Cesiulis, H. Electrodeposited Tungsten-Rich Ni-W, Co-W and Fe-W Cathodes for Efficient Hydrogen Evolution in Alkaline Medium. *Electrochim. Acta* **2019**, *318*, 597–606. [CrossRef]
101. Silkin, S.A.; Gotelyak, A.V.; Tsyntsaru, N.I.; Dikusar, A.I. Electrodeposition of Alloys of the Iron Group Metals with Tungsten from Citrate and Gluconate Solutions: Size Effect of Microhardness. *Surf. Eng. Appl. Electrochem.* **2017**, *53*, 7–14. [CrossRef]
102. Belevskii, S.S.; Gotelyak, A.V.; Silkin, S.A.; Dikusar, A.I. Macroscopic Size Effect on the Microhardness of Electroplated Iron Group Metal—Tungsten Alloy Coatings: Impact of Electrode Potential and Oxygen-Containing Impurities. *Surf. Eng. Appl. Electrochem.* **2019**, *55*, 46–52. [CrossRef]
103. Belevskii, S.S.; Danilchuk, V.V.; Gotelyak, A.V.; Lelis, M.; Yushchenko, S.P.; Dikusar, A.I. Electrodeposition of Fe–W Alloys from Citrate Bath: Impact of Anode Material. *Surf. Eng. Appl. Electrochem.* **2020**, *56*, 1–12. [CrossRef]
104. Myrzak, V.; Gotelyak, A.V.; Dikusar, A.I. Size Effects in the Surface Properties of Electroplated Alloys between Iron Group Metals and Tungsten. *Surf. Eng. Appl. Electrochem.* **2021**, *57*, 409–418. [CrossRef]
105. Gamburg, Y.D.; Zakharov, E.N. Electrodeposition of Ternary Fe–W–H Alloys. *Surf. Eng. Appl. Electrochem.* **2019**, *55*, 402–409. [CrossRef]
106. Belevskii, S.S.; Bobanova, Z.I.; Buravets, V.A.; Gotelyak, A.V.; Danil'chuk, V.V.; Silkin, S.A.; Dikusar, A.I. Electrodeposition of Co–W Coatings from Boron Gluconate Electrolyte with a Soluble Tungsten Anode. *Russ. J. Appl. Chem.* **2016**, *89*, 1427–1433. [CrossRef]
107. Ludwig, P.; Kac, M.; Kopia, A.; Falkus, J.; Dubiel, B. Microstructure and Properties of Electrodeposited Nanocrystalline Ni-Co-Fe Coatings. *Materials* **2021**, *14*, 3886. [CrossRef] [PubMed]
108. Chou, M.-C.; Chu, C.-F.; Wu, S.-T. Phase Transformations of Electroplated Amorphous Iron–Tungsten–Carbon Film. *Mater. Chem. Phys.* **2003**, *78*, 59–66. [CrossRef]
109. Mulone, A.; Nicolenco, A.; Hoffmann, V.; Klement, U.; Tsyntsaru, N.; Cesiulis, H. In-Depth Characterization of as-Deposited and Annealed Fe-W Coatings Electrodeposited from Glycolate-Citrate Plating Bath. *Electrochim. Acta* **2018**, *261*, 167–177. [CrossRef]
110. Park, J.-H.; Kim, J.-I.; Shinohara, Y.; Hagio, T.; Umehara, N.; Ichino, R. Super Hardening of Fe W Alloy Plating by Phase Transformation of Amorphous to Metal Carbides-Dispersed Nanocrystalline Alloys and Application as Promising Alternative for Hard Chromium Plating. *Surf. Coat. Technol.* **2024**, *477*, 130388. [CrossRef]
111. Köse, M.; Tan, S.; Algül, H.; Alp, A.; Akbulut, H.; Uysal, M. Effect of Different Heat Treatment Temperatures on Fe W Alloy Electrodeposits: Tribological and Electrochemical Analysis. *Surf. Coat. Technol.* **2024**, *485*, 130852. [CrossRef]
112. Marvel, C.J.; Cantwell, P.R.; Harmer, M.P. The Critical Influence of Carbon on the Thermal Stability of Nanocrystalline Ni–W Alloys. *Scripta Mater.* **2015**, *96*, 45–48. [CrossRef]
113. Marvel, C.J.; Yin, D.; Cantwell, P.R.; Harmer, M.P. The Influence of Oxygen Contamination on the Thermal Stability and Hardness of Nanocrystalline Ni–W Alloys. *Mater. Sci. Eng. A* **2016**, *664*, 49–57. [CrossRef]
114. Mulone, A.; Nicolenco, A.; Fornell, J.; Pellicer, E.; Tsyntsaru, N.; Cesiulis, H.; Sort, J.; Klement, U. Enhanced Mechanical Properties and Microstructural Modifications in Electrodeposited Fe-W Alloys through Controlled Heat Treatments. *Surf. Coat. Technol.* **2018**, *350*, 20–30. [CrossRef]
115. Herzer, G. Grain Size Dependence of Coercivity and Permeability in Nanocrystalline Ferromagnets. *IEEE Trans. Magn.* **1990**, *26*, 1397–1402. [CrossRef]
116. Zhang, Y.; Ivey, D.G. Electrodeposition of Nanocrystalline CoFe Soft Magnetic Thin Films from Citrate-Stabilized Baths. *Mater. Chem. Phys.* **2018**, *204*, 171–178. [CrossRef]
117. Spasojevic, M.; Plazinic, M.; Lukovic, M.; Maricic, A.; Spasojevic, M. The Effect of Annealing and Frequency of the External Magnetic Field on Magnetic Properties of Nanostructured Electrodeposit of the Ni86,0Fe9,8W1,3Cu2,9 Alloy. *Mater. Chem. Phys.* **2020**, *254*, 123513. [CrossRef]
118. Mehrizi, S.; Molaei, M.J.; Sohi, M.H. An Investigation on Magnetic Properties and Electrical Resistivity of Nanocrystalline CoFeNi Thick Films Synthesized through Stabilized Bath. *J. Mater. Res. Technol.* **2022**, *21*, 2547–2554. [CrossRef]
119. Hernández, S.C.; Yoo, B.Y.; Stefanescu, E.; Khizroev, S.; Myung, N.V. Electrodeposition of Iron–Palladium Thin Films. *Electrochim. Acta* **2008**, *53*, 5621–5627. [CrossRef]
120. Faltas, M.; Pillars, J.; Soule, L.; Meyerson, M.L.; Rodriguez, M.A.; Valdez, N.R.; Oglesby, S.; Jackson, N.; El-Kady, I. Electrodeposited NiFeCo + Tb and Dy for Enhanced Magnetostrictive Properties and Soft Magnetism. *Thin Solid Films* **2024**, *800*, 140396. [CrossRef]

121. Wang, Q.; Li, M.; Weng, L.; Huang, W. Dual-Mode Sensing Based on Oriented Magnetostrictive Films by Electrodeposition. *IEEE Sensors J.* **2024**, *24*, 23667–23675. [CrossRef]
122. Barbano, E.P.; Carlos, I.A.; Vallés, E. Electrochemical Synthesis of Fe-W and Fe-W-P Magnetic Amorphous Films and Fe-W Nanowires. *Surf. Coat. Technol.* **2017**, *324*, 80–84. [CrossRef]
123. Ren, Q.Q.; Fan, J.L.; Han, Y.; Gong, H.R. Structural, Thermodynamic, Mechanical, and Magnetic Properties of FeW System. *J. Appl. Phys.* **2014**, *116*, 093909. [CrossRef]
124. Mundotiya, B.M.; Dinulovic, D.; Rissing, L.; Wurz, M.C. Fabrication and Characterization of a Ni-Fe-W Core Microtransformer for High-Frequency Power Applications. *Sens. Actuators A Phys.* **2017**, *267*, 42–47. [CrossRef]
125. Kanakaraj, M.; Ramalingam, H.B. Effect of Electrolytic Bath Temperature on Magnetic and Structural Properties of Electrodeposited Ni Fe W Nanocrystalline Thin Films. *Orient. J. Chem.* **2017**, *33*, 2899–2904. [CrossRef]
126. Ved', M.; Yermolenko, I.; Sachanova, Y.; Sakhnenko, N. Refractory Metals Influence on the Properties of Fe-Co-Mo(W) Electrolytic Alloys. *Mater. Today Proc.* **2019**, *6*, 121–128. [CrossRef]
127. Kannan, R.; Ganesan, S.; Selvakumari, T.M. An Investigation on Effects of Annealing on Magnetic Properties of Ni-Fe-W-S Electrodeposited Coatings in Tri Sodium Citrate Bath. *Trans. A Sci.* **2013**, *37*, 181–187. [CrossRef]
128. Restriction Proposal on Chromium (VI) to Cover More Substances. Available online: https://echa.europa.eu/-/restriction-proposal-on-chromium-vi-to-cover-more-substances (accessed on 12 September 2024).
129. Weston, D.P.; Harris, S.J.; Shipway, P.H.; Weston, N.J.; Yap, G.N. Establishing Relationships between Bath Chemistry, Electrodeposition, and Microstructure of Co–W Alloy Coatings Produced from a Gluconate Bath. *Electrochim. Acta* **2010**, *55*, 5695–5708. [CrossRef]
130. Leyland, A.; Matthews, A. On the Significance of the H/E Ratio in Wear Control: A Nanocomposite Coating Approach to Optimized Tribological Behaviour. *Wear* **2000**, *246*, 1–11. [CrossRef]
131. Rupert, T.J.; Schuh, C.A. Sliding Wear of Nanocrystalline Ni–W: Structural Evolution and the Apparent Breakdown of Archard Scaling. *Acta Mater.* **2010**, *58*, 4137–4148. [CrossRef]
132. Su, F.; Huang, P. Microstructure and Tribological Property of Nanocrystalline Co–W Alloy Coating Produced by Dual-Pulse Electrodeposition. *Mater. Chem. Phys.* **2012**, *134*, 350–359. [CrossRef]
133. Bobanova, Z.I.; Dikusar, A.I.; Cesiulis, H.; Celis, J.-P.; Tsyntsaru, N.I.; Prosycevas, I. Micromechanical and Tribological Properties of Nanocrystalline Coatings of Iron-Tungsten Alloys Electrodeposited from Citrate-Ammonia Solutions. *Russ. J. Electrochem.* **2009**, *45*, 895–901. [CrossRef]
134. Park, J.-H.; Hagio, T.; Kamimoto, Y.; Ichino, R. The Effect of Bath pH on Electrodeposition and Corrosion Properties of Ternary Fe-W-Zn Alloy Platings. *J. Solid State Electrochem.* **2021**, *25*, 1901–1913. [CrossRef]
135. Zouch, F.; Antar, Z.; Bahri, A.; Elleuch, K.; Ürgen, M. Tribological Study of Fe–W–P Electrodeposited Coating on 316 L Stainless Steel. *J. Tribol.* **2018**, *140*, 011301. [CrossRef]
136. Tian, L.; Li, J.; Xing, H.; Yue, L.; Li, Z.; Wang, Y. Study of the Process of Preparing Amorphous Fe–W(La) Alloy Plating by Induced Co-Deposition. *Mat. Express* **2023**, *13*, 1764–1771. [CrossRef]
137. Nicolenco, A.; Tsyntsaru, N.; Matijošius, T.; Asadauskas, S.; Cesiulis, H. Wear Resistance of Electrodeposited Fe-W Alloy Coatings under Dry Conditions and in the Presence of Rapeseed Oil. *Green Tribol.* **2018**, *1*, 16–23. [CrossRef]
138. Bajwa, R.S.; Khan, Z.; Bakolas, V.; Braun, W. Water-Lubricated Ni-Based Composite (Ni–Al_2O_3, Ni–SiC and Ni–ZrO_2) Thin Film Coatings for Industrial Applications. *Acta Metall. Sin.* **2016**, *29*, 8–16. [CrossRef]
139. Liang, J.; Wu, X.-W.; Ling, Y.; Yu, S.; Zhang, Z. Trilaminar Structure Hydrophobic Graphene Oxide Decorated Organosilane Composite Coatings for Corrosion Protection. *Surf. Coat. Technol.* **2018**, *339*, 65–77. [CrossRef]
140. Vainoris, M.; Cesiulis, H.; Tsyntsaru, N. Metal Foam Electrode as a Cathode for Copper Electrowinning. *Coatings* **2020**, *10*, 822. [CrossRef]
141. Herraiz-Cardona, I.; Ortega, E.; Antón, J.G.; Pérez-Herranz, V. Assessment of the Roughness Factor Effect and the Intrinsic Catalytic Activity for Hydrogen Evolution Reaction on Ni-Based Electrodeposits. *Int. J. Hydrogen Energy* **2011**, *36*, 9428–9438. [CrossRef]
142. Vainoris, M.; Tsyntsaru, N.; Cesiulis, H. Modified Electrodeposited Cobalt Foam Coatings as Sensors for Detection of Free Chlorine in Water. *Coatings* **2019**, *9*, 306. [CrossRef]
143. Zhu, L.; Zhang, R.; Lv, W.; Wei, M.; Wang, W. One-Pot Hydrothermal Synthesis of Fe, W Co-Doped Ni_3S_2/NiS on Ni Foam for Bifunctional Oxygen Evolution and Urea Oxidation Reactions. *Int. J. Electrochem. Sci.* **2022**, *17*, 221043. [CrossRef]
144. Wang, S.; Yuan, D.; Sun, S.; Huang, S.; Wu, Y.; Zhang, L.; Dou, S.X.; Liu, H.K.; Dou, Y.; Xu, J. Iron, Tungsten Dual-Doped Nickel Sulfide as Efficient Bifunctional Catalyst for Overall Water Splitting. *Small* **2024**, *25*, 2311770. [CrossRef] [PubMed]
145. Lačnjevac, Č.; Jaksic, M.M. Synergetic Electrocatalytic Effects of D-Metals on the Hydrogen Evolution Reaction in Industrially Important Electrochemical Processes. *J. Res. Inst. Catal. Hokkaido Univ.* **1983**, *31*, 7–33.
146. Tharamani, C.N.; Beera, P.; Jayaram, V.; Begum, N.S.; Mayanna, S.M. Studies on Electrodeposition of Fe–W Alloys for Fuel Cell Applications. *Appl. Surf. Sci.* **2006**, *253*, 2031–2037. [CrossRef]

147. Bersirova, O.L.; Bilyk, S.V.; Kublanovs'kyi, V.S. Electrochemical Synthesis of Fe–W Nanostructural Electrocatalytic Coatings. *Mater. Sci.* **2018**, *53*, 732–738. [CrossRef]
148. Zhang, P.; Tan, W.; He, H.; Fu, Z. Binder-Free Quaternary Ni–Fe–W–Mo Alloy as a Highly Efficient Electrocatalyst for Oxygen Evolution Reaction. *J. Alloys Compd.* **2021**, *853*, 157265. [CrossRef]
149. Bachvarov, V.; Arnaudova, M.; Lefterova, E.; Rashkov, R. Study of Oxygen Evolution Reaction on Iron Group-Based Electrodeposited Multicomponent Catalysts in Alkaline Media. Part I: Influence of the Composition. *J. Chem. Technol. Metall.* **2022**, *57*, 910–918.
150. Zhang, J.; Ling, Y.; Gao, W.; Wang, S.; Li, J. Enhanced Photoelectrochemical Water Splitting on Novel Nanoflake WO3 Electrodes by Dealloying of Amorphous Fe–W Alloys. *J. Mater. Chem. A* **2013**, *1*, 10677. [CrossRef]
151. Ved', M.; Sakhnenko, N.; Nenastina, T.; Volobuyev, M.; Yermolenko, I. Corrosion and Mechanical Properties of Nanostructure Electrolytic Co-W and Fe-Co-W Alloys. *Mater. Today Proc.* **2022**, *50*, 463–469. [CrossRef]
152. Stojić, D.L.; Cekić, B.D.; Maksić, A.D.; Kaninski, M.P.M.; Miljanić, Š.S. Intermetallics as Cathode Materials in the Electrolytic Hydrogen Production. *Int. J. Hydrogen Energy* **2005**, *30*, 21–28. [CrossRef]
153. Shen, Y.; Wu, P.; Wang, C.; Yuan, W.; Yang, W.; Shang, X. Electrodeposition of Amorphous Ni–Fe–Mo Composite as a Binder-Free and High-Performance Electrocatalyst for Hydrogen Generation from Alkaline Water Electrolysis. *Int. J. Hydrogen Energy* **2023**, *48*, 33130–33138. [CrossRef]
154. Zhou, D.; Li, P.; Xu, W.; Jawaid, S.; Mohammed-Ibrahim, J.; Liu, W.; Kuang, Y.; Sun, X. Recent Advances in Non-Precious Metal-Based Electrodes for Alkaline Water Electrolysis. *ChemNanoMat* **2020**, *6*, 336–355. [CrossRef]
155. Vernickaitė, E.; Bersirova, O.; Cesiulis, H.; Tsyntsaru, N. Design of Highly Active Electrodes for Hydrogen Evolution Reaction Based on Mo-Rich Alloys Electrodeposited from Ammonium Acetate Bath. *Coatings* **2019**, *9*, 85. [CrossRef]
156. Bachvarov, V.; Lefterova, E.; Rashkov, R. Electrodeposited NiFeCo and NiFeCoP Alloy Cathodes for Hydrogen Evolution Reaction in Alkaline Medium. *Int. J. Hydrogen Energy* **2016**, *41*, 12762–12771. [CrossRef]
157. Huang, Y.; Wu, Y.; Zhang, Z.; Yang, L.; Zang, Q. Rapid Electrodeposited of Self-Supporting Ni-Fe-Mo Film on Ni Foam as Affordable Electrocatalysts for Oxygen Evolution Reaction. *Electrochimica Acta* **2021**, *390*, 138754. [CrossRef]
158. Huang, X.; Liu, K.; Vrijburg, W.L.; Ouyang, X.; Iulian Dugulan, A.; Liu, Y.; Tiny Verhoeven, M.W.G.M.; Kosinov, N.A.; Pidko, E.A.; Hensen, E.J.M. Hydrogenation of Levulinic Acid to γ-Valerolactone over Fe-Re/TiO_2 Catalysts. *Appl. Catal. B Environ.* **2020**, *278*, 119314. [CrossRef]
159. Carroll, Z.L.; Haché, M.J.R.; Wang, B.; Chen, L.; Wu, S.; Erb, U.; Thorpe, S.; Zou, Y. Electrodeposited NiFeCoMoW High-Entropy Alloys with Nanoscale Amorphous Structure as Effective Hydrogen Evolution Electrocatalysts. *ACS Appl. Energy Mater.* **2024**, *7*, 8412–8422. [CrossRef]
160. Wu, K.; Liu, C.; Li, Q.; Huo, J.; Li, M.; Chang, C.; Sun, Y. Magnetocaloric Effect of Fe25Co25Ni25Mo5P10B10 High-Entropy Bulk Metallic Glass. *J. Magn. Magn. Mater.* **2019**, *489*, 165404. [CrossRef]
161. Lee, C.; Song, G.; Gao, M.C.; Feng, R.; Chen, P.; Brechtl, J.; Chen, Y.; An, K.; Guo, W.; Poplawsky, J.D.; et al. Lattice Distortion in a Strong and Ductile Refractory High-Entropy Alloy. *Acta Mater.* **2018**, *160*, 158–172. [CrossRef]
162. George, E.P.; Raabe, D.; Ritchie, R.O. High-Entropy Alloys. *Nat. Rev. Mater.* **2019**, *4*, 515–534. [CrossRef]
163. Bian, H.; Wang, R.; Zhang, K.; Zheng, H.; Wen, M.; Li, Z.; Li, Z.; Wang, G.; Xie, G.; Liu, X.; et al. Facile Electrodeposition Synthesis and Super Performance of Nano-Porous Ni-Fe-Cu-Co-W High Entropy Alloy Electrocatalyst. *Surf. Coat. Technol.* **2023**, *459*, 129407. [CrossRef]
164. Dehestani, M.; Sharafi, S.; Khayati, G.R. The Effect of Pulse Current Density on the Microstructure, Magnetic, Mechanical, and Corrosion Properties of High-Entropy Alloy Coating Fe–Co–Ni–Mo–W, Achieved through Electro Co-Deposition. *Intermetallics* **2022**, *147*, 107610. [CrossRef]
165. Temiz, M.; Kara, A.Y.G.; Erdemir, D.; Dincer, I. More Efficient Way of Clean Hydrogen Production: The Synergetic Roles of Magnetic Effects and Effective Catalysts. *Fuel* **2024**, *376*, 132708. [CrossRef]
166. Xiong, Z.; Ning, X.; Zhou, F.; Yang, B.; Tu, Y.; Jin, J.; Lu, W.; Liu, Z. Environment-Friendly Magnetic Fe–Ce–W Catalyst for the Selective Catalytic Reduction of NOx with NH_3: Influence of Citric Acid Content on Its Activity-Structure Relationship. *RSC Adv.* **2018**, *8*, 21915–21925. [CrossRef] [PubMed]
167. Xiao, S.; Ma, L.; An, W.; Shen, P. Stability Optimization of Iron-Based Catalysts: Application of Cr-Enriched Alloy Steel Catalyst in Ozonation Pharmaceutical Wastewater. *Sep. Purif. Technol.* **2025**, *354*, 129007. [CrossRef]
168. Humphreys, J.; Lan, R.; Tao, S. Development and Recent Progress on Ammonia Synthesis Catalysts for Haber–Bosch Process. *Adv. Energy Sustain. Res.* **2021**, *2*, 2000043. [CrossRef]
169. Liu, H. Ammonia Synthesis Catalyst 100 Years: Practice, Enlightenment and Challenge. *Chin. J. Catal.* **2014**, *35*, 1619–1640. [CrossRef]
170. Fan, H.; Ma, Y.; Chen, W.; Tang, Y.; Li, L.; Wang, J. Facile One-Step Electrodeposition of Two-Dimensional Nickel-Iron Bimetallic Sulfides for Efficient Electrocatalytic Oxygen Evolution. *J. Alloys Compd.* **2022**, *894*, 162533. [CrossRef]

171. Kırlıoğlu, A.C.; Ölmez, B.; Rahbarshendi, F.; Buldu-Akturk, M.; Yürüm, A.; Alkan Gürsel, S.; Yarar Kaplan, B. Scalable Nano-Sized Fe-N-C Catalysts for Fuel Cells: Evaluating the Impact of Iron Precursors and CeO_2 Addition. *Mater. Res. Bull.* **2024**, *179*, 112952. [CrossRef]
172. Zhao, L.; Wang, Y.; Jin, S.; An, N.; Yan, M.; Zhang, X.; Hong, Z.; Yang, S. Rational Electrochemical Design of Hierarchical Microarchitectures for SERS Sensing Applications. *Nat. Synth.* **2024**, *3*, 867–877. [CrossRef]
173. Jamal, A.; Haq, G.-U.; Hussain, S.; Gul, M.; Saifullah, M.; Anjum, M.A.R. Enhanced Tribological Properties of Electrodeposited Fe–W Alloy Coatings through Carburization. *Langmuir* **2023**, *39*, 16328–16335. [CrossRef] [PubMed]
174. Li, H.; Zhang, C.; Xiang, W.; Amin, M.A.; Na, J.; Wang, S.; Yu, J.; Yamauchi, Y. Efficient Electrocatalysis for Oxygen Evolution: W-Doped NiFe Nanosheets with Oxygen Vacancies Constructed by Facile Electrodeposition and Corrosion. *Chem. Eng. J.* **2023**, *452*, 139104. [CrossRef]
175. Rajendiran, R.; Chinnadurai, D.; Chen, K.; Selvaraj, A.R.; Prabakar, K.; Li, O.L. Electrodeposited Trimetallic NiFeW Hydroxide Electrocatalysts for Efficient Water Oxidation. *ChemSusChem* **2021**, *14*, 1324–1335. [CrossRef] [PubMed]
176. Capelli, S.; Cattaneo, S.; Stucchi, M.; Villa, A.; Prati, L. Iron as Modifier of Pd and Pt-Based Catalysts for Sustainable and Green Processes. *Inorganica Chim. Acta* **2022**, *535*, 120856. [CrossRef]
177. Hosseini, J.; Abdolmaleki, M.; Pouretedal, H.R.; Keshavarz, M.H. Electrocatalytic Activity of Porous Nanostructured Fe/Pt-Fe Electrode for Methanol Electrooxidation in Alkaline Media. *Chin. J. Catal.* **2015**, *36*, 1029–1034. [CrossRef]

materials

MDPI

Article

Composites Based on Electrodeposited WO_3 and TiO_2 Nanoparticles for Photoelectrochemical Water Splitting

Ramunas Levinas [1,2,*], Elizabeth Podlaha [3,*], Natalia Tsyntsaru [1,4] and Henrikas Cesiulis [1]

1 Faculty of Chemistry and Geosciences, Vilnius University, Naugarduko Str. 24, LT-03225 Vilnius, Lithuania; natalia.tintaru@chf.vu.lt (N.T.); henrikas.cesiulis@chf.vu.lt (H.C.)
2 State Research Institute, Center for Physical Sciences and Technology (FTMC), Saulėtekio Ave. 3, LT-10257 Vilnius, Lithuania
3 Department of Chemical and Biomolecular Engineering, Clarkson University, 8 Clarkson Ave., Potsdam, NY 13699, USA
4 Institute of Applied Physics, Moldova State University, MD-2028 Chisinau, Moldova
* Correspondence: ramunas.levinas@ftmc.lt (R.L.); epodlaha@clarkson.edu (E.P.)

Abstract: Photoelectrochemically active WO_3 films were fabricated by electrodeposition from an acidic (pH 2), hydrogen-peroxide-containing electrolyte at −0.5 V vs. SCE. WO_3-TiO_2 composites were then synthesized under the same conditions, but with 0.2 g/L of anatase TiO_2 nanoparticles (⌀ 36 nm), mechanically suspended in the solution by stirring. After synthesis, the films were annealed at 400 °C. Structural characterization by XRD showed that the WO_3 films exhibit the crystalline structure of a non-stoichiometric hydrate, whereas, in WO_3-TiO_2, the WO_3 phase was monoclinic. The oxidation of tungsten, as revealed by XPS, was W^{6+} for both materials. Ti was found to exist mainly as Ti^{4+} in the composite, with a weak Ti^{3+} signal. The efficiency of the WO_3 films and composites as an oxygen evolution reaction (OER) photo-electrocatalyst was examined. The composite would generate approximately three times larger steady-state photocurrents at 1.2 V vs. SCE in a neutral 0.5 M Na_2SO_4 electrolyte compared to WO_3 alone. The surface recombination of photogenerated electron–hole pairs was characterized by intensity-modulated photocurrent spectroscopy (IMPS). Photogenerated charge transfer efficiencies were calculated from the spectra, and at 1.2 V vs. SCE, were 86.6% for WO_3 and 62% for WO_3-TiO_2. Therefore, the composite films suffered from relatively more surface recombination but generated larger photocurrents, which resulted in overall improved photoactivity.

Keywords: tungsten trioxide titania composite; electrodeposition; photoelectrochemical water splitting; IMPS of OER

1. Introduction

The adverse environmental impact of using fossil fuels for energy needs and generating H_2 for chemical processing has led to significant research toward finding alternative and sustainable energy carriers, such as hydrogen (H_2), with H_2 produced by solar water splitting is more advantageous compared to other forms of green hydrogen [1–3]. The photoelectrochemical (PEC) water-splitting process includes light harvesting, photogenerated charge separation, and surface catalysis, and is an effective strategy as an alternative to fossil fuels. The reduction of water to produce hydrogen is coupled with the more kinetically hindered oxygen evolution reaction, which is typically carried out on metal oxide photoanodes [4,5]. Although TiO_2 is recognized as an effective inexpensive photocatalyst in the UV range, [6,7], the addition of tungsten trioxide (WO_3) is widely known to lower the band gap (~2.75 eV) and hence, allows absorption to occur within the visible range of the solar spectrum, [8–11]. However, WO_3 alone suffers from high recombination of photon-generated electron–hole pairs [12] and is susceptible to photocorrosion caused by peroxo-species formed on the surface during the photoelectrochemical reaction [13]. Tungsten trioxide doped by TiO_2 shows a red-shifted optical response in the visible light range due to a band gap decrease from a lowering of the conduction band edge, and can act as a photocatalyst. Momeni and Nazari [14] reported a change in band gap from 3.23 eV for bare TiO_2 to 2.78 eV for WO_3-doped TiO_2 nanotubes. Also, doping TiO_2 with cations like tungsten (W^{6+}) introduces energy levels in TiO_2 just below the conduction band that acts as an electron trap increasing the lifetime of charge carriers, and thus retarding recombination [15].

Heterojunctions of WO_3/TiO_2 layers have been used to increase the efficiency of the photocatalysts [16–20] by aligning the band structures of the two coupled semiconductors to encourage migration of photogenerated electrons (e^-) and holes (h^+) in separate directions across the heterojunction boundary, which reduces the electron–hole recombination. The highly oxidative valence band of WO_3, arising from the O 2p orbitals, also helps to drive oxidation reactions, such as PEC oxygen evolution or degradation of dyes. WO_3/TiO_2 composites have been reported to perform as stable and efficient photoanodes in water splitting [21–23]. Numerous reports of dye degradation with tungsten doped TiO_2 or heterojunctions are reported through the reactions of a photogenerated superoxide [24–28]. One such heterojunction unique to electrodeposition is the work presented by Martin et al. [29]. They electrodeposited WO_3 onto a metallic Ti electrode then heat treated it to create a $WO_3/TiO_2/Ti$ layered material that exhibited high photocurrent values toward the photoelectrocatalytic degradation of propyl paraben. An alternative strategy to reduce recombination rates is to apply a potential or bias and photogenerated electrons are drawn away from the electrode–electrolyte interface through the external surface. This electrically enhanced photocatalysis was observed for electrochemically prepared TiO_2/WO_3 layers on stainless steel for the oxidation of organics [30].

Nanostructuring is also important for generating quantum confinement effects that alter the electron and hole transport behaviors, shifting of the electronic band structure, and the large surface-to-volume ratio can significantly increase the surface reaction sites [31]. However, it can also introduce defects or additional grain boundaries that can promote recombination [32].

As mentioned, synergistic effects, as a result of combining different photoactive materials, provide a promising pathway to improve the photoelectrochemical performance in heterojunction devices for solar water splitting [33]. To this end, a nanocomposite WO_3-TiO_2 film was electrodeposited with a WO_3 matrix incorporating nanoparticles of TiO_2, in contrast to layered materials or those with a matrix of TiO_2. The electrodeposition of WO_3 films followed the approach by Pauporté [34], where the electrochemical deposition mechanism is multistaged and occurs through the reduction of a peroxy-tungstate species as outlined in Equations (1) to (3) [35–37].

$$2WO_4^{2-} + 4H_2O_2 \rightarrow W_2O_{11}^{2-} + 2OH^- + 3H_2O \tag{1}$$

$$W_2O_{11}^{2-} + (2+3x)H^+ + 3xe^- \rightarrow 2H_xWO_3 + \frac{(2+x)}{2}H_2O + \frac{(8-x)}{4}O_2 \tag{2}$$

$$H_xWO_3 \rightarrow WO_3 + xH^+ + xe^- \tag{3}$$

In this study, the approach is modified by adding TiO_2 nanoparticles to the sodium tungstate–hydrogen peroxide electrolyte. The electrodeposition is then carried out to simultaneously electrodeposit a WO_3 film and capture TiO_2 nanoparticles, resulting in a one-step synthesis of a composite with improved photoactivity for water splitting. Moreover, the use of an advanced non-stationary photoelectrochemical technique (IMPS) is demonstrated to probe charge carrier photogeneration in the space-charge layer and to elucidate important kinetic constants for charge transfer and recombination.

2. Materials and Methods

2.1. Composite Electrodeposition

A schematic representation of the cell and process used to deposit WO_3 and WO_3-TiO_2 films is shown in Figure 1. All used chemicals were of ACS reagent grade. The electrolyte to electrodeposit WO_3 films was prepared from 0.025 M $Na_2WO_4 \cdot 2H_2O$ (Fisher Chemical, Hampton, NH, USA), 0.075% (v/v) H_2O_2 (Fisher Chemical, Hampton, NH, USA) which was 1.25 mL of 30% hydrogen peroxide, and the pH was adjusted to 1.4 with sulfuric acid (Fisher Cchemical, Hampton, NH, USA). To electrodeposit composite WO_3-TiO_2 films, the same electrolyte composition was used, but additionally, a particle loading of 0.2 g/L of anatase TiO_2 (Thermo Fisher Scientific, Waltham, MA, USA) of average ⌀ 36 nm was added. The nanoparticles were mechanically suspended in the electrolyte by using a magnetic stir bar. Electrodeposition was carried out in a three-electrode cell, with a stainless-steel (type 316) working electrode (plates of 2 cm × 2 cm dimensions were used as the substrate, with a circular working area of 1.226 cm^2), a Pt counter electrode, and an SCE reference electrode. The substrate was positioned in a holder parallel to the bottom of the electrodeposition bath. Deposition occurred during an applied potential of −0.5 V vs. SCE for 5 min.

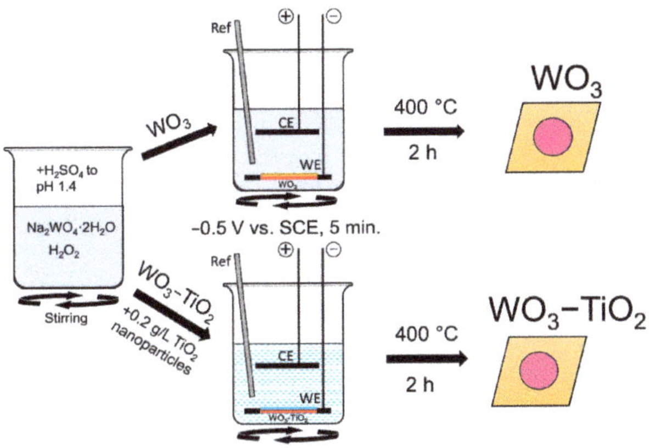

Figure 1. Schematic representation of material synthesis and annealing process.

2.2. Structural and Surface Morphology Characterization

The as-deposited WO_3 and WO_3-TiO_2 films were thermally annealed at 400 °C for two hours, with a 2 °C min^{-1} temperature ramp-up rate. The surface morphology was characterized by a Hitachi TM 3000 scanning electron microscope (Hitachi Ltd., Tokyo, Japan). XRD diffraction patterns were obtained with a Rigaku MiniFlex II x-ray diffractometer (Rigaku Corporation, Tokyo, Japan).

The XPS analyses were carried out with a Kratos Axis Supra spectrometer (Kratos Analytical Limited, Manchester, UK), using a monochromatic Al K(alpha) source (15 mA, 15 kV). Survey scan analyses were carried out on the area of 300 × 700 μm at a pass energy of 160 eV. High-resolution analyses were also carried out on the area of 300 × 700 μm, but at a pass energy of 20 eV. The XPS signal due to adventitious carbon located at 284.8 eV was used as a binding energy (BE) reference.

2.3. PEC Characterization

The photoelectrochemical response of the resulting deposits was examined with linear sweep voltammetry, chronoamperometry, and IMPS, using a ZAHNER Zennium CIMPS-QE/IPCE system (Zahner-Elektrik, Kronach, Germany). The photoelectrochemical cell included a quartz window for front-illumination of the films, a platinum counter electrode, and an SCE reference electrode. A neutral 0.5 M Na_2SO_4 electrolyte was used. The light source was a variable-intensity LED, emitting monochromatic 365 nm UV light.

A typical characterization procedure of an electrochemically deposited and annealed WO_3 or WO_3-TiO_2 film consisted of these steps:

Voltammetry experiments were conducted from 0.2 V to 1.6 V (vs. SCE) at 5 mV s^{-1}, with a 4-s on/off illumination pulse (2 s on/2 s off). Chronoamperometry at a constant potential of 1.2 V used an illumination pulse of 120 s (60 s on/60 s off). Light intensity was set to 30 mW cm^{-2}. IMPS was measured at steady state potentials: 0.6 V, 0.8 V, 1.0 V, 1.2 V (100 mW cm^{-2} light intensity, 10% amplitude modulation, 10 kHz to 0.1 Hz).

3. Results and Discussion

3.1. Structure and Surface Morphology

Figure 2 shows the XRD patterns of electrodeposited WO_3 and WO_3-TiO_2 films that had been annealed at 400 °C for 2 h. The XRD pattern of the annealed stainless steel substrate is also included. It contains two peaks characteristic of austenite iron (JCPDS No. 00-023-0298) at 43.5° and 50.7°, and the substrate peaks also emerge in the diffractograms of the deposited films. The diffractograms of WO_3 and WO_3-TiO_2 reveal the radically different phase structure of the films. WO_3-TiO_2 displays all the characteristic peaks of monoclinic WO_3 [38,39]. However, there are a couple of relatively low-intensity peaks that cannot be attributed to this phase, in particular, at 27.6° and 49.5°. The analysis of WO_3 reveals the origin of these peaks. The diffractogram of WO_3 contains a very intense peak at ~ 24.1°, which may correspond to the (200) face of monoclinic WO_3. There is also a broad hump at a slightly lower 2Θ of 23.3°, and two small peaks at 27.8° and 49.5°. These reasonably correspond to the unidentified peaks from WO_3-TiO_2 and have been attributed to hydrated forms of WO_3 in the literature [40,41]. It is claimed that the formation of $WO_3 \cdot xH_2O$ can occur through tuning of the intermediate peroxo-complex and thus is related to the H_2O_2 content in the solution [40]. The electrodeposition of a hydrated tungsten oxide has also recently been reported under conditions very similar to those used in this study [42]. Therefore, the formation of monoclinic WO_3 when electrodepositing the composite must have been related to an equilibrium shift in the intermediate peroxo-complex caused by the interaction of solution species with nanoparticulate TiO_2, possibly solvation. The crystallite sizes of WO_3, as approximated by the Scherer equation, are 41 ± 9 nm for WO_3 and 27.4 ± 3.3 nm for WO_3-TiO_2. Regarding the presence of TiO_2, in the diffractogram of the composite film, a very faint hump at ~25° can perhaps be distinguished, which could be related to the main (101) peak of anatase TiO_2.

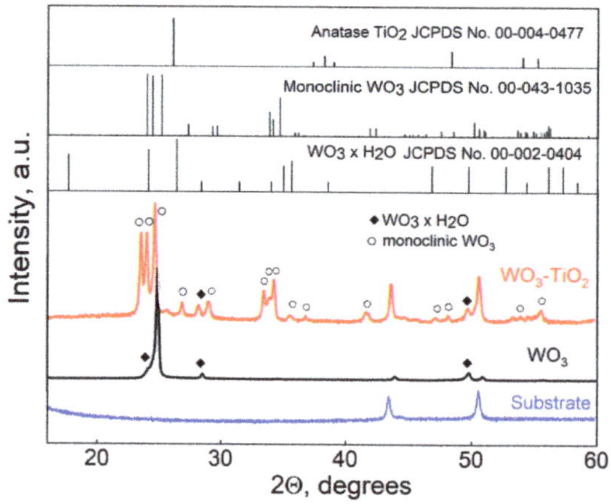

Figure 2. XRD patterns of stainless steel substrate and electrodeposited WO_3 and WO_3-TiO_2 composite films, after annealing at 400 °C.

The visual appearance and SEM surface morphology images of WO_3 and WO_3-TiO_2 films are shown in Figure 3. The WO_3 film (Figure 3a) is inhomogeneously polychromic, which is indicative of a compact thin film of slightly varying thicknesses. The surface morphology, as observed by SEM, is smooth and fine-grained, as was also observed by Pauporté when first reporting on WO_3 electrodeposition by this method [34]. For both films, signs of emerging cracks can be seen. It was noted earlier [43,44] that cracking and delamination occur if the layer is thicker (i.e., after longer deposition times). Accordingly, the WO_3-TiO_2 film is already cracked (Figure 3b), which may indicate a larger thickness. Scattered agglomerates of white particles, which are most likely TiO_2, can be observed on the surface. Visually, the film appears "cloudy" because TiO_2 particles had settled at the surface of the film during deposition. These particles were not always well-adhered to the surface. If the electrodeposition time was prolonged, they would detach from the surface when immersed in an electrolyte for further characterization. For this reason, the PEC characterization was carried out on films that had been deposited for 5 min.

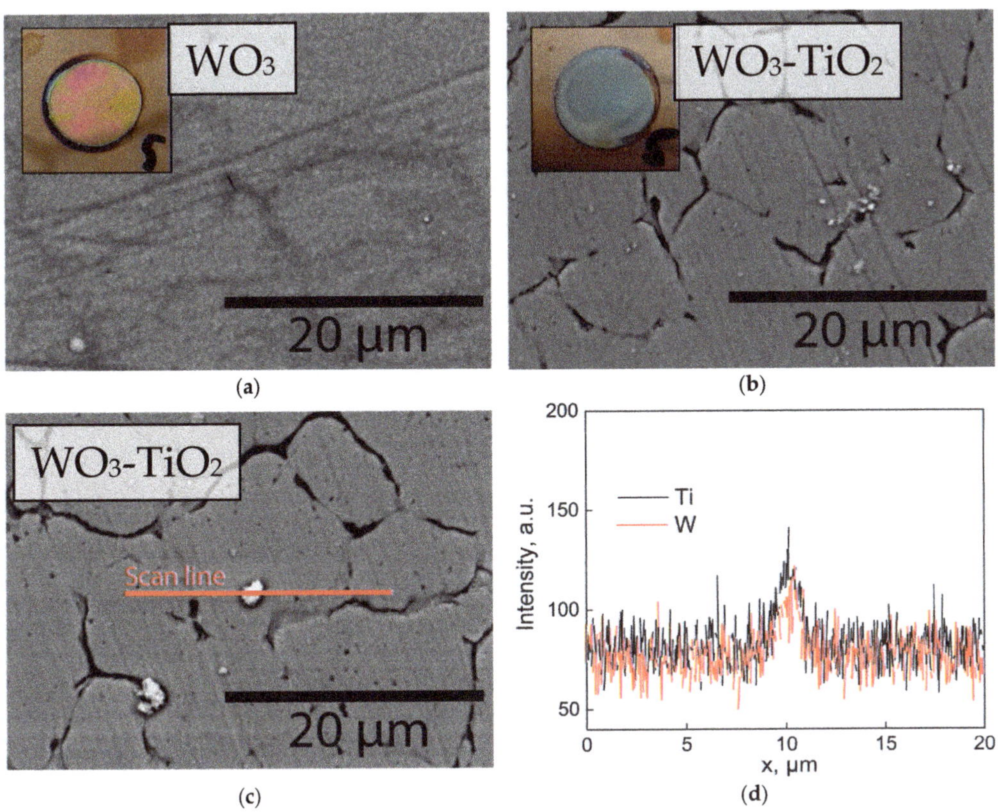

Figure 3. SEM surface morphology images of (**a**) WO_3; (**b**) WO_3-TiO_2 films electrodeposited for 5 min, (**c**) SEM image of EDX line scan on a WO_3-TiO_2 film, (**d**) EDX line scan data portraying titanium and tungsten signals.

The results of an EDX line scan (Figure 3c,d) show that a TiO_2 nanoparticle agglomeration was observed on the surface. Titanium was detected, albeit with a very weak signal. It is also interesting to note that the tungsten signal increases along with titanium. This may simply be a topographical effect (i.e., a stronger signal obtained from a point closer to the detector), but it could also indicate that WO_3 had electrodeposited onto TiO_2.

The near-surface electronic structure of the WO_3 and WO_3-TiO_2 films was investigated by XPS. It was found that for both films, the core-level structure of W 4f deconvolutes into a single doublet with W $4f_{7/2}$ and W $4f_{5/2}$ peaks at binding energies of 35.46 eV and 37.6 eV, respectively (Figure 4a,b). The position of these peaks, and the spin-orbit splitting energy of 2.14 eV, indicate the total dominance of the W^{6+} oxidation state [45,46]. Interestingly, the O 1s core-level spectra reveal that oxygen can be found in at least two oxidation states (Figure 4c,d). The O 1s peak seen here at 530.22 eV for WO_3 and 530.48 eV for WO_3-TiO_2 is commonly attributed to lattice oxygen (i.e., the W-O bond) [47].

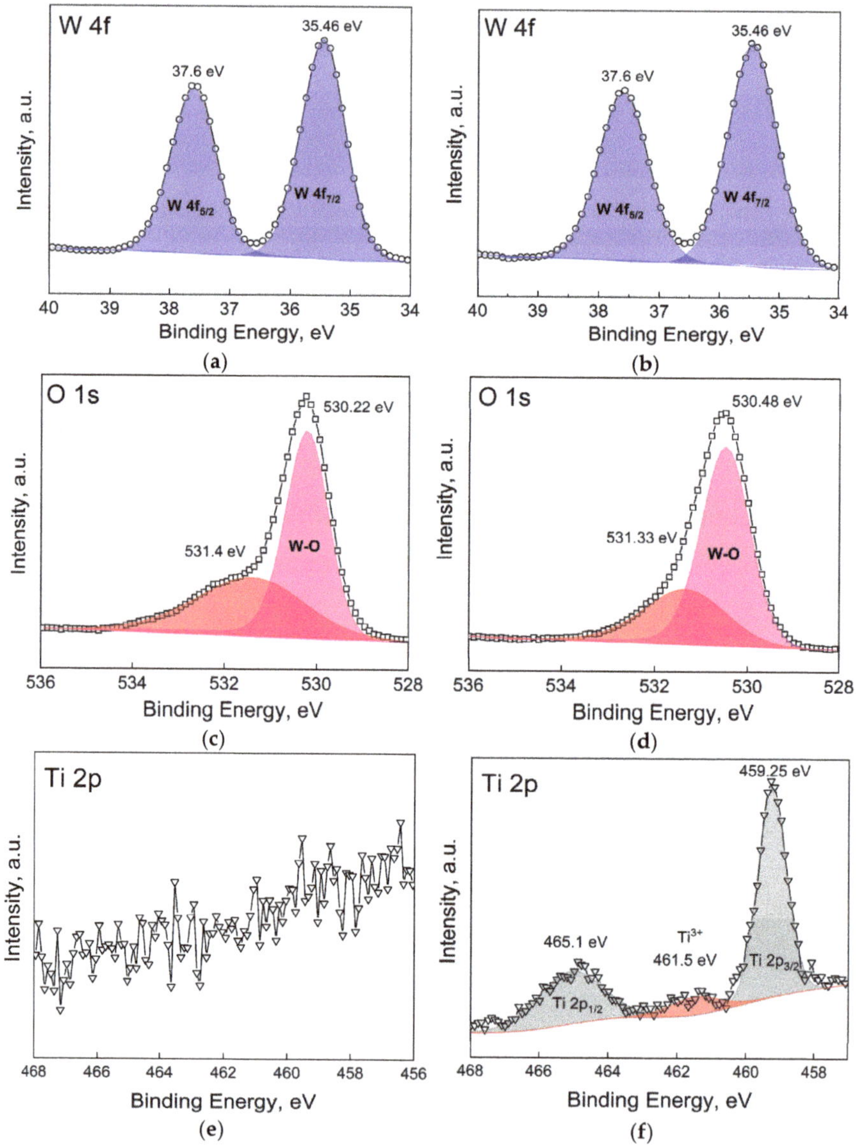

Figure 4. Deconvoluted core-level XPS spectra of W 4f, O 1s, and Ti 2p of a WO_3 film (**a**,**c**,**e**) and WO_3-TiO_2 film (**b**,**d**,**f**).

The origin of the second much broader peak at ~531.4 eV is slightly more complicated to explain. First and foremost, this cannot be Ti-related O because (1) the lattice oxygen peak energy is expected to be around 529–530 eV [48] and (2) the same peak also exists for WO_3 without TiO_2. A couple of other possibilities that are often reported are oxygen from non-stoichiometric WO_{3-x} or -OH groups incorporated within the lattice, which can result in peaks in the O 1s spectra at approximately 532.5 eV [47,49]. However, these energies also do not match. A very recent work by T. J. Frankcombe and Y. Liu describes what is most likely the cause of these peaks [50]. They found that a peak at this energy (commonly ascribed in the literature to oxygen vacancies) is instead caused by O 1s electrons from water molecules that are chemisorbed to the surface or surface oxygen passivated by hydrogen. This agrees well with the W 4f spectra, which show 100% dominance of W^{6+} and no trace of reduced oxidation states that would be expected with a large number of oxygen vacancies. The existence of chemisorbed water may also be related to the observation of hydrates in XRD characterization (Figure 2). Lastly, no Ti is found on WO_3, and only a very weak Ti 2p signal is obtained from the WO_3-TiO_2 film (Figure 4e,f). The core level spectrum deconvolutes into one doublet with peaks at 459.25 eV and 465.1 eV, corresponding to Ti^{4+} from anatase TiO_2 [51]. An indistinct hump at 461.5 eV may also be observed. It is on the verge of being noise, but a peak at this position has been attributed to Ti^{3+} [52]. The reduction of titanium dioxide nanoparticles is feasible under cathodic conditions in an acidic electrolyte, so the signal is attributed to Ti^{3+} here as well.

3.2. Photoelectrochemical Characterization

Initially, the PEC properties of WO_3 and WO_3-TiO_2 films were characterized by carrying out a potential scan with intermittent on/off light pulses (Figure 5a). Two differences are immediately apparent: (1) the WO_3-TiO_2 film generates larger photocurrents throughout the entire potential range and (2) the photocurrent onset potential of the composite is more cathodic (~0.2 V vs. SCE) in comparison to just WO_3 (~0.35 V vs. SCE). As expected, the magnitude of the photocurrent increases with the applied potential as the space-charge layer, where the photoexcitation of charge carriers occurs and becomes wider due to the increasing strength of the applied electrical field. It is also worth noting that both WO_3 and WO_3-TiO_2 films show an appreciable amount of surface hole recombination, which is evident from the initial photocurrent overshoot that occurs at the instant light is turned on.

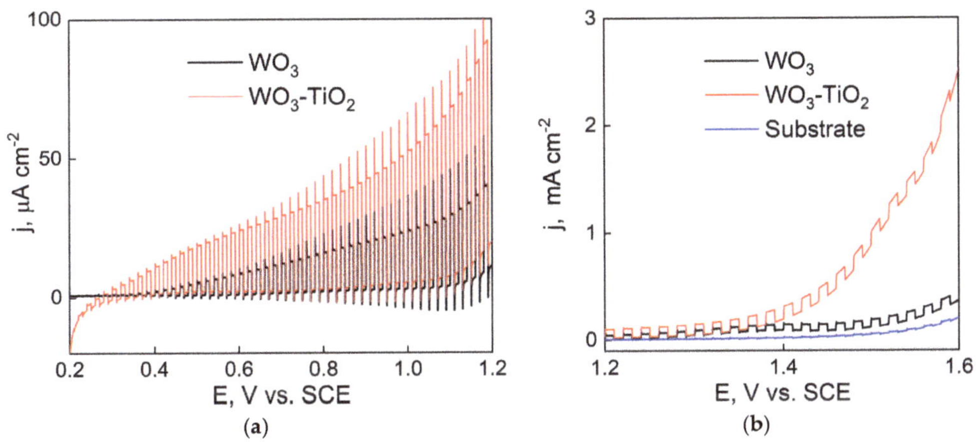

Figure 5. Chopped UV LSV measurements at low potentials (**a**) and high potentials (**b**) in 0.5 M Na_2SO_4 of WO_3 and WO_3-TiO_2 films deposited within 5 min; 5 mV s^{-1}, 30 mW cm^{-2} light intensity.

An additional difference between the activities of these films is noted at higher applied potentials of $\geq$~1 V vs. SCE, which roughly corresponds to 1.23 V vs. RHE in neutral pH (Figure 5b). In this case, particularly for WO_3-TiO_2, the electrochemical current rises sharply and is most probably attributable to the onset of electrochemical water splitting/oxygen evolution (OER). A similar effect, although with much lower current densities, is observed for the annealed stainless steel substrate. Thus, the composite films exhibit better electrochemical OER activity. This may simply be an effect of increased surface area [53] or the improved OER mechanistic kinetics of WO_3 [54].

Because the profile of the photocurrent pulse provides useful information about the material photoactivity and surface recombination [55], potentiostatic photocurrent pulses were obtained at 1.2 V vs. SCE and with a longer on/off cycle duration of 120 s (Figure 6). This made it possible to observe the initial photocurrent overshoot immediately upon turning on the light, and the subsequent decay over several pulses.

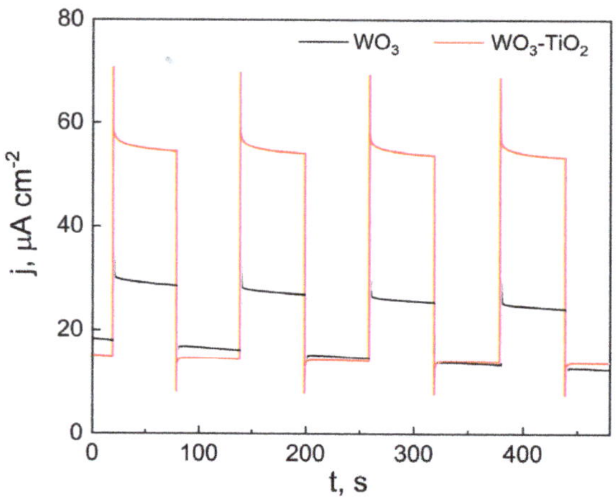

Figure 6. UV on/off pulses in 0.5 M Na_2SO_4 of WO_3 and WO_3-TiO_2 films; 365 nm LED, 30 mW cm^{-2} light intensity.

At the moment the light is turned on, the photocurrent density ($j_{t=0}$) is proportional to the maximum amount of photogenerated electron–hole pairs. For a photoanode, the electrons move to the back contact and are registered as photocurrent, and the holes move toward the electrode–electrolyte interface, where they are either lost to surface recombination or are transferred into the electrolyte to carry out an oxidative photoelectrochemical reaction. After this, the photocurrent decays until a steady state value (j_{ss} is assumed to be the photocurrent at the end of the 60 s pulse) is reached. In this case, it is observed that the WO_3-TiO_2 film displays slightly more relative surface recombination than WO_3. For example, for the WO_3 film, the $j_{ss}/j_{t=0}$ ratio is 0.8; whereas, for the comparable composite film, it is lower at 0.77. However, the WO_3-TiO_2 film also generates larger photocurrents overall (~40 µA cm^{-2} compared to just under 12 µA cm^{-2} for WO_3). A similar result was observed by Castro et al., who found that a hydrothermally synthesized WO_3 composite with 40 wt% of TiO_2 exhibited larger photocurrents than WO_3 [21]. Similarly, WO_3/TiO_2 heterostructures on reduced graphene oxide were shown to generate significantly larger photocurrents than the plain films alone [22].

Lastly, similar to the initial jump, there is an "overshoot" to the downside when illumination is turned off. This occurs because of the recombination of holes trapped on the surface with electrons—a phenomenon similar to the decay of the photocurrent.

3.3. IMPS Study

IMPS is a non-stationary photoelectrochemical analysis method, modeled by Ponomarev and Peter [56], to determine the features of the space charge region relative to the reaction rate constants. A sinusoidal modulation is applied to the intensity of illumination, and the working electrode under investigation is kept at potentiostatic conditions. In this case, the (photo)current response is proportional entirely to the incident photon flux, i.e., conversion efficiency. It must be noted that here, the light intensity was increased in comparison to previous experiments (30 mW cm^{-2} to 100 mW cm^{-2}). This is because the signal for IMPS is not obtained from the entire photocurrent, but from the light-perturbation-induced photocurrent, which corresponds to the photocurrent generated by the amplitude of light intensity modulation (10%). If this light intensity is too low (e.g., 3 mW cm^{-2} as 10% of 30 mW cm^{-2}), the spectra can be noisy. However, photogenerated charge carrier recombination is generally larger at higher light intensities [57], so care should be taken to find the optimal measurement conditions for a particular system.

IMPS Nyquist spectra obtained for WO_3 are shown in Figure 7, where H represents the complex transfer function of the resulting current measured by controlling the light intensity, normalized to the geometric electrode area. The first quadrant reflects the response from charge-transfer kinetic and surface recombination. The low-frequency intercept H'_{LF} (*left side*) when $H''\rightarrow 0$ represents the differential steady-state photocurrent increase due to a differential increase in the light intensity and should be zero if no photoelectrochemical reactions occur. ω_{LF}, the radial frequency of the highest point of the recombination semicircle, is a product of first-order rate constants of photogenerated charge transfer and recombination and is inversely related to the recombination time constant.

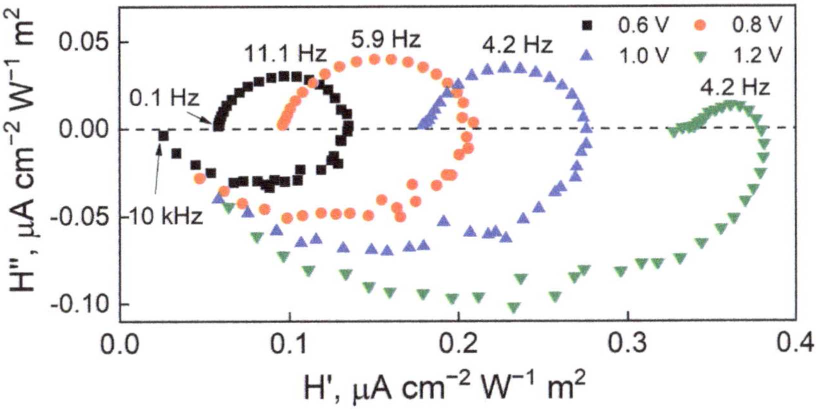

Figure 7. Nyquist plots for IMPS of a WO_3 film that had been electrodeposited for 5 min, obtained in 0.5 M Na_2SO_4, 365 nm LED, 100 mW cm^{-2} light intensity with 10% modulation amplitude.

Figure 7 for WO_3 shows a non-zero H'_{LF}, suggesting that an electrochemical reaction occurs at all applied potentials. In Figure 8 for WO_3-TiO_2 at 0.6 and 0.8 V, this intercept is nearly zero and thus, electron–hole generation and surface recombination primarily occur. A significant shift in this intercept occurs at 1 V where a significant amount of OER is expected to commence driven by the applied potential. The intercept when the imaginary part is zero at a higher frequency (*right side*) represents the amplitude of the flux of charge carriers to the surface. This intercept occurs at a value of H′ that can be expressed as Equation (4).

$$H' = j_h \frac{C_H}{C_H + C_{SC}} \quad (4)$$

where j_h is the hole current amplitude, C_H and C_{SC} are the Helmholtz and space charge layer capacitances, respectively.

Since the Helmholtz layer size is typically smaller than the width of the depletion, the space charge layer, then it is expected that the Helmholtz capacitance is higher, $C_H > C_{SC}$ [58]. Then, this intercept equals j_h, i.e., the hole current. At higher frequencies relative to this intercept, the influence of the capacitance of the semiconductor space charge dominates and the response moves to the lower quadrant, with a negative imaginary transfer function. The discussed parameters that had been obtained from the spectra are presented in Table 1. The overall trend is that, as the applied potential is raised, H'_{LF} increases because more photogenerated holes are transferred as surface recombination is suppressed. Moreover, the IMPS spectra are not normalized to the amplitude of the photogenerated hole current, and therefore, larger IMPS spectra magnitudes of WO_3-TiO_2 suggest higher conversion efficiencies overall, corresponding well to LSV and potentiostatic pulse data.

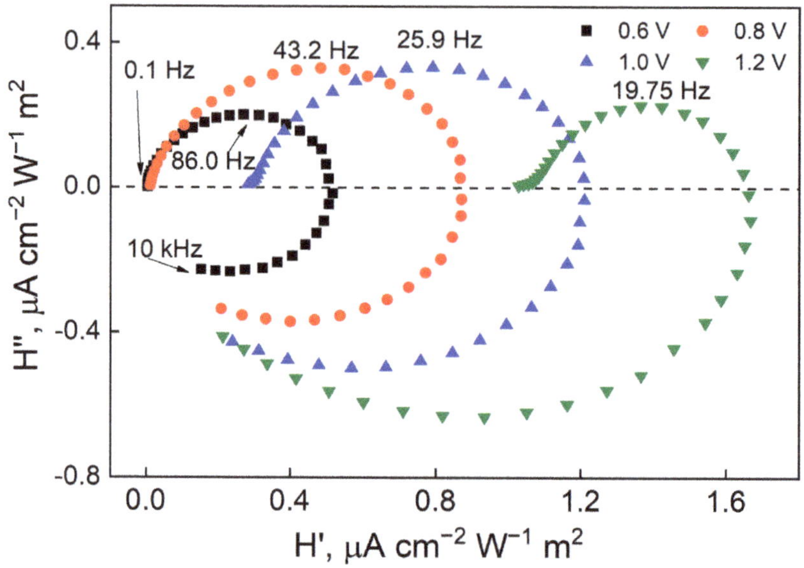

Figure 8. Nyquist plots for IMPS of a WO_3-TiO_2 film that had been electrodeposited for 5 min, obtained in 0.5 M Na_2SO_4, 365 nm LED, 100 mW cm^{-2} light intensity with 10% modulation amplitude.

Table 1. Parameters, obtained from interpretation of IMPS Nyquist plots.

	WO_3			WO_3-TiO_2		
E, V	**ω_{LF}, Rad s^{-1}**	**H'_{LF}, μA cm^{-2} W^{-1} m^2**	**j_h, μA cm^{-2} W^{-1} m^2**	**ω_{LF}, Rad s^{-1}**	**H'_{LF}, μA cm^{-2} W^{-1} m^2**	**j_h, μA cm^{-2} W^{-1} m^2**
0.6	69.6	0.058	0.135	546.2	0.004	0.511
0.8	37.0	0.097	0.208	271.3	0.010	0.867
1.0	26.4	0.180	0.276	162.7	0.284	1.208
1.2	26.4	0.330	0.378	124.0	1.028	1.653

In IMPS spectra, the high-frequency semicircle occurs because of the RC time constant of the electrochemical cell, i.e., the series resistance of the substrate and the total capacitance. Conversely, the low-frequency semicircle describes the recombination and charge transfer kinetics. Kinetically, these processes are described by first order rate constants: k_{tr}—the rate constant of charge transfer across this same interface and k_{rec}—the rate of electron–hole recombination at the semiconductor/electrolyte interface. The spectra shown in Figures 5 and 6 exhibit some characteristic tendencies. Most notably—the radius of the low frequency semicircle decreases as the applied potential is increased. In IMPS, the value of the low frequency intercept H'_{LF} corresponds to the DC photocurrent, scaled by the 10% modulation, and is expressed as Equation (5).

$$H'_{LF} = j_h \frac{k_{tr}}{k_{tr} + k_{rec}} \tag{5}$$

Here, $k_{tr}/k_{tr} + k_{rec}$ corresponds to η_{trans}—the fraction of hole flux from the space charge region that is involved in the charge transfer reaction, or the transfer efficiency.

The transfer efficiency increases with applied potential as photogenerated charge carriers can participate in an electrochemical reaction. Therefore, electron–hole recombination is significant at less positive potentials (the low frequency intercept value is closer to 0), because there is no electrochemical reaction, but at 1.2 V, it decreases considerably with the occurrence of OER.

Several important parameters can be calculated from IMPS spectra. Firstly, the transfer efficiency is an indicator of how many of the holes that had arrived at the semiconductor/electrolyte interface that were transferred into the electrolyte to provide the photoelectrochemical reaction. The data for WO_3 and WO_3-TiO_2 films are shown in Figure 9. As expected, the efficiencies are larger with increasing applied potential. Plain WO_3 films have larger photogenerated transfer efficiency over the entire measured potential range, while the TiO_2-containing composites have total recombination at lower potentials, but improve rapidly at higher potentials.

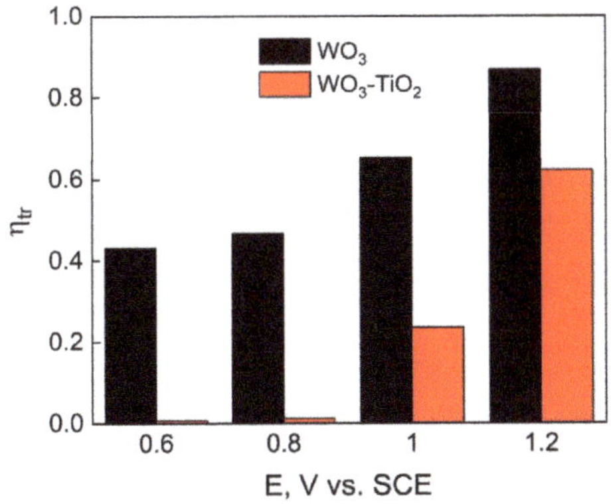

Figure 9. Transfer efficiency values, obtained from IMPS spectra for WO_3 and WO_3-TiO_2 films.

To fully discern reaction kinetics, the rate constants should be considered. Both k_{tr} and k_{rec} can be calculated from two characteristic points of the IMPS spectrum: the low frequency intercept H'_{LF} (Equation (5)) and ω_{LF}, the radial frequency of the highest point of the recombination semicircle, which is described by Equation (6).

$$\omega_{LF} = k_{tr} + k_{rec} \tag{6}$$

The spectra were analyzed and the rate constants are presented in Figure 10 in relation to applied potential. Firstly, it can be seen that for both measured photoelectrodes, k_{rec} decreases with applied potential (Figure 10a), which is expected from the added potential bias as has been discussed earlier.

The k_{rec} values of WO_3-TiO_2 are larger, showing faster recombination kinetics. However, when considering the kinetic transfer constant, it is apparent that k_{tr} increases significantly for the composite films at higher potentials, compared to WO_3 (Figure 10b). Thus, there is a higher OER reaction rate on the WO_3-TiO_2 under a biased photoelectrochemical environment. At an applied potential of 1.2 V vs. SCE, the k_{tr} of the composite is three times larger than that of the respective WO_3 film, consistent with the voltammetry result in Figure 5 and the chronoamperometry in Figure 6.

While the addition of discrete TiO_2 nanoparticles to WO_3 generated a higher photocurrent compared to WO_3 alone, it had the negative effect of increasing the recombination rate constant, particularly near open circuit conditions at small applied potentials. The increased recombination rate may be due to the introduction of grain boundaries and local potential perturbation by the particles. The recombination rate constant decreased with applied potential, and in the applied potential range for conventional electrolysis, the WO_3-TiO_2 composite greatly exceeded the performance of the WO_3. At an even larger applied potential exceeding 1.2 V vs. SCE, evident in the polarization curve in Figure 5, the WO_3-TiO_2 composite OER current density greatly exceeded WO_3 even without light; thus, the TiO_2 addition to WO_3 improved its electrocatalytic performance. Georgieva et al. [30] also found an enhanced photocatalysis for TiO_2/WO_3 layers on stainless steel compared to WO_3 but attributed the enhancement to reduced recombination. The results presented here are in contrast to this reasoning, elucidated by the additional evidence from IMPS. The enhancement in our work is due to larger charge transfer kinetics and not suppressed recombination.

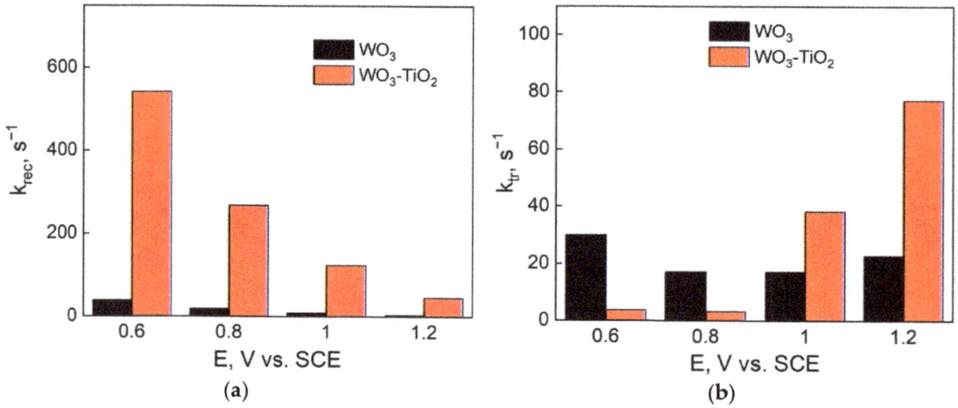

Figure 10. Values of (**a**) k_{rec} and (**b**) k_{tr} for WO_3 and WO_3-TiO_2 films that had been electrodeposited for 5 min.

4. Conclusions

In this study, WO_3 and composite WO_3-TiO_2 films were prepared by electrodeposition from an acidic peroxo-tungstate precursor solution containing suspended anatase TiO_2 nanoparticles for the formation of the composite. The annealed films were found to be crystalline, with WO_3-TiO_2 having a distinct monoclinic WO_3 structure, but plain WO_3 displaying characteristics of hydrate formation. XPS showed the dominance of W^{6+} in both samples. In addition, the inclusion of Ti^{4+} and a small amount of Ti^{3+} was proven by deconvoluting the Ti 2p core level spectra. The films prepared in this study were found to be photoelectrochemically active, and WO_3-TiO_2 generated significantly larger photocurrent magnitudes compared to the WO_3 alone. However, the WO_3-TiO_2 composite photocurrents possessed considerable recombination at and near open circuit conditions. While the overall transfer efficiency is higher for WO_3, applying a bias potential resulted in a significant increase in the OER current density of the WO_3-TiO_2 composite: it is three times higher at an applied potential of 1.2 V vs. SCE, and an associated expected decrease in the electron–hole recombination rate. The implications of these results show that electrodeposition of a WO_3-TiO_2 composite from an electrolyte containing a peroxo-tungstate precursor and TiO_2 nanoparticles result in significant changes in the material's structure, morphology, and consequently, PEC activity. The synthesis of composites is facile, they exhibit improved PEC performance as photoanodes when compared to WO_3, but also suffer from higher photogenerated electron–hole recombination at lower applied potentials.

Author Contributions: Conceptualization, R.L., E.P., N.T. and H.C.; methodology, R.L. and E.P.; formal analysis, R.L.; investigation, R.L., E.P. and N.T.; resources, E.P. and H.C.; data curation, R.L., E.P., N.T. and H.C.; writing—original draft preparation, R.L. and E.P.; writing—review and editing, R.L., E.P., N.T. and H.C.; visualization, R.L.; supervision, E.P. and H.C.; project administration, H.C.; funding acquisition, H.C. All authors have read and agreed to the published version of the manuscript.

Funding: This research has received funding from the European Union's Horizon 2020 research and innovation program under the Marie Sklodowska-Curie grant agreement No. 778357-SMARTELECTRODES. It has also received funding from the Research Council of Lithuania (LMTLT), agreement No. S-PD-22-5 (TICAL).

Institutional Review Board Statement: Not applicable.

Informed Consent Statement: Not applicable.

Data Availability Statement: Dataset available on request from the authors.

Conflicts of Interest: The authors declare no conflicts of interest.

References

1. Turner, J.A. Sustainable Hydrogen Production. *Science* **2004**, *305*, 972–974. [CrossRef] [PubMed]
2. Bard, A.J.; Fox, M.A. Artificial Photosynthesis: Solar Splitting of Water to Hydrogen and Oxygen. *Acc. Chem. Res.* **1995**, *28*, 141–145. [CrossRef]
3. Walter, M.G.; Warren, E.L.; McKone, J.R.; Boettcher, S.W.; Mi, Q.; Santori, E.A.; Lewis, N.S. Solar Water Splitting Cells. *Chem. Rev.* **2010**, *110*, 6446–6473. [CrossRef]
4. Corby, S.; Rao, R.R.; Steier, L.; Durrant, J.R. The Kinetics of Metal Oxide Photoanodes from Charge Generation to Catalysis. *Nat. Rev. Mater.* **2021**, *6*, 1136–1155. [CrossRef]
5. Holmes-Gentle, I.; Tembhurne, S.; Suter, C.; Haussener, S. Kilowatt-Scale Solar Hydrogen Production System Using a Concentrated Integrated Photoelectrochemical Device. *Nat. Energy* **2023**, *8*, 586–596. [CrossRef]
6. Fujishima, A.; Honda, K. Electrochemical Photolysis of Water at a Semiconductor Electrode. *Nature* **1972**, *238*, 37–38. [CrossRef]
7. Thompson, T.L.; Yates, J.T. Surface Science Studies of the Photoactivation of TiO_2 New Photochemical Processes. *Chem. Rev.* **2006**, *106*, 4428–4453. [CrossRef] [PubMed]
8. Yagi, M.; Maruyama, S.; Sone, K.; Nagai, K.; Norimatsu, T. Preparation and Photoelectrocatalytic Activity of a Nano-Structured WO_3 Platelet Film. *J. Solid State Chem.* **2008**, *181*, 175–182. [CrossRef]
9. Amano, F.; Li, D.; Ohtani, B. Tungsten(VI) Oxide Flake-Wall Film Electrodes for Photoelectrochemical Oxygen Evolution from Water. *ECS Trans.* **2010**, *28*, 127–133. [CrossRef]
10. Bamwenda, G.R.; Arakawa, H. The Visible Light Induced Photocatalytic Activity of Tungsten Trioxide Powders. *Appl. Catal. A Gen.* **2001**, *210*, 181–191. [CrossRef]
11. González-Borrero, P.P.; Sato, F.; Medina, A.N.; Baesso, M.L.; Bento, A.C.; Baldissera, G.; Persson, C.; Niklasson, G.A.; Granqvist, C.G.; Ferreira Da Silva, A. Optical Band-Gap Determination of Nanostructured WO_3 Film. *Appl. Phys. Lett.* **2010**, *96*, 061909. [CrossRef]
12. Wu, H.; Xu, M.; Da, P.; Li, W.; Jia, D.; Zheng, G. WO_3–Reduced Graphene Oxide Composites with Enhanced Charge Transfer for Photoelectrochemical Conversion. *Phys. Chem. Chem. Phys.* **2013**, *15*, 16138. [CrossRef] [PubMed]
13. Wang, G.; Ling, Y.; Wang, H.; Yang, X.; Wang, C.; Zhang, J.Z.; Li, Y. Hydrogen-Treated WO_3 Nanoflakes Show Enhanced Photostability. *Energy Environ. Sci.* **2012**, *5*, 6180. [CrossRef]
14. Momeni, M.M.; Nazari, Z. Easy Synthesis of Titania–Tungsten Trioxide Nanocomposite Films by Anodising Method for Solar Water Splitting. *Mater. Sci. Technol.* **2016**, *32*, 855–862. [CrossRef]
15. Zhang, X.; Liu, F.; Huang, Q.-L.; Zhou, G.; Wang, Z.-S. Dye-Sensitized W-Doped TiO_2 Solar Cells with a Tunable Conduction Band and Suppressed Charge Recombination. *J. Phys. Chem. C* **2011**, *115*, 12665–12671. [CrossRef]
16. Lv, K.; Li, J.; Qing, X.; Li, W.; Chen, Q. Synthesis and Photo-Degradation Application of WO_3/TiO_2 Hollow Spheres. *J. Hazard. Mater.* **2011**, *189*, 329–335. [CrossRef]
17. Sotelo-Vazquez, C.; Quesada-Cabrera, R.; Ling, M.; Scanlon, D.O.; Kafizas, A.; Thakur, P.K.; Lee, T.; Taylor, A.; Watson, G.W.; Palgrave, R.G.; et al. Evidence and Effect of Photogenerated Charge Transfer for Enhanced Photocatalysis in WO_3/TiO_2 Heterojunction Films: A Computational and Experimental Study. *Adv. Funct. Mater.* **2017**, *27*, 1605413. [CrossRef]
18. Zhang, L.; Guo, J.; Hao, B.; Ma, H. WO_3/TiO_2 Heterojunction Photocatalyst Prepared by Reactive Magnetron Sputtering for Rhodamine B Dye Degradation. *Opt. Mater.* **2022**, *133*, 113035. [CrossRef]
19. Gao, L.; Gan, W.; Qiu, Z.; Zhan, X.; Qiang, T.; Li, J. Preparation of Heterostructured WO_3/TiO_2 Catalysts from Wood Fibers and Its Versatile Photodegradation Abilities. *Sci. Rep.* **2017**, *7*, 1102. [CrossRef]
20. Székely, I.; Kovács, Z.; Rusu, M.; Gyulavári, T.; Todea, M.; Focșan, M.; Baia, M.; Pap, Z. Tungsten Oxide Morphology-Dependent $Au/TiO_2/WO_3$ Heterostructures with Applications in Heterogenous Photocatalysis and Surface-Enhanced Raman Spectroscopy. *Catalysts* **2023**, *13*, 1015. [CrossRef]
21. Castro, I.A.; Byzynski, G.; Dawson, M.; Ribeiro, C. Charge Transfer Mechanism of WO_3/TiO_2 Heterostructure for Photoelectrochemical Water Splitting. *J. Photochem. Photobiol. A Chem.* **2017**, *339*, 95–102. [CrossRef]
22. Yoon, D.-H.; Biswas, M.R.U.D.; Sakthisabarimoorthi, A. Composite Nanostructures of Black TiO_2/WO_3 on rGO Nanosheets for Photoelectrochemical Water Splitting. *Diam. Relat. Mater.* **2022**, *129*, 109363. [CrossRef]
23. Liu, K.-I.; Perng, T.-P. Fabrication of Flower-Like WO_3/TiO_2 Core–Shell Nanoplates by Atomic Layer Deposition for Improved Photoelectrochemical Water-Splitting Activity. *ACS Appl. Energy Mater.* **2020**, *3*, 4238–4244. [CrossRef]

24. Yadav, M.; Yadav, A.; Fernandes, R.; Popat, Y.; Orlandi, M.; Dashora, A.; Kothari, D.C.; Miotello, A.; Ahuja, B.L.; Patel, N. Tungsten-Doped TiO_2/Reduced Graphene Oxide Nano-Composite Photocatalyst for Degradation of Phenol: A System to Reduce Surface and Bulk Electron-Hole Recombination. *J. Environ. Manag.* **2017**, *203*, 364–374. [CrossRef] [PubMed]
25. Ki, S.J.; Park, Y.-K.; Kim, J.-S.; Lee, W.-J.; Lee, H.; Jung, S.-C. Facile Preparation of Tungsten Oxide Doped TiO_2 Photocatalysts Using Liquid Phase Plasma Process for Enhanced Degradation of Diethyl Phthalate. *Chem. Eng. J.* **2019**, *377*, 120087. [CrossRef]
26. Balta, Z.; Simsek, E.B. Insights into the Photocatalytic Behavior of Carbon-Rich Shungite-Based WO_3/TiO_2 Catalysts for Enhanced Dye and Pharmaceutical Degradation. *New Carbon. Mater.* **2020**, *35*, 371–383. [CrossRef]
27. Patil, S.M.; Deshmukh, S.P.; More, K.V.; Shevale, V.B.; Mullani, S.B.; Dhodamani, A.G.; Delekar, S.D. Sulfated TiO_2/WO_3 Nanocomposite: An Efficient Photocatalyst for Degradation of Congo Red and Methyl Red Dyes under Visible Light Irradiation. *Mater. Chem. Phys.* **2019**, *225*, 247–255. [CrossRef]
28. Hepel, M.; Hazelton, S. Photoelectrocatalytic Degradation of Diazo Dyes on Nanostructured WO_3 Electrodes. *Electrochim. Acta* **2005**, *50*, 5278–5291. [CrossRef]
29. Martins, A.S.; Cordeiro-Junior, P.J.M.; Bessegato, G.G.; Carneiro, J.F.; Zanoni, M.V.B.; Lanza, M.R.D.V. Electrodeposition of WO_3 on Ti Substrate and the Influence of Interfacial Oxide Layer Generated in Situ: A Photoelectrocatalytic Degradation of Propyl Paraben. *Appl. Surf. Sci.* **2019**, *464*, 664–672. [CrossRef]
30. Georgieva, J.; Armyanov, S.; Valova, E.; Philippidis, N.; Poulios, I.; Sotiropoulos, S. Photoelectrocatalytic Activity of Electrosynthesised Tungsten Trioxide- Titanium Dioxide Bi-Layer Coatings for the Photooxidation of Organics. *J. Adv. Oxid. Technol.* **2008**, *11*, 300–307. [CrossRef]
31. Wang, X.; Li, Z.; Shi, J.; Yu, Y. One-Dimensional Titanium Dioxide Nanomaterials: Nanowires, Nanorods, and Nanobelts. *Chem. Rev.* **2014**, *114*, 9346–9384. [CrossRef] [PubMed]
32. Sherkar, T.S.; Momblona, C.; Gil-Escrig, L.; Ávila, J.; Sessolo, M.; Bolink, H.J.; Koster, L.J.A. Recombination in Perovskite Solar Cells: Significance of Grain Boundaries, Interface Traps, and Defect Ions. *ACS Energy Lett.* **2017**, *2*, 1214–1222. [CrossRef]
33. Rodríguez-Gutiérrez, I.; Djatoubai, E.; Su, J.; Vega-Poot, A.; Rodríguez-Gattorno, G.; Souza, F.L.; Oskam, G. An Intensity-Modulated Photocurrent Spectroscopy Study of the Charge Carrier Dynamics of WO_3/$BiVO_4$ Heterojunction Systems. *Sol. Energy Mater. Sol. Cells* **2020**, *208*, 110378. [CrossRef]
34. Pauporté, T. A Simplified Method for WO_3 Electrodeposition. *J. Electrochem. Soc.* **2002**, *149*, C539. [CrossRef]
35. Meulenkamp, E.A. Mechanism of WO_3 Electrodeposition from Peroxy-Tungstate Solution. *J. Electrochem. Soc.* **1997**, *144*, 1664–1671. [CrossRef]
36. Wang, Y.; Meng, Z.; Chen, H.; Li, T.; Zheng, D.; Xu, Q.; Wang, H.; Liu, X.Y.; Guo, W. Pulsed Electrochemical Deposition of Porous WO_3 on Silver Networks for Highly Flexible Electrochromic Devices. *J. Mater. Chem. C* **2019**, *7*, 1966–1973. [CrossRef]
37. Kim, C.-H.; Kim, Y.-S.; Choi, J.-Y.; Lee, I.-S.; Cha, B.-C.; Kim, D.-W.; Lee, J. Enhancement of Electrochromic Properties Using Nanostructured Amorphous Tungsten Trioxide Thin Films. *RSC Adv.* **2022**, *12*, 35320–35327. [CrossRef]
38. Boruah, P.J.; Khanikar, R.R.; Bailung, H. Synthesis and Characterization of Oxygen Vacancy Induced Narrow Bandgap Tungsten Oxide (WO_{3-x}) Nanoparticles by Plasma Discharge in Liquid and Its Photocatalytic Activity. *Plasma Chem. Plasma Process* **2020**, *40*, 1019–1036. [CrossRef]
39. Liu, Y.; Wygant, B.R.; Kawashima, K.; Mabayoje, O.; Hong, T.E.; Lee, S.-G.; Lin, J.; Kim, J.-H.; Yubuta, K.; Li, W.; et al. Facet Effect on the Photoelectrochemical Performance of a WO_3/$BiVO_4$ Heterojunction Photoanode. *Appl. Catal. B Environ.* **2019**, *245*, 227–239. [CrossRef]
40. Székely, I.; Kovács, G.; Baia, L.; Danciu, V.; Pap, Z. Synthesis of Shape-Tailored WO_3 Micro-/Nanocrystals and the Photocatalytic Activity of WO_3/TiO_2 Composites. *Materials* **2016**, *9*, 258. [CrossRef]
41. Sun, H.; Song, F.; Zhou, C.; Wan, X.; Jin, Y.; Dai, Y.; Zheng, J.; Yao, S.; Yang, Y. Lattice-Water-Induced Acid Sites in Tungsten Oxide Hydrate for Catalyzing Fructose Dehydration. *Catal. Commun.* **2021**, *149*, 106254. [CrossRef]
42. Zhou, Z.; Chen, Z.; Ma, D.; Wang, J. Porous $WO_3{\cdot}2H_2O$ Film with Large Optical Modulation and High Coloration Efficiency for Electrochromic Smart Window. *Sol. Energy Mater. Sol. Cells* **2023**, *253*, 112226. [CrossRef]
43. Martín, A.J.; Maffiotte, C.; Chaparro, A.M. Mechanisms for the Growth of Thin Films of WO_3 and Bronzes from Suspensions of WO_3 Nanoparticles. *ECS Trans.* **2015**, *64*, 43–56. [CrossRef]
44. Mineo, G.; Ruffino, F.; Mirabella, S.; Bruno, E. Investigation of WO_3 Electrodeposition Leading to Nanostructured Thin Films. *Nanomaterials* **2020**, *10*, 1493. [CrossRef]
45. Yang, B.; Miao, P.; Cui, J. Characteristics of Amorphous WO_3 Thin Films as Anode Materials for Lithium-Ion Batteries. *J. Mater. Sci: Mater. Electron.* **2020**, *31*, 11071–11076. [CrossRef]
46. Nguyen, H.T.T.; Truong, T.H.; Nguyen, T.D.; Dang, V.T.; Vu, T.V.; Nguyen, S.T.; Cu, X.P.; Nguyen, T.T.O. Ni-Doped WO_3 Flakes-Based Sensor for Fast and Selective Detection of H_2S. *J. Mater. Sci: Mater. Electron.* **2020**, *31*, 12783–12795. [CrossRef]
47. Vasilopoulou, M.; Soultati, A.; Georgiadou, D.G.; Stergiopoulos, T.; Palilis, L.C.; Kennou, S.; Stathopoulos, N.A.; Davazoglou, D.; Argitis, P. Hydrogenated Under-Stoichiometric Tungsten Oxide Anode Interlayers for Efficient and Stable Organic Photovoltaics. *J. Mater. Chem. A* **2014**, *2*, 1738–1749. [CrossRef]
48. Wei, Z.; Zhao, M.; Yang, Z.; Duan, X.; Jiang, G.; Li, G.; Zhang, F.; Hao, Z. Oxygen Vacancy-Engineered Titanium-Based Perovskite for Boosting H_2O Activation and Lower-Temperature Hydrolysis of Organic Sulfur. *Proc. Natl. Acad. Sci. USA* **2023**, *120*, e2217148120. [CrossRef] [PubMed]

49. Malikov, I.F.; Lyadov, N.M.; Salakhov, M.K.; Tagirov, L.R. Anion Doping of Tungsten Oxide with Nitrogen: Reactive Magnetron Synthesis, Crystal Structure, Valence Composition, and Optical Properties. *Crystals* **2024**, *14*, 109. [CrossRef]
50. Frankcombe, T.J.; Liu, Y. Interpretation of Oxygen 1s X-ray Photoelectron Spectroscopy of ZnO. *Chem. Mater.* **2023**, *35*, 5468–5474. [CrossRef]
51. Xu, Y.; Wu, S.; Wan, P.; Sun, J.; Hood, Z.D. Introducing Ti^{3+} Defects Based on Lattice Distortion for Enhanced Visible Light Photoreactivity in TiO_2 Microspheres. *RSC Adv.* **2017**, *7*, 32461–32467. [CrossRef]
52. Li, J.; Li, W.; Dong, H.; Li, N.; Guo, X.; Wang, L. Enhanced Performance in Hybrid Perovskite Solar Cell by Modification with Spinel Lithium Titanate. *J. Mater. Chem. A* **2015**, *3*, 8882–8889. [CrossRef]
53. Sun, S.; Li, H.; Xu, Z.J. Impact of Surface Area in Evaluation of Catalyst Activity. *Joule* **2018**, *2*, 1024–1027. [CrossRef]
54. Diao, J.; Yuan, W.; Qiu, Y.; Cheng, L.; Guo, X. A Hierarchical Oxygen Vacancy-Rich WO_3 with "Nanowire-Array-on-Nanosheet-Array" Structure for Highly Efficient Oxygen Evolution Reaction. *J. Mater. Chem. A* **2019**, *7*, 6730–6739. [CrossRef]
55. Le Formal, F.; Sivula, K.; Grätzel, M. The Transient Photocurrent and Photovoltage Behavior of a Hematite Photoanode under Working Conditions and the Influence of Surface Treatments. *J. Phys. Chem. C* **2012**, *116*, 26707–26720. [CrossRef]
56. Ponomarev, E.A.; Peter, L.M. A Generalized Theory of Intensity Modulated Photocurrent Spectroscopy (IMPS). *J. Electroanal. Chem.* **1995**, *396*, 219–226. [CrossRef]
57. Grigioni, I.; Polo, A.; Nomellini, C.; Vigni, L.; Poma, A.; Dozzi, M.V.; Selli, E. Nature of Charge Carrier Recombination in $CuWO_4$ Photoanodes for Photoelectrochemical Water Splitting. *ACS Appl. Energy Mater.* **2023**, *6*, 10020–10029. [CrossRef]
58. Nozik, A.J. Photoelectrochemistry: Applications to Solar Energy Conversion. *Annu. Rev. Phys. Chem.* **1978**, *29*, 189–222. [CrossRef]

materials

MDPI

Article

Synergetic Effect and Phase Engineering by Formation of $Ti_3C_2T_x$ Modified 2H/1T-$MoSe_2$ Composites for Enhanced HER

Lei Xiao [1], Qichao Yang [2], Xiangyang Zhu [2], Yang Wei [3,*] and Jing Wang [1,*]

1 School of Integrated Circuits, Tsinghua University, Beijing 100084, China; xiaolei@mail.tsinghua.edu.cn
2 School of Materials Science and Engineering, University of Science and Technology Beijing, Beijing 100083, China; 18866986789@163.com (Q.Y.); 202310192378@mail.scut.edu.cn (X.Z.)
3 Department of Physics and Tsinghua-Foxconn Nanotechnology Research Center, Tsinghua University, Beijing 100084, China
* Correspondence: weiyang@tsinghua.edu.cn (Y.W.); wang_j@mail.tsinghua.edu.cn (J.W.)

Abstract: The typical semi conductivity and few active sites of hydrogen evolution of 2H $MoSe_2$ severely restrict its electrocatalytic hydrogen evolution performance. At the same time, the 1T $MoSe_2$ has metal conductivity and plentiful hydrogen evolution sites, making it feasible to optimize the electrocatalytic hydrogen evolution behavior of $MoSe_2$ using phase engineering. In this study, we, through a simple one-step hydrothermal method, composed 1T/2H $MoSe_2$, and then used newly emerging transition metal carbides with several atomic-layer thicknesses $Ti_3C_2T_x$ to improve the conductivity of a $MoSe_2$-based electrocatalyst. Finally, $MoSe_2$@$Ti_3C_2T_x$ was successfully synthesized, according to the control of the additional amount of $Ti_3C_2T_x$, to form a proper $MoSe_2$/$Ti_3C_2T_x$ heterostructure with a better electrochemical HER performance. As obtained $MoSe_2$@4 mg-$Ti_3C_2T_x$ achieved a low overpotential, a small Tafel slope and this work offers additional insight into broadened $MoSe_2$ and MXenes-based catalyst's electrochemical application.

Keywords: $MoSe_2$@$Ti_3C_2T_x$ heterostructures; hydrogen evolution reaction; phase structure

1. Introduction

Hydrogen energy is one of the most prevalent potential candidates for renewable clean energy with high energy capacity to replace fossil fuel [1–3]. However, up to now, the most used industrial hydrogen production is still electrochemical water splitting [4–6]. Electrochemical HER is a cathode double-electron-transfer reaction that begins with the Volmer step, in which electrons are transferred to the cathode surface, bind to H+ at an unoccupied active site, and produce adsorbed H*. Currently, noble metals are the most feasible catalysts for electrochemical water splitting to reduce overpotential and simultaneously facilitate reaction kinetics of hydrogen evolution reaction (HER) [7–9]. Their scarcity and high price significantly restrict further commercial development and large-scale applications of these noble metals for water-splitting. Hence, finding cheap and earth-abundant electrocatalytic electrodes is essential to accelerating the commercial application of hydrogen energy.

Two-dimensional transition-metal dichalcogenides (2D TMDs) have many promoting properties for HER; 2D TMD's atomically thin-based structure increased reaction surface area and available active site [10], and different transition metals have various d orbitals. Thus, exhibiting different catalytic performance makes it feasible to improve the HER performance by doping different metal elements [11,12]. However, excellent catalytic activity and low cost compared to noble metals give it considerable potential in HER. $MoSe_2$, as a member of the 2D transition-metal dichalcogenide (TMDC), has received widespread attention due to its corrosion stability [13], chemically tunable electronic properties [14–18]. $MoSe_2$ can present as a octahedral coordination 1T (tetragonal symmetry, one layer per repeat unit, octahedral coordination) phase with metallic properties and a triangular coordination 2H (hexagonal symmetry, two lavers per repeat unit, trigonal prismatic coordination) phase with semiconductor properties. However, $MoSe_2$ electrochemical HER performances mainly depend upon its nanostructure and Crystalline phase, but for 2H phase $MoSe_2$, lack of active sites and weak conductivity restrict its performance in electrochemical HER [19–21]. Researchers use commonly used regulatory methods to improve their electrochemical performance such as strain engineering [22], surface/interface engineering [23], alloy engineering [24,25], defect engineering [26], and compounds with other materials [27–30]. However, designing and fabricating excellent $MoSe_2$-based composite catalysts for HER is of great fundamental importance to promoting TMDC-catalyst use of clean energy.

In this study, we, through a simple one-step hydrothermal method, use an emerging type of 2D transition-metal carbide $Ti_3C_2T_x$ (T = -F, -OH, -O terminal group) to improve the conductivity of $MoSe_2$-based electrocatalysts, $Ti_3C_2T_x$. A typical MXene material is a typically a layered material with a sandwich structure consisting of three alternating layers of Ti atoms and two alternating layers of C atoms. In addition, due to the introduction of a small number of residual chemical elements during the preparation process of acid-solution etching, the surface adsorbs rich terminal groups such as -F, -OH, -O, etc. As a conductor can improve the HER performance of $MoSe_2$ [31], the addition amount of $Ti_3C_2T_x$ must be controlled accordingly to form a proper $MoSe_2/Ti_3C_2T_x$ heterostructure which has a better electrochemical HER performance. A small Tafel slope (79 mV dec^{-1}) and low overpotential (−185.29 mV) were achieved with the obtained $MoSe_2$@4 mg-$Ti_3C_2T_x$. However, the $Ti_3C_2T_x$ improves the performance of 1H/2H nanoflower-like $MoSe_2$ in HER and is still not researched. This work utilizes $Ti_3C_2T_x$ as a composite substrate, and controls the $MoSe_2$ two-phase ratio. Both of these simultaneously enhance catalytic activity, which has significance for applying TMDC materials in the HER catalytic field.

2. Experiment Section

2.1. Synthesis of $Ti_3C_2T_x$ MXene Nanosheets

Firstly, via a relatively safe method, we combined a 5 wt % LiF powder and a 1 mol/HCl solution-etching in a glove box inert atmosphere, using a magnetic force-driving stirred reactor for the reaction. After heating in an oil bath for 96 h, AlF_3 is generated from the bulk-layered MXene phase, and Ti_3AlC_2 raw material is used to remove Al and generate $Ti_3C_2T_x$. Then, ultrasonic dispersion and cleaning are carried out to obtain the required few-layer $Ti_3C_2T_x$ thin sheets. The material was then dried in a vacuum-drying oven for 12 h. The product was then ground into powder in an agate mortar, collected, and vacuumed, successfully preparing layered $Ti_3C_2T_x$.

2.2. Carbon Cloth Preprocess for Hydrophilic Treatment

According to the volume ratio of 4:1, we mixed 10% HNO_3 and 20% H_2SO_4; soaked the carbon cloth (1 cm × 1 cm) at room temperature for 12 h; took it out, and put it in ultrasonic ethanol solution for 30 min; then, used DI water wash for 30 s; and repeated 3 times to remove the residual acid. It must be sealed and stored in deionized water.

2.3. Preparation of 1T/2H Phase $MoSe_2$@$Ti_3C_2T_x$ on CC

Figure 1 shows that Se powder (1.0 mmol) was dissolved into 15% hydrazine hydrate. Magnetic solution was stirred and kept for 24 h. At the same time, a few $Ti_3C_2T_x$ powders were dispersed with ultrasound (600 W) in 20 mL deionized water for 30 min; then, 1 mmol of $Na_2MoO_4 \cdot 2H_2O$ was added to the dispersion and stirred for 45 min. After that, we added the reduced Se ion solution to the dispersion and sealed it in a 50 mL high-pressure reactor, using a carbon cloth (1 cm × 1 cm) as the loaded substrate, and then heated it to 180 °C for insulation for 24 h. After cooling, the sample was washed and dried in a vacuum environment of 60 °C. During the process, attention should be paid to preventing oxidation.

Figure 1. Schematic diagram of hydrothermal preparation of 1T/2H $MoSe_2$@$Ti_3C_2T_x$.

2.4. Characterization

The morphology of the sample was characterized using field emission scanning electron microscopy (FEI, HONG KONG, SEM, FEI Nova 400), and X-ray diffraction (XRD) patterns were recorded using Cu Ka radiation on Rigaku D/Max 4000 V to determine sample-composition information. Raman spectral information was obtained using a confocal microscope (HORIBA FRANCE LabRAM HR Evolution), using a 532 nm laser at low power. XPS (Thermo ESCALAB 250XI) determined the chemical configuration.

2.5. Electrochemical Measurement

The electrocatalytic HER measurement of the obtained products was conducted using a standard three-electrode electrochemical workstation consisting of an electrode counter electrode (Pt), a reference electrode Ag/AgCl, and a stainless-steel working electrode. All the electrocatalytic HER measurements were performed in 1 mol/L H^+ solution. Electrode potential was converted into reversible hydrogen electrode potential (vs. RHE), and solution IR compensation was performed. The LSV curves of the prepared material were measured at different scanning rates in an inert constant-room-temperature environment to characterize its catalytic hydrogen evolution performance. Electrochemical impedance spectroscopy and scanning CV curves were also measured to obtain the electrochemical surface area.

3. Result and Discussion

Characterization of $Ti_3C_2T_x$

First, the $Ti_3C_2T_x$ was characterized using a Scanning Electron Microscope shown in Figure 2a $Ti_3C_2T_x$ which has prominent layered characteristics, which allows for a large contact area with $MoSe_2$ and CC. However, due to introducing some terminal groups on the surface of Ti_3C_2 during the preparation process, the Raman spectrum was used to characterize its molecular vibration. As shown in Figure 2b, in addition to its inherent interlayer vibration mode E_g peak, there were three apparent peaks at 235 cm^{-1}, 373 cm^{-1}, and 610 cm^{-1}, which, respectively, represent the vibration modes of terminal groups such as F and OH, which are the same as previously reported [32]. To determine the phase-composition structure of the $Ti_3C_2T_x$, we carried out XRD, as shown in Figure 2c. $Ti_3C_2T_x$ had prominent peaks near 7.3°, 34.9°, 35.2° and 63°, which, respectively, correspond to the (002), (001), (104), (105) crystal planes of Ti_3C_2 [33], and the sharp peak represents its good crystallinity. However, XPS was used further to determine the element composition and valence state of $Ti_3C_2T_x$. As shown in Figure 2d, the XPS full spectra of $Ti_3C_2T_x$ showed distinctive peaks of Ti and OH F, O. The Ti 2p XPS spectra after deconvoluting is shown in Figure 2e, where there are double peaks at lower binding energies of 455.6 eV and higher binding energies of 465 eV were obtained, confirming the existence of oxygen-containing terminal groups, which further confirmed our previous Raman results. The C XPS spectra after deconvoluting are shown in Figure 2f; double peaks at lower binding energies of 284.8 eV and higher binding energies of 281.8 eV were obtained after Gaussian Lorentz fitting. The stoichiometric ratio of Ti and C is about 100:68. However, The $Ti_3C_2T_x$ we obtained was rich in oxygen, hydroxyl, F, and other terminal groups.

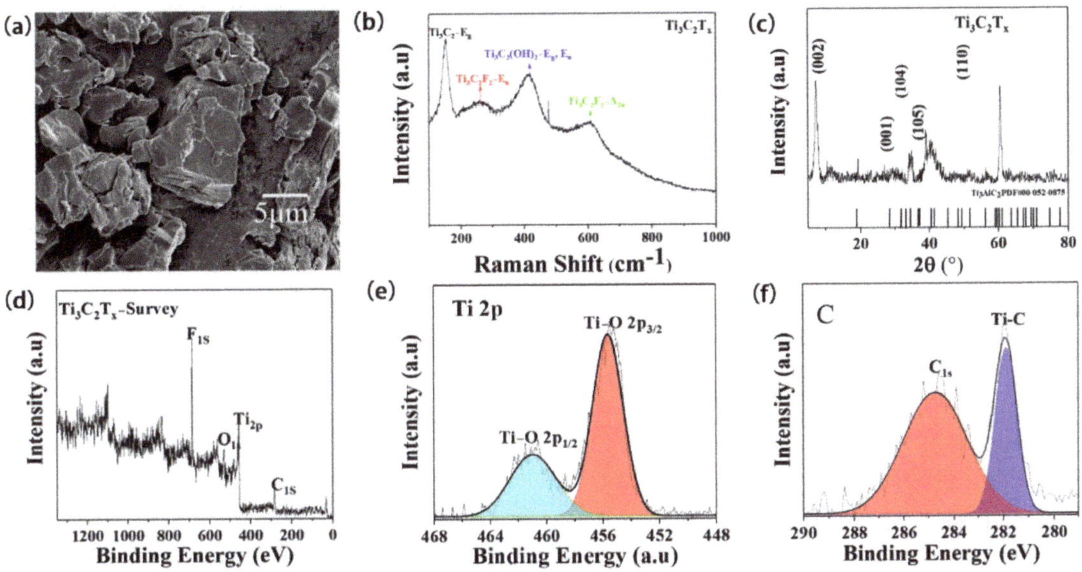

Figure 2. (**a**) SEM image; (**b**) Raman Spectrum; (**c**) XRD; and (**d**–**f**) XPS of $Ti_3C_2T_x$.

$MoSe_2$@$Ti_3C_2T_x$ was successfully synthesized on the Carbon Clothes with a facile one-step hydrothermal. During the preparation of the samples, the $Ti_3C_2T_x$ played a key role, and SEM was used to observe the influence of the $Ti_3C_2T_x$ on the surface morphology features of the synthesized samples. As shown in Figure 3a–c, with the increase of $Ti_3C_2T_x$ content, more and more $MoSe_2$ became attached to the carbon cloth substrate. This may be due to the intense attraction between terminal groups (-F, -OH, -O) on $Ti_3C_2T_x$ and carbon fiber cloth substrate. However, the surface of $MoSe_2$@4 Mg $Ti_3C_2T_x$ had the most evenly distributed $MoSe_2$ nanoflowers. A high-magnification scanning electron microscope was used to analyze its surface morphology further to study the $MoSe_2$@4 Mg $Ti_3C_2T_x$. High-magnification scanning microscope photographs of $MoSe_2$ nanoflowers are shown in Figure 3d–f. The size distribution of a single nanoflower was between 30 and 60 nm.

Figure 3. SEM images of synthesized $MoSe_2@Ti_3C_2T_x/CC$ samples; (**a**) $MoSe_2@2\ mgTi_3C_2T_x/CC$; (**b**) $MoSe_2@4\ mgTi_3C_2T_x/CC$; (**c**) $MoSe_2@6\ mgTi_3C_2T_x/CC$; and $MoSe_2$ nanoflower on the samples of $MoSe_2@4\ mgTi_3C_2T_x/CC$ at (**d**) ×5000, (**e**) ×100,000, (**f**) ×200,000.

The Raman spectrum of all the samples show a characteristic peak at ≈285.4 cm^{-1}. In accurate compliance to the in-plane ($E^1{}_{2g}$) mode, two other peaks at ≈239.4 cm^{-1} and ≈342.5 cm^{-1} on both sides of the main peak were found, (Figure 4a). Corresponding to A_{1g} and $B^1{}_{2g}$, the characteristic Raman manifested as 2H $MoSe_2$ [34–37]. Furthermore, there are two inconspicuous peaks at ≈194 cm^{-1} and ≈376 cm^{-1}, which correspond to the characteristic Raman mode of 1T $MoSe_2$ [38,39]. However, with the increase of $Ti_3C_2T_x$, there was no apparent shift in each peak position, and no characteristic peak of $Ti_3C_2T_x$ was found in each sample; this may be due to the low amount of $Ti_3C_2T_x$ or the adsorption of $Ti_3C_2T_x$ on the carbon fiber cloth substrate. Raman spectra show that the $MoSe_2$ we synthesized were 1T/2H multiple phases. To ascertain the existing phase composition and structural information of the as-synthesized samples XRD was carried out. The results are shown in Figure 4b; three prominent peaks were found which were located around 13.7°, 31.4° and 58°, respectively. Representing the (002), (110) and (008) crystal planes of $MoSe_2$, due to the presence of surface structural defects, the (002) and (100) peaks of $MoSe_2$ broaden in the XRD pattern [40]. Compared with the 2H-phase $MoSe_2$ (JCPDS No 29–0914), standard XRD pattern means all the sample peaks of the (100) diffraction peaks shift right to a larger angle and the (002) crystal plane shifts left to a smaller angle, which is in good agreement with the result demonstrating the presence of 1T-phase $MoSe_2$ in all synthesized samples [41,42]. With the increase in the amount of $Ti_3C_2T_x$, a prominent peak corresponding to the (110) crystal plane of Ti_3C_2 was observed at 60.5° in $MoSe_2@4$ mg. At $Ti_3C_2T_x$, $MoSe_2@6$ mg $Ti_3C_2T_x$, another peak was observed near 7.3° and in $MoSe_2@6$ mg $Ti_3C_2T_x$, corresponding to the (002) crystal plane of Ti_3C_2. Therefore, we reached the preliminary conclusion that $Ti_3C_2T_x$ did not decompose or participate in the reaction in the $MoSe_2@Ti_3C_2T_x$ synthesis. Next, XPS survey spectra illustrate that C, Ti, Mo, and Se elements are uniformly distributed in all samples. Next, the XPS spectra of $MoSe_2@Ti_3C_2T_x$ were further investigated for the presence of the 1T phase in the samples. Figure 4c shows the XPS spectrum in the Mo region. Two peaks at 54.7 and 53.7 eV correspond to 1T-phase $MoSe_2$; double peaks at 228.3 and 231.4 eV correspond to Mo within the 1T phase, whereas 55.2 and 54.3 eV are assigned to the 2H phase $MoSe_2$. The other pair at 229.3 eV and 232.5 eV can be assigned to the $3d_{5/2}$ and $3d_{3/2}$ orbitals of Mo within the 2H phase [43,44]. However, the 1T and 2H contents of each sample were different. The 1T-phase content in

$MoSe_2$@2 mg $Ti_3C_2T_x$ was 64.4%; for $MoSe_2$@4 mg $Ti_3C_2T_x$, it was 66.9%; and the 1T-phase content in $MoSe_2$@6 mg $Ti_3C_2T_x$ was 54.58%. However, according to the previous report, temperature plays an important role in the formation of 1T $MoSe_2$ [45]. Theoretically, there should be little difference in the 1T $MoSe_2$ content of the sample. However, the addition of $Ti_3C_2T_x$ may have affected the synthesis of 1T-phase $MoSe_2$, but the terminal groups of $Ti_3C_2T_x$ may prevent the formation of $MoSe_2$. All the XPS results verified that 1T and 2H phases coexist in the samples we synthesized.

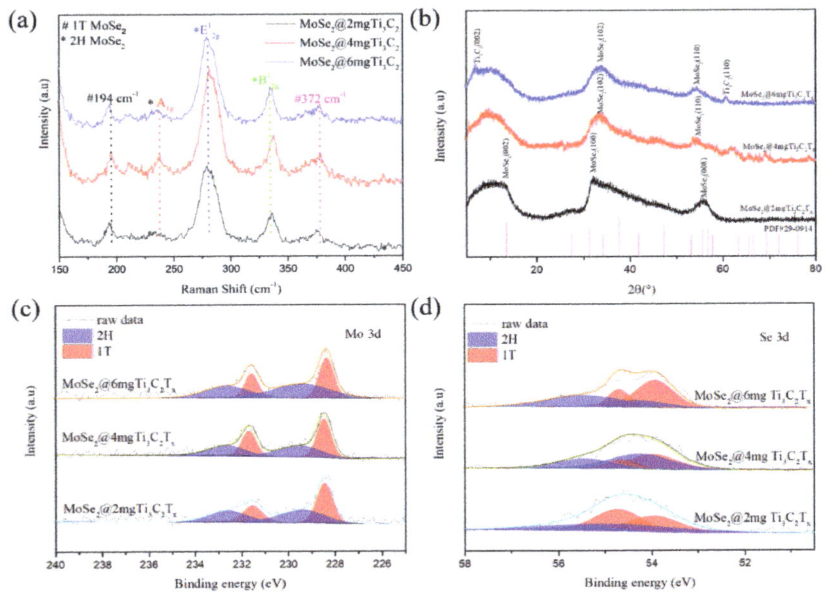

Figure 4. $MoSe_2$ synthesized with different $Ti_3C_2T_x$-added amounts; (**a**) Raman spectrometer shifts; (**b**) XRD pattern; (**c**) Mo 3d XPS spectra; and (**d**) Se 3d XPS spectra.

We first tested the LSV curves for bare CC and MXene on CC, as shown in Figure 5. Dropping a small amount of MXene onto the carbon cloth can significantly improve the catalytic performance of the carbon cloth by about 2 mg. This may be due to the good conductivity of MXene. In addition, to determine the stability of MXene on the carbon cloth, we used a similar method in the paper 1000 CV cycles, and the performance of the sample did not change significantly. This may be due to the strong adsorption ability of MXene as a nanosheet and CC substrate, which did not considerably detach during the catalytic process. The HER electrochemical catalytic properties of $MoSe_2$ obtained with a diverse mass of $Ti_3C_2T_x$ were investigated in Figure 6a. According to XPS, the polarization curves after IR calibration are our work's best value when the mass of $Ti_3C_2T_x$ was 4 mg. The overpotential of η@10 mA/cm^2 was 184. 4 mV. Still, with the increase of the $Ti_3C_2T_x$, the overpotential of @6 mg $Ti_3C_2T_x$ drops to 244.8 mV. This is due to its low 1T-phase content. However, compared with the sample without $Ti_3C_2T_x$, using different masses of $Ti_3C_2T_x$ suggests that the existence of $Ti_3C_2T_x$ plays an important role in the majorization of HER electrocatalytic performance. The Tafel slope reflecting chemical reactivity in the HER further confirms the above results. Derived from their corresponding polarization curves, the samples prepared under different conditions are displayed in Figure 6b. The Tafel slope significantly decreased (79 mV dec^{-1}), indicating a faster HER process in the sample. To understand the difference in the HER catalytic performance of the samples we synthesized varied $Ti_3C_2T_x$ amounts in the Heyrovsky reaction rate-resolving stage. Furthermore, we analyzed CV measurement results to evaluate the effective electrochemical activity surface area (ECSA) by measuring C_{dl} stemming.

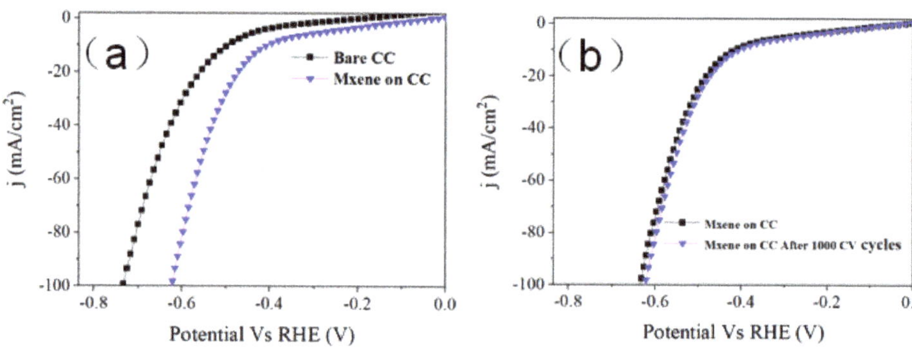

Figure 5. (**a**) LSV of Bare CC and MXene on CC; (**b**) LSV of MXene on CC and MXene on CC after 1000 CV cycles.

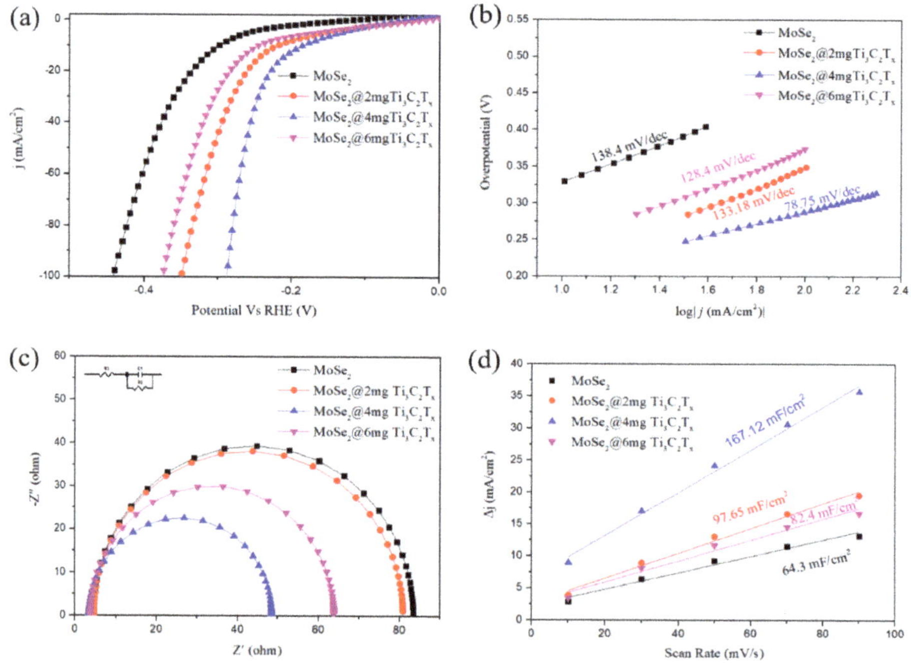

Figure 6. Electrochemical properties of $MoSe_2$@ $Ti_3C_2T_x$ synthesized with diverse amount of $Ti_3C_2T_x$. (**a**) Polarization curves after IR calibration; (**b**) corresponding Tafel plots of the samples stemming from (**a**); (**c**) Nyquist plots of the samples; and (**d**) C_{dl} current at different scan rates.

Samples with different component structures' capacitive current and scan rates are presented in Figure 6c. $MoSe_2$@4 mg-$Ti_3C_2T_x$ also manifested larger C_{dl} (167.12 mF cm^{-2}). Because of the existence of surface structural defects, the composite phase structure exposes more active site, which improves the performance of electrocatalytic HER. EIS tests were also conducted further to understand the electrochemical behavior in the HER process. Figure 6d presents the Nyquist plots of tested samples, displaying stable-system semicircle features. The charge-transfer resistance (R_{ct}) can be obtained by calculating and fitting the Nyquist plot of Figure 6c. The R_{ct} values $MoSe_2$@4 mg-$Ti_3C_2T_x$ (23 Ω), $MoSe_2$@2 mg-$Ti_3C_2T_x$ (39 Ω), $MoSe_2$@6 mg-$Ti_3C_2T_x$ (30 Ω), and $MoSe_2$ (40 Ω) results as mentioned above expound that the addition of $Ti_3C_2T_x$ is critical to enhancing HER catalytic activity. From work obtained samples, $MoSe_2$@4 mg$Ti_3C_2T_x$ that exhibit the best HER electrocatalytic performance.

The double plate capacitance extracted in the nonfaradaic-zone CV cycle is proportional to the electrochemical activation area of the catalytic electrode. Figure 7 shows the cyclic voltammetry curve of the sample for different $MoSe_2$@$Ti_3C_2T_x$. The sample was subjected to cyclic voltammetry testing. The test solution was consistent with the previous electrochemical workstation and still used a 0.5 mol/L H_2SO_4 solution. The scanning rates were 10, 30, 50, 70, and 90 mV/s. In addition, to ensure the stability of the sample, the last one was selected for 100 cycles at each rate and plotted. As shown in Figure 7a, the $MoSe_2$ sample electrode without adding $Ti_3C_2T_x$ showed the smallest capacitive behavior, followed by the electrode sample with the added 2 mg $Ti_3C_2T_x$. As the content of $Ti_3C_2T_x$ increases, the area of the CV curve gradually increases. The sample with a dosage of 4 mg $Ti_3C_2T_x$ has the largest CV area. When the dosage of $Ti_3C_2T_x$ is further increased to 6 mg, the geometric area of the CV curve shows a significant reduction trend. This indicates that the doping amount of $Ti_3C_2T_x$ has an optimal value. When the doping amount of $Ti_3C_2T_x$ reaches an optimal value, its catalytic activity will exceed the contribution of the carbon cloth matrix, becoming the dominant role instead. However, its performance will gradually decline after exceeding a specific optimization interval. Based on the previous XPS characterization results, this is due to the addition of a large amount of $Ti_3C_2T_x$, which is affected by the adsorption of terminal groups -F, -OH, etc., on the surface of $Ti_3C_2T_x$. The solution's environment during the preparation of 1T phase is disrupted, thereby affecting the generation of 1T-phase $MoSe_2$, and reducing the content of 1T phase in the sample. In addition, due to the reduction of $MoSe_2$ content in the 1T phase, the number of hydrogen evolution catalytic active sites introduced by the disordered arrangement of atoms at the interface of 1T and 2H phase in the sample due to different structures, will also be reduced. This further affects its catalytic activity and reduces the CV area at different scanning rates.

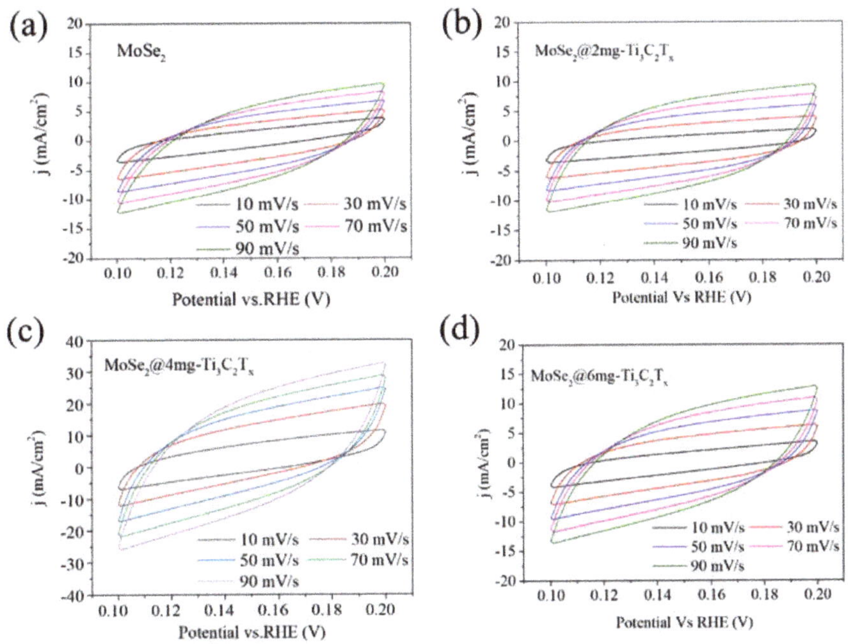

Figure 7. CV scanning curve in the scanning range of 0.1 V–0.2 V for different samples of (**a**) $MoSe_2$, (**b**) $MoSe_2$@2 mg-$Ti_3C_2T_x$, (**c**) $MoSe_2$@4 mg-$Ti_3C_2T_x$, and (**d**) $MoSe_2$@6 mg-$Ti_3C_2T_x$.

We investigated the catalytic performance of $Ti_3C_2T_x$ during the in situ growth of $MoSe_2$ electrodes on carbon cloth. The results showed that adding $Ti_3C_2T_x$ optimized the electrocatalytic hydrogen evolution performance of $MoSe_2$ electrodes to varying degrees. Based on characterization methods, it is speculated that adding $Ti_3C_2T_x$ changed the in situ growth process of $MoSe_2$ on carbon cloth. Due to the introduction of active groups, such as F during the preparation of $Ti_3C_2T_x$, it is more inclined to adhere to the carbon cloth. On the other hand, as a two-dimensional material, it is easy to synthesize heterostructures with $MoSe_2$ based on the interlayer van der Waals force. In addition, its natural good conductivity can act as a good electronic channel between $MoSe_2$ and the carbon cloth. Therefore, $MoSe_2$ not only exposes more electrocatalytic active sites but also greatly improves the adhesion between the catalyst and the carbon cloth, so it can effectively optimize the electrocatalytic performance of $MoSe_2$. Based on the successful preparation of 1T/2H impure-phase $MoSe_2$, we investigated the effect of $Ti_3C_2T_x$ addition on the electrocatalytic performance of $MoSe_2$. For samples with a low amount of $Ti_3C_2T_x$ addition, $Ti_3C_2T_x$ cannot perform its expected conductivity, and the optimization of $MoSe_2$ electrocatalytic performance is limited. However, when more $Ti_3C_2T_x$ is added, the terminal groups adsorbed on the surface of $Ti_3C_2T_x$ will affect the formation of 1T-phase $MoSe_2$. With the increase of $Ti_3C_2T_x$ addition, the sample with 4 mg $Ti_3C_2T_x$ exhibited good electrocatalytic hydrogen evolution performance.

As shown in Table 1, the enhanced catalytic performance is attributed to a larger density of exposed catalytically active edge sites and a higher electrical conductivity. Other studies on improved catalytic activity of $MoSe_2$ have shown that the structure of the high-content 1T phase with defects is instrumental in enhancing HER performance [46].

Table 1. Electrochemical performance (j = 10 mA cm^{-2}) comparison of different $MoSe_2$-based composite samples.

Sample	Overpotential (V vs. RHE)	Tafel Slope (mV dec^{-1})	References
2H $MoSe_2$	0.300	82	[47]
$MoSe_2$-amorphous CNT	0.254	49	[48]
C@$MoSe_2$	0.270	72	[49]
Graphene-carbon nanotube aerogel-$MoSe_2$ hybrid	0.228	68	[50]
N-doped RGO/$MoSe_2$ composites	0.229	78	[51]
$MoSe_2/Ti_3C_2T_x$	0.185	79	This work

To study the stability of the sample, we conducted experiments on the stability of the samples $MoSe_2$@$Ti_3C_2T_x$. After the CV test, the following experiments were carried out: the SEM of samples shown in Figure 8a, and the sample Raman spectroscopy shown in Figure 8b. There were no shifts for the Raman peak of 1T and 2H $MoSe_2$ after the test, and the main crystal planes remain the same as before the CV test as shown in Figure 8c, and the half-peak width did not change significantly, indicating that the test did not significantly change its crystal structure. We used XPS to illustrate whether $MoSe_2$ transitioned from 1T to 2H after the test, since phase 2H is a thermodynamically stable phase. As shown in the figure, the chemical elements and valence states that the full spectrum of XPS did not change significantly, and in spectroscopy, we used the same method as in the paper to split the Mo 3d and Se 3d peaks, respectively. However, the phase content in $MoSe_2$@2 mg $Ti_3C_2T_x$ after the CV test decreased from 64.4% to 63.3%. This reduction is tolerable since the samples were dried before testing.

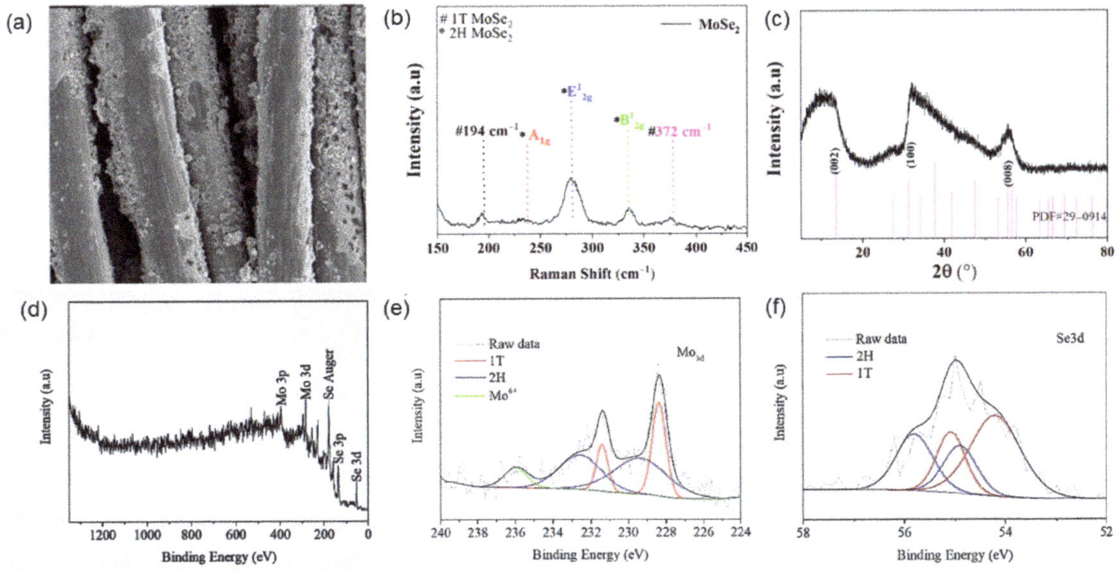

Figure 8. Characterization of MoSe2@2 mg $Ti_3C_2T_x$ after CV test samples of (**a**) SEM, (**b**) Raman spectrum, (**c**) XRD spectrum, (**d**) XPS full spectrum, (**e**) XPS Mo 3d peak, and (**f**) XPS Se 3d peak.

In addition, the surface morphology of the samples after different cycles was observed. Figure 9a–c show the initial morphology of the sample without cycling. It can be seen that a large number of $MoSe_2$ clusters uniformly adhere to the surface of the sample, and there is also a small amount of $MoSe_2$ powder attached to the surface of the carbon cloth fiber. Still the contribution to hydrogen evolution performance is extremely limited due to the small contact area with the carbon cloth. A clear tissue network can be seen with significant adhesion to the substrate at high magnification. Figure 9d–f shows the surface morphology of the sample after 500 cycles, and $MoSe_2$ powder on the surface of the carbon cloth has obviously fallen off. Still, the $MoSe_2$ clusters adsorbed on the surface have not obviously fallen off. Still the cluster morphology has changed greatly, the $MoSe_2$ on the surface has changed from loose nanoflower clusters to "chocolate" like whole blocks attached to the surface of the carbon cloth, and small $MoSe_2$ particles can still be seen at local magnification. The particles are closely bound to the carbon cloth substrate. In Figure 9g–i, a large number of $MoSe_2$ nanoparticles are still uniformly adhered to the surface of the sample after 1000 cycles. However, the difference is that the particle size is significantly reduced under local magnification, with a small amount of detachment locally. Overall, the majority of $MoSe_2$ has a relatively tight binding force. It is worth noting that Nafion and other reagents were not used to promote the adhesion of $MoSe_2$ on the substrate during this experimental testing process. The small amount of $MoSe_2$ attached to the surface of the carbon cloth itself has a limited contribution to the catalytic performance of the catalyst due to its small interface with the carbon cloth, and its detachment will not significantly impact the catalyst performance.

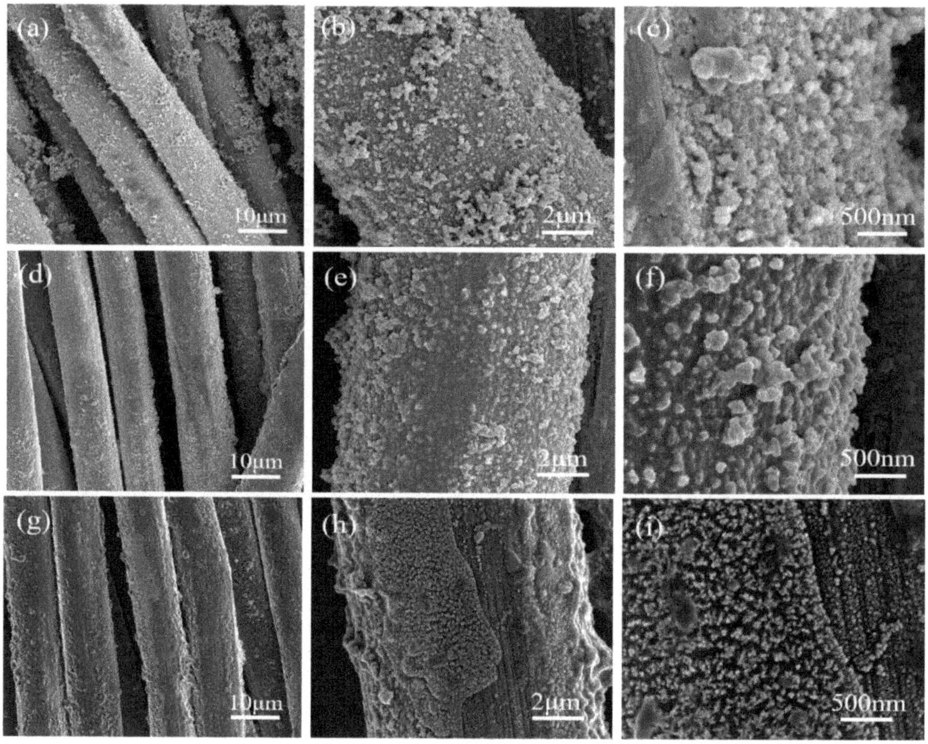

Figure 9. (**a–c**) Carbon cloth in situ growth $MoSe_2$@4 mg $Ti_3C_2T_x$ initial surface morphology of sample (**d–f**) after 500 Cycles, and microscopic morphology of the sample at different magnification times, and (**g–i**) after 1000 cycles, and microscopic morphology of the sample at different magnification times.

4. Conclusions

In this work, the two-dimensional $Ti_3C_2T_x$ was added to form the heterostructure with 1T/2H $MoSe_2$ to improve the contact with the conductive carbon fiber cloth substrate. In addition, $Ti_3C_2T_x$ itself, as a good conductor, can provide fast channels and more active sites for electron transfer during electrochemical reactions. According to the control, the addition of a mass of $Ti_3C_2T_x$ to form proper $MoSe_2/Ti_3C_2T_x$ heterostructures with low overpotentials (−185.29 mV) and small Tafel slopes (79 mV dec^{-1}) are achieved with the obtained $MoSe_2$@4 mg-$Ti_3C_2T_x$, Moreover, phase engineering synergistic effects and remaking heterostructures using $MoSe_2$ with other materials can also be used for $MoSe_2$ future development in the field of electrochemical catalysis. However, how the terminal groups of $Ti_3C_2T_x$ influence the formation of the 1T $MoSe_2$ is still not clear. More comparative experiments need to be carried out with further $Ti_3C_2T_x$ addition.

Author Contributions: Conceptualization, Y.W. and J.W.; Methodology, Y.W. and J.W.; Software, Q.Y.; Formal analysis, L.X.; investigation, Q.Y.; Data curation, X.Z.; Writing—original draft, L.X.; writing-review and editing, Y.W. and J.W. All authors have read and agreed to the published version of the manuscript.

Funding: This research received no external funding.

Informed Consent Statement: Not applicable.

Data Availability Statement: Not applicable.

Conflicts of Interest: The authors declare no conflict of interest.

References

1. Seh, Z.W.; Kibsgaard, J.; Dickens, C.F.; Chorkendorff, I.B.; Nørskov, J.K.; Jaramillo, T.F. Combining theory and experiment in electrocatalysis: Insights into materials design. *Science* **2017**, *355*, eaad4998. [CrossRef] [PubMed]
2. Schlapbach, L.; Züttel, A. Hydrogen-storage materials for mobile applications. *Nature* **2001**, *414*, 353–358. [CrossRef] [PubMed]
3. Wang, X.; Maeda, K.; Thomas, A.; Takanabe, K.; Xin, G.; Carlsson, J.M.; Domen, K.; Antonietti, M. A metal-free polymeric photocatalyst for hydrogen production from water under visible light. *Nat. Mater.* **2009**, *8*, 76–80. [CrossRef] [PubMed]
4. Bodhankar, P.M.; Sarawade, P.B.; Kumar, P.; Vinu, A.; Kulkarni, A.P.; Lokhande, C.D.; Dhawale, D.S. Nanostructured Metal Phosphide Based Catalysts for Electrochemical Water Splitting: A Review. *Small* **2022**, *18*, 2107572. [CrossRef]
5. Luo, Y.; Zhang, Z.; Chhowalla, M.; Liu, B. Recent Advances in Design of Electrocatalysts for High-Current-Density Water Splitting. *Adv. Mater.* **2021**, *34*, 2108133. [CrossRef]
6. Cabán-Acevedo, M.; Stone, M.L.; Schmidt, J.R.; Thomas, J.G.; Ding, Q.; Chang, H.C.; Tsai, M.L.; He, J.H.; Jin, S. Efficient hydrogen evolution catalysis using ternary pyrite-type cobalt phosphosulphide. *Nat. Mater.* **2015**, *14*, 1245–1251. [CrossRef]
7. Mahmood, J.; Li, F.; Jung, S.M.; Okyay, M.S.; Ahmad, I.; Kim, S.J.; Park, N.; Jeong, H.Y.; Baek, J.B. An efficient and pH-universal ruthenium-based catalyst for the hydrogen evolution reaction. *Nat. Nanotechnol.* **2017**, *12*, 441–446. [CrossRef]
8. Han, X.; Wu, X.; Deng, Y.; Liu, J.; Lu, J.; Zhong, C.; Hu, W. Ultrafine Pt Nanoparticle-Decorated Pyrite-Type CoS_2 Nanosheet Arrays Coated on Carbon Cloth as a Bifunctional Electrode for Overall Water Splitting. *Adv. Energy Mater.* **2018**, *8*, 1800935. [CrossRef]
9. Zhao, Z.; Liu, H.; Gao, W.; Xue, W.; Liu, Z.; Huang, J.; Pan, X.; Huang, Y. Surface-Engineered PtNi-O Nanostructure with Record-High Performance for Electrocatalytic Hydrogen Evolution Reaction. *J. Am. Chem. Soc.* **2018**, *140*, 9046–9050. [CrossRef]
10. Jaramillo, T.F.; Jørgensen, K.P.; Bonde, J.; Nielsen, J.H.; Horch, S.; Chorkendorff, I. Identification of Active Edge Sites for Electrochemical H_2 Evolution from MoS_2 Nanocatalysts. *Science* **2007**, *317*, 100–102. [CrossRef]
11. Fu, Q.; Han, J.; Wang, X.; Xu, P.; Yao, T.; Zhong, J.; Zhong, W.; Liu, S.; Gao, T.; Zhang, Z.; et al. 2D Transition Metal Dichalcogenides: Design, Modulation, and Challenges in Electrocatalysis. *Adv. Mater.* **2021**, *33*, 1907818. [CrossRef] [PubMed]
12. Voiry, D.; Yang, J.; Chhowalla, M. Recent Strategies for Improving the Catalytic Activity of 2D TMD Nanosheets Toward the Hydrogen Evolution Reaction. *Adv. Mater.* **2016**, *28*, 6197–6206. [CrossRef] [PubMed]
13. Zhang, Z.; Wang, J.; Li, Y.; Zhang, S.; Xiao, L.; Wang, J.; Qi, J. Hydrothermal preparation of MoX_2 (X = S, Se, Te)/TaS_2 hybrid materials on carbon cloth as efficient electrocatalyst for hydrogen evolution reaction. *Int. J. Hydrogen Energy* **2023**, *48*, 4207–4219. [CrossRef]
14. Deng, S.; Zhong, Y.; Zeng, Y.; Wang, Y.; Yao, Z.; Yang, F.; Lin, S.; Wang, X.; Lu, X.; Xia, X.; et al. Directional Construction of Vertical Nitrogen-Doped 1T-2H $MoSe_2$/Graphene Shell/Core Nanoflake Arrays for Efficient Hydrogen Evolution Reaction. *Adv. Mater.* **2017**, *29*, 1700748. [CrossRef]
15. Xu, C.; Peng, S.; Tan, C.; Ang, H.; Tan, H.; Zhang, H.; Yan, Q. Ultrathin S-doped $MoSe_2$ nanosheets for efficient hydrogen evolution. *J. Mater. Chem. A* **2014**, *2*, 5597–5601. [CrossRef]
16. Najafi, L.; Bellani, S.; Oropesa-Nuñez, R.; Ansaldo, A.; Prato, M.; Del Rio Castillo, A.E.; Bonaccorso, F. Doped-$MoSe_2$ Nanoflakes/3d Metal Oxide–Hydr(Oxy)Oxides Hybrid Catalysts for pH-Universal Electrochemical Hydrogen Evolution Reaction. *Adv. Energy Mater.* **2018**, *8*, 1801764. [CrossRef]
17. Chen, X.; Qiu, Y.; Liu, G.; Zheng, W.; Feng, W.; Gao, F.; Cao, W.; Fu, Y.; Hu, W.; Hu, P. Tuning electrochemical catalytic activity of defective 2D terrace $MoSe_2$ heterogeneous catalyst via cobalt doping. *J. Mater. Chem. A* **2017**, *5*, 11357–11363. [CrossRef]
18. Qian, J.; Wang, T.; Xia, B.; Xi, P.; Gao, D. Zn-doped $MoSe_2$ nanosheets as high-performance electrocatalysts for hydrogen evolution reaction in acid media. *Electrochim. Acta* **2019**, *296*, 701–708. [CrossRef]
19. Ambrosi, A.; Sofer, Z.; Pumera, M. 2H → 1T phase transition and hydrogen evolution activity of MoS_2, $MoSe_2$, WS_2 and WSe_2 strongly depends on the MX_2 composition. *Chem. Commun.* **2015**, *51*, 8450–8453. [CrossRef]
20. Deng, S.; Yang, F.; Zhang, Q.; Zhong, Y.; Zeng, Y.; Lin, S.; Wang, X.; Lu, X.; Wang, C.Z.; Gu, L.; et al. Phase Modulation of (1T-2H)-$MoSe_2$/TiC-C Shell/Core Arrays via Nitrogen Doping for Highly Efficient Hydrogen Evolution Reaction. *Adv. Mater.* **2018**, *30*, 1802223. [CrossRef]
21. Jiang, M.; Zhang, J.; Wu, M.; Jian, W.; Xue, H.; Ng, T.W.; Lee, C.S.; Xu, J. Synthesis of 1T-$MoSe_2$ ultrathin nanosheets with an expanded interlayer spacing of 1.17 nm for efficient hydrogen evolution reaction. *J. Mater. Chem. A* **2016**, *4*, 14949–14953. [CrossRef]
22. Wang, L.; Zeng, Z.; Gao, W.; Maxson, T.; Raciti, D.; Giroux, M.; Pan, X.; Wang, C.; Greeley, J. Tunable intrinsic strain in two-dimensional transition metal electrocatalysts. *Science* **2019**, *363*, 870–874. [CrossRef] [PubMed]
23. Wang, P.; Wang, B. Designing Self-Supported Electrocatalysts for Electrochemical Water Splitting: Surface/Interface Engineering toward Enhanced Electrocatalytic Performance. *ACS Appl. Mater. Interfaces* **2021**, *13*, 59593–59617. [CrossRef] [PubMed]
24. Lin, Z.; Lin, B.; Wang, Z.; Chen, S.; Wang, C.; Dong, M.; Gao, Q.; Shao, Q.; Ding, T.; Liu, H.; et al. Facile Preparation of 1T/2H-$Mo(S_{1-x}Se_x)_2$ Nanoparticles for Boosting Hydrogen Evolution Reaction. *ChemCatChem* **2019**, *11*, 2217–2222. [CrossRef]
25. Gong, Q.; Cheng, L.; Liu, C.; Zhang, M.; Feng, Q.; Ye, H.; Zeng, M.; Xie, L.; Liu, Z.; Li, Y. Ultrathin $MoS_{2(1-x)}Se_{2x}$ Alloy Nanoflakes For Electrocatalytic Hydrogen Evolution Reaction. *ACS Catal.* **2015**, *5*, 2213–2219. [CrossRef]
26. Kwak, I.H.; Kwon, I.S.; Lee, J.H.; Lim, Y.R.; Park, J. Chalcogen-vacancy group VI transition metal dichalcogenide nanosheets for electrochemical and photoelectrochemical hydrogen evolution. *J. Mater. Chem. C* **2021**, *9*, 101–109. [CrossRef]

27. Zhou, X.; Liu, Y.; Ju, H.; Pan, B.; Zhu, J.; Ding, T.; Wang, C.; Yang, Q. Design and Epitaxial Growth of $MoSe_2$–NiSe Vertical Heteronanostructures with Electronic Modulation for Enhanced Hydrogen Evolution Reaction. *Chem. Mater.* **2016**, *28*, 1838–1846. [CrossRef]
28. Tang, X.; Zhang, J.Y.; Mei, B.; Zhang, X.; Liu, Y.; Wang, J.; Li, W. Synthesis of hollow $CoSe_2/MoSe_2$ nanospheres for efficient hydrazine-assisted hydrogen evolution. *Chem. Eng. J.* **2021**, *404*, 126529. [CrossRef]
29. Inta, H.R.; Ghosh, S.; Mondal, A.; Tudu, G.; Koppisetti, H.V.; Mahalingam, V. $Ni_{0.85}Se/MoSe_2$ Interfacial Structure: An Efficient Electrocatalyst for Alkaline Hydrogen Evolution Reaction. *ACS Appl. Energy Mater.* **2021**, *4*, 2828–2837. [CrossRef]
30. Zhang, Y.; Zhang, S.; He, Y.; Li, H.; He, T.; Shi, H.; Ma, X.; Yang, Q.; Chen, L.; Chen, J. Self-supporting $MoSe_2/CoSe_2$@CFP electrocatalyst electrode for high-efficiency HER under alkaline solution. *J. Solid State Chem.* **2021**, *298*, 122108. [CrossRef]
31. Mahmood, A.; Khan, S.; Rahayu, F.; Shah, A.U.H.A.; Rahman, A.U.; Muhammad, T.; Khan, A.; Ullah, N. Factors Effecting and Structural Engineering of Molybdenum Nitride-Based Electrocatalyst for Overall Water Splitting: A Critical Review. *Energy Technol.* **2023**, *11*, 2194–4288. [CrossRef]
32. Zhang, S.; Ying, H.; Huang, P.; Yang, T.; Han, W.Q. Hierarchical utilization of raw $Ti_3C_2T_x$ MXene for fast preparation of various $Ti_3C_2T_x$ MXene derivatives. *Nano Res.* **2022**, *15*, 2746–2755. [CrossRef]
33. Chen, B.; Feng, A.; Liu, K.; Wu, J.; Yu, Y.; Song, L. High-performance capacitive deionization using 3D porous $Ti_3C_2T_x$ with improved conductivity. *J. Electroanal. Chem.* **2021**, *895*, 115515. [CrossRef]
34. Qi, Z.; Wang, S.; Li, Y.; Wang, L.; Zhao, L.; Ge, Q.; Zhang, J.Z. Scavenging activity and reaction mechanism of $Ti_3C_2T_x$ MXene as a novel free radical scavenger. *Ceram. Int.* **2021**, *47*, 16555–16561. [CrossRef]
35. Alhabeb, M.; Maleski, K.; Anasori, B.; Lelyukh, P.; Clark, L.; Sin, S.; Gogotsi, Y. Guidelines for Synthesis and Processing of Two-Dimensional Titanium Carbide ($Ti_3C_2T_x$ MXene). *Chem. Mater.* **2017**, *29*, 7633–7644. [CrossRef]
36. Li, C.; Zhu, L.; Li, H.; Li, H.; Wu, Z.; Liang, C.; Zhu, X.; Sun, Y. Dual surfactants applied in synthesis of $MoSe_2$ for high-efficiency hydrogen evolution reaction. *J. Alloys Compd.* **2021**, *863*, 158092. [CrossRef]
37. Truong, Q.D.; Nakayasu, Y.; Nguyen, Q.T.; Nguyen, D.N.; Nguyen, C.T.; Devaraju, M.K.; Rangappa, D.; Nayuki, K.; Sasaki, Y.; Tran, P.D.; et al. Defect-rich exfoliated $MoSe_2$ nanosheets by supercritical fluid process as an attractive catalyst for hydrogen evolution in water. *Appl. Surf. Sci.* **2020**, *505*, 144537. [CrossRef]
38. Zhao, X.; Zhao, Y.; Huang, B.; Cai, W.; Sui, J.; Yang, Z.; Wang, H.E. $MoSe_2$ nanoplatelets with enriched active edge sites for superior sodium-ion storage and enhanced alkaline hydrogen evolution activity. *Chem. Eng. J.* **2020**, *382*, 123047. [CrossRef]
39. Ullah, N.; Xie, M.; Hussain, S.; Yaseen, W.; Shah, S.A.; Yusuf, B.A.; Oluigbo, C.J.; Rasheed, H.U.; Xu, Y.; Xie, J. Simultaneous synthesis of bimetallic@3D graphene electrocatalyst for HER and OER. *Front. Mater. Sci.* **2021**, *15*, 305–315. [CrossRef]
40. Setayeshgar, S.; Karimipour, M.; Molaei, M.; Moghadam, M.R.; Khazraei, S. Synthesis of scalable 1T/2H–$MoSe_2$ nanosheets with a new source of Se in basic media and study of their HER activity. *Int. J. Hydrogen Energy* **2020**, *45*, 6090–6101. [CrossRef]
41. Yin, Y.; Zhang, Y.; Gao, T.; Yao, T.; Zhang, X.; Han, J.; Wang, X.; Zhang, Z.; Xu, P.; Zhang, P.; et al. Synergistic Phase and Disorder Engineering in 1T-$MoSe_2$ Nanosheets for Enhanced Hydrogen-Evolution Reaction. *Adv. Mater.* **2017**, *29*, 1700311. [CrossRef] [PubMed]
42. Shi, H.; Zhang, H.; Li, M.; Wang, Y.; Wang, D. Nanoflower-like 1T/2H mixed-phase $MoSe_2$ as an efficient electrocatalyst for hydrogen evolution. *J. Alloys Compd.* **2021**, *878*, 160381. [CrossRef]
43. Zhang, J.; Chen, Y.; Liu, M.; Du, K.; Zhou, Y.; Li, Y.; Wang, Z.; Zhang, J. 1T@2H-$MoSe_2$ nanosheets directly arrayed on Ti plate: An efficient electrocatalytic electrode for hydrogen evolution reaction. *Nano Res.* **2018**, *11*, 4587–4598. [CrossRef]
44. Li, N.; Wu, J.; Lu, Y.; Zhao, Z.; Zhang, H.; Li, X.; Zheng, Y.Z.; Tao, X. Stable multiphasic 1T/2H $MoSe_2$ nanosheets integrated with 1D sulfide semiconductor for drastically enhanced visible-light photocatalytic hydrogen evolution. *Appl. Catal. B Environ.* **2018**, *238*, 27–37. [CrossRef]
45. Li, N.; Zhang, Y.; Jia, M.; Lv, X.; Li, X.; Li, R.; Ding, X.; Zheng, Y.Z.; Tao, X. 1T/2H $MoSe_2$-on-MXene heterostructure as bifunctional electrocatalyst for efficient overall water splitting. *Electrochim. Acta* **2019**, *326*, 134976. [CrossRef]
46. Zhang, L.; Sun, L.; Huang, Y.; Sun, Y.; Hu, T.; Xu, K.; Ma, F. Hydrothermal synthesis of N-doped RGO/$MoSe_2$ composites and enhanced electro-catalytic hydrogen evolution. *J. Mater. Sci.* **2017**, *52*, 13561–13571. [CrossRef]
47. Vikraman, D.; Hussain, S.; Prasanna, K.; Karuppasamy, K.; Jung, J.; Kim, H.S. Facile method to synthesis hybrid phase 1T@2H $MoSe_2$ nanostructures for rechargeable lithium ion batteries. *J. Electroanal. Chem.* **2019**, *833*, 333–339. [CrossRef]
48. Mao, Z.; Wang, C.; Lu, H.; Tang, K.; Li, Q.; Yan, C.; Wang, X. Boron-modified electron transfer in metallic 1T $MoSe_2$ for enhanced inherent activity on per-catalytic site toward hydrogen evolution. *Adv. Mater. Interfaces* **2020**, *7*, 1901560. [CrossRef]
49. Maity, S.; Das, B.; Samanta, M.; Das, B.K.; Ghosh, S.; Chattopadhyay, K.K. $MoSe_2$-amorphous CNT hierarchical hybrid core–shell structure for efficient hydrogen evolution reaction. *ACS Appl. Energ. Mater.* **2020**, *3*, 5067–5076. [CrossRef]
50. Ren, X.; Yao, Y.; Ren, P.; Wang, Y.; Peng, Y. Facile sol-gel synthesis of C@$MoSe_2$ coreshell composites as advanced hydrogen evolution reaction catalyst. *Mater. Lett.* **2019**, *238*, 286–289. [CrossRef]
51. Shi, Y.; Gao, W.; Lu, H.; Huang, Y.; Zuo, L.; Fan, W.; Liu, T. Carbon-nanotube-incorporated graphene scroll-sheet conjoined aerogels for efficient hydrogen evolution reaction. *ACS Sustain. Chem. Eng.* **2017**, *5*, 6994–7002. [CrossRef]

materials

MDPI

Article

Controlling the Cooling Rate of Hydrothermal Synthesis to Enhance the Supercapacitive Properties of β-Nickel Hydroxide Electrode Materials

Yang-Ming Lu * and Sheng-Huai Hong *

Department of Electrical Engineering, National University of Tainan, Tainan 7005, Taiwan
* Correspondence: ymlu@mail.nutn.edu.tw (Y.-M.L.); x8558x@gmail.com (S.-H.H.)

Abstract: The demand for power storage devices with good quality, fast charging and high energy density is becoming more and more urgent in today's electronic technology. For batteries and traditional capacitors, it is an insurmountable challenge to combine fast charging and discharging, large capacitance and long-life properties. The characteristics of supercapacitors can meet all the above requirements at the same time. In this study, a simple one-step hydrothermal method was successfully used to grow β-nickel hydroxide nanocone particles directly on the 3D foamed nickel substrate as a working electrode material for supercapacitors. After growing β-nickel hydroxide crystals on 3D foamed nickel substrate, by controlling the cooling rate, a well-crystalized β-nickel hydroxide with good capacitance characteristics can be obtained. Cyclic voltammetry (CV), galvanostatic charge–discharge (GCD) and electrochemical impedance spectroscopy (EIS) were used to analyze the capacitance characteristics of the β-nickel hydroxide electrode. The research results show that the specific capacitance value of the β-$Ni(OH)_2$/3D nickel foam electrode material prepared at the cooling rate of 10 °C/h can reach 539 F/g with the charge–discharge test at a current density of 3 A/g. After 1000 continuous charge and discharge cycles, the material still retains 94.1% of the specific capacitance value.

Keywords: β-nickel hydroxide; supercapacitors; electrochemical materials; electrode materials; capacitance characteristics

1. Introduction

Energy issues are always a topic of concern in the development of modern society. Non-renewable fossil fuels will run out one day, and their pollution of the environment and waste of resources cannot be ignored. Renewable energy is receiving more and more attention because it is expected to take into account both environmental protection and sustainable development. Part of the renewable energy is used directly, and the excess energy is stored in the form of electric energy. Therefore, energy storage devices with high energy-storage efficiency and environmental friendliness have attracted great attention all over the world. However, renewable energy is heavily influenced by the environment, and there are still great challenges in power generation and electric energy storage technologies. Therefore, it is particularly important to develop energy storage devices with fast charging and discharging capabilities [1].

Supercapacitors have the characteristics of high power density, long cycle life, fast charge and discharge rate, wide operating temperature and low maintenance cost [2–4]. Their energy density is higher than that of traditional capacitors, and their power density is higher than that of lithium-ion batteries. Supercapacitors are a new type of energy storage device between lithium-ion batteries and traditional capacitors [5,6] and have great potential to replace them. Because of their excellent power density and charge–discharge rate, supercapacitors are widely used in various fields such as biomedicine [7], automobile [8], aerospace [9], military [10] and so on. In supercapacitor devices, the electrochemical performance of electrodes plays an important role in energy storage. Excellent electrode materials need to be safe and have high conductivity, high specific surface area and good stability. Therefore, knowing how to design electrode materials that are environmentally friendly, easy to prepare and have excellent performance is very important for the power storage of supercapacitors.

Although transition metal oxides or hydroxides have high capacitance, they are limited by the kinetic properties of redox reactions, and their stability is not good when charging and discharging at high rates [11–14]. To solve this problem, the active capacitor material is directly grown on the three-dimensional structure of the charge collector, which can effectively shorten the transmission and diffusion paths of electrons and ions, respectively, and improve the reactivity and utilization between the electrolyte and the electrode material [15–17]. Porous three-dimensional network metals are a very promising substrate, and their high surface area makes them attractive in the research of various energy storage devices such as fuel cells [18,19], lithium-ion batteries [20] and supercapacitors [21–23]. Three-dimensional substrates include foamed titanium, foamed copper, foamed iron, foamed cobalt and foamed nickel. Because of the excellent physical and chemical properties of foamed nickel mesh, it is the most widely used 3D mesh metal substrate [24].

It can be understood from the literature that ruthenium oxide has a very high theoretical specific capacitance (1400–2000 F/g) [25–27], but ruthenium is a noble metal with a high cost and scarce output, which is not conducive to commercialization. Therefore, the use of transition metal oxides as electrode materials is one of the current research directions for supercapacitors. Table 1 presents a comprehensive comparison of different transition metal oxides as supercapacitor electrode materials. Among them, the nickel-based electrode material [28] stands out among many metal materials and has attracted attention because of its relatively low price and excellent electrochemical performance. The theoretical specific capacitance of nickel hydroxide $Ni(OH)_2$ is as high as 2082 F/g [29], and it is easy to prepare with a simple method. Therefore, in consideration of cost, quantity and electrochemical performance, nickel hydroxide is quite suitable as an electrode material for supercapacitors. Nickel hydroxide exists in two crystal forms, α and β. There is ion intercalation in α-$Ni(OH)_2$, and the water molecules in it increase the carrier mobility during the discharge process and quickly replenish the missing carriers on the reaction surface. Therefore, α-$Ni(OH)_2$ has a higher specific capacitance value than β-$Ni(OH)_2$. However, α-$Ni(OH)_2$ undergoes an aging phenomenon during the charging and discharging process, so its cycle life and stability are worse than those of β-$Ni(OH)_2$. However, there is no ion intercalation in the β-$Ni(OH)_2$ structure, resulting in a slower carrier transport rate and a smaller specific capacitance than α-$Ni(OH)_2$, but its cycle life and stability are better [30]. Jingyao Qi et al. [31] obtained nickel hydroxide crystalline materials with two different preparation methods. The crystallinity characteristics were judged from the characteristic diffraction peak width and shape of the XRD. According to the analysis results, it can be reported that nickel hydroxide with poor crystallinity has poor redox reversibility, capacitance value and cycle life. From these research results, it can be seen that the crystallinity of nickel hydroxide has a considerable influence on the electrochemical performance of the electrode.

Table 1. Comparison of different materials as supercapacitor electrode materials.

Electrode Material	Preparation Method	GCD Current Density (A/g)	Specific Capacitance Value (F/g)	Reference in This Work
MnO_2/graphene	CVD, hydrothermal	0.2	333.4	[30]
MnCo−layered double hydroxide	Hydrothermal	1	1581.3	[32]
Nickel–cobalt hydroxide	CVD, hydrothermal	10	2023.5	[33]
$rGO - Fe_3O_4$	Hydrothermal (with rGO)	1.25	186.6	[34]
NiO	Electrospun	1	763	[35]
Cu/Cu_xO	Thermal reduction and oxidation with MOF	3	400	[36]
Graphene foam − ZnO	Solid-state (SS) synthesis	1	448	[37]
ZnO/MnO_x	Electrodeposition, chemical bath	1	556	[38]
$Ni(OH)_2$/graphene	CVD, hydrothermal	3	539	This work

2. Experimental Section

A 3D nickel foam was used as the substrate to grow nickel hydroxide on it using a hydrothermal method. It was first cut into a 1 cm × 1 cm test piece. All test pieces were sequentially cleaned in acetone, isopropanol, hydrochloric acid and deionized water with an ultrasonic oscillator for 15 min to completely clean their surfaces. A simple one-step hydrothermal method was used to directly grow nickel hydroxide crystals on the surface of the 3D nickel foam test piece (substrate). The hydrothermal reaction solution was obtained by adding 2.25 mmol of nickel nitrate ($(Ni(NO_3)_2 \cdot 6H_2O)$) and 4.50 mmol of urea (urea) into a beaker filled with 45 mL of deionized water.

All the reaction liquids were mixed with ultrasonic vibration for 40 min to obtain a light-green and clear mixed solution. This mixed solution and the pre-treated 3D nickel foam substrate were placed into a Teflon lining of a 75 mL high-pressure hydrothermal kettle (Figure 1). The kettle was heated (Figure 2) to 200 °C and kept at this high temperature for hydrothermal reaction for 6 h. The program controller was set to regulate the cooling rate of the high-temperature furnace. There were three different cooling rates of 100 °C/h, 25 °C/h and 10 °C/h used for the test pieces to cool down to room temperature. These three test pieces were named N_{100}, N_{25} and N_{10}, respectively.

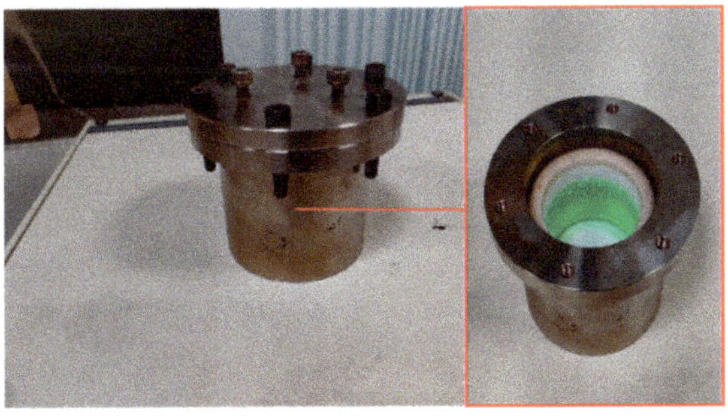

Figure 1. Exterior and interior of high-pressure hydrothermal kettle.

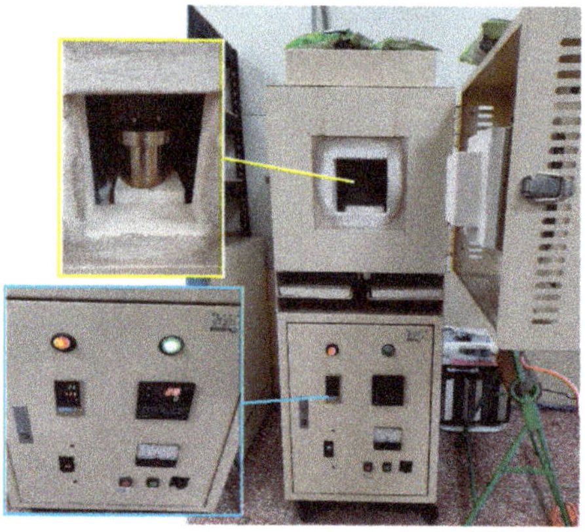

Figure 2. Exterior of high-temperature furnace and program controller.

Sun et al. used urea as an OH^- source to grow nickel hydroxide crystals using a hydrothermal method. In that paper, they proposed a possible mechanism for the growth of nickel hydroxide crystals [39]. The high-temperature hydrothermal growth of the nickel hydroxide crystals used in our study also has a similar growth mechanism. When the solution in the hydrothermal reaction is heated to a high temperature, a combination reaction of the Ni^{2+} and OH^- in the aqueous solution takes place on the surface of the 3D nickel foam and starts to nuclate $Ni(OH)_2$ nanocrystals because those nuclei have high surface energy and are thermodynamically unstable. The dipole interaction between the nano-nuclei enables them to gather and form nanocrystals. Finally, those $Ni(OH)_2$ nanocrystals continue to aggregate and begin to crystallize. The crystal starts to grow along the c-axis direction and gradually grows into a nanocone crystalline appearance. The crystalline appearance of the crystal is determined by the state of the solution, such as pH value, solution concentration, temperature, etc.

In this experiment, a D8 Advance ECO X-ray diffractometer was used to analyze the crystal structure of nickel hydroxide, and a HITACHI SU8000 (bought from Hitachi, Japan) high-resolution scanning electron microscope (HRSEM) was used to observe the surface morphology of the test piece. A three-electrode electrochemical system was used to measure the electrochemical properties of electrode materials (nickel hydroxide on 3D nickel foam substrates) prepared using a hydrothermal method. The three-electrode electrochemical system is composed of the following components: a 1 M KOH solution as electrolyte, 1 cm × 1 cm platinum sheet as counter electrode, Ag/AgCl electrode (keep in saturated KCl solution when not in use) as reference electrode and the well-prepared $Ni(OH)_2$/3D nickel foam as working electrode.

The Admiral Squidstat Solo electrochemical instrument was used to perform cyclic voltammetry (CV), galvanostatic charge–discharge (GCD) and electrochemical impedance spectroscopy (EIS) to analyze the electrochemical properties of electrode material.

The specific capacitance value of the working electrode can be calculated by bringing the integral result of the CV measurement curve into the following formula [40]:

$$C_s = \frac{\int idV}{2\,m(\Delta V)(V_s)} \tag{1}$$

where C_s is the specific capacitance value (F/g), $\int idV$ is the integral value of the curve, m is the mass of the electrode material (g), ΔV is the charge and discharge potential window (V) and V_s is the scan rate (V/s).

The specific capacitance value of the working electrode can also be calculated by bringing the GCD analysis results into the following formula [41]:

$$C_s = \frac{(\Delta t)(I)}{m(\Delta V)} \tag{2}$$

where $\triangle$t is the total discharge time (s), I (A) is the charge and discharge current, m is the mass of the electrode material (g) and ΔV is the charge and discharge potential window (V).

3. Results and Discussion

After completing the hydrothermal reaction for the test pieces, the temperature was cooled to room temperature at three different cooling rates. The effect of the cooling rate on the crystallization characteristics of nickel hydroxide was investigated. Figure 1 displays the XRD patterns of N_{100}, N_{25} and N_{10}. Among them, the two more obvious characteristic peaks at 44.50° and 51.84° are the signals of the foamed nickel substrate.

The diffraction peak signals at 19.25°, 33.06°, 38.54°, 52.10°, 59.05° and 62.72° respectively correspond to (001), (100), (101) of β-phase nickel hydroxide (102), (110) and (111) crystal faces (JCPDS 14-0117), as shown in Figure 3.

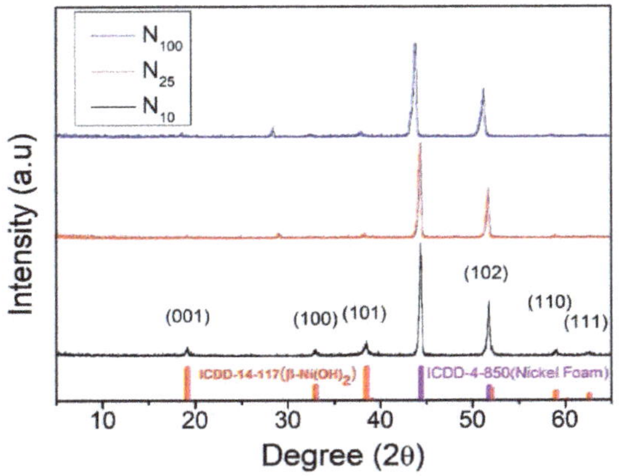

Figure 3. XRD patterns of nickel hydroxide cooled to room temperature at different rates.

This XRD analysis result (Figure 3) confirms that the nickel hydroxide prepared using the hydrothermal method in this experiment has a β-$Ni(OH)_2$ crystal form. In addition, the signal intensity of the diffraction peak of β-$Ni(OH)_2$ at 52.10° is the highest one, but the diffraction peak of 3D foamed nickel substrate is quite close to this peak, which is easily confused and makes it difficult to distinguish between them. The intensity of the signal peak located at 38.54° (101) is the second-strongest diffraction peak and occurred without the interference of the substrate material diffraction peak signal. Therefore, this experiment uses this diffraction peak as the main basis for subsequent analysis of the crystallization characteristics of β-$Ni(OH)_2$. The (101) peak intensities of the test pieces cooled down to room temperature using different cooling rates after the hydrothermal synthesis method are significantly different. The (101) peak intensities of N_{100}, N_{25} and N_{10} are 33, 73 and 253, respectively. The (101) diffraction peak of N_{10} is particularly obvious as shown in Figure 4. It can be deduced that as the cooling rate slows down, more time is provided for atoms to diffuse, which results in a more ordered crystal structure. The better the crystallization of the working electrode material, the more complete the redox reaction can be during the charge and discharge process of the supercapacitor. As a result, the specific capacitance value can be increased.

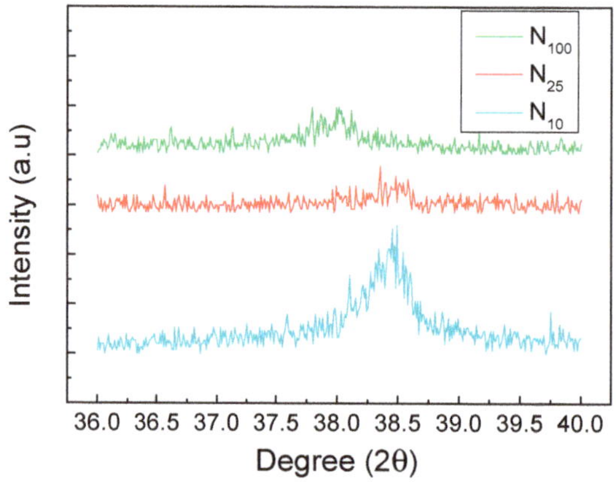

Figure 4. Enlarged XRD pattern of nickel hydroxide (101) diffraction peak.

The surface microstructures of β-nickel hydroxide are shown in Figure 5. From Figure 5a–c, it can be observed that the β-nickel hydroxide films prepared at different cooling rates are all composed of cone-shaped particles ranging in size from 500 nm to 1 μm, and there is not much change in morphology. Figure 6a–c displays the CV curves of N_{100}, N_{25} and N_{10} at scan rates from 1 mV/s to 25 mV/s. It can be observed that with the increase in the scan rate, the peak position of the oxidation peak gradually shifts to a high potential, while the peak position of the reduction peak gradually shifts to a low potential. This phenomenon is caused by the fact that the speed at which ions in the electrolyte migrate to the electrode surface and react with the material cannot keep up with the migration speed of electrons.

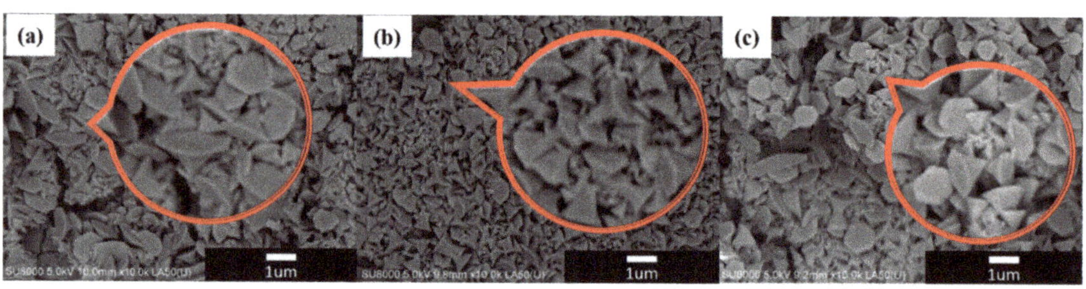

Figure 5. The SEM images of nickel hydroxide with different cooling rates (**a**) N_{100}, (**b**) N_{25} and (**c**) N_{10}. Red area is zoom-in figure of the area where it's pointed to.

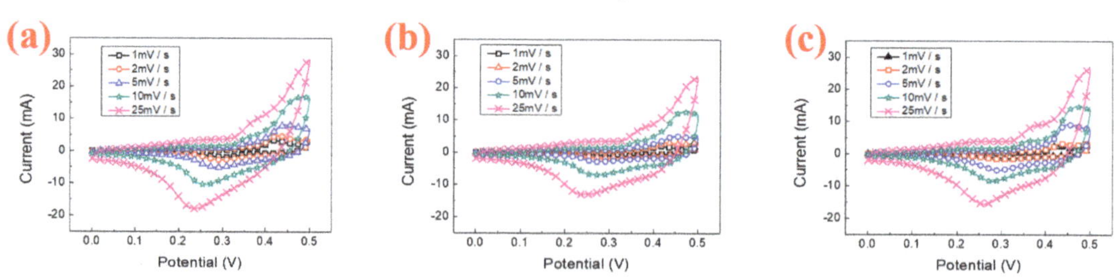

Figure 6. The CV curves of nickel hydroxide prepared at different cooling rates (**a**) N_{100}, (**b**) N_{25} and (**c**) N_{10}.

From the CV curve in Figure 7 (at a scan rate of 2 mV/s), the specific capacitance value was calculated according to Equation (2). The specific capacitance values of N_{100}, N_{25} and N_{10} are 257.6, 281.4 and 378.9 F/g, respectively. The specific capacitance values increase as the cooling rate becomes slower, which is consistent with the inference results of XRD analysis. Because the crystallization of N10 is more obvious and orderly, it can carry out a more complete redox reaction during the charging and discharging process so that the peak current value increases, resulting in an increase in the specific capacitance value. In addition to the improvement in the specific capacitance value, the redox stability of the material can also be inferred from the ratio I_{pb}/I_{pa} of the oxidation peak current (I_{pb}) to the reduction peak current (I_{pb}). The I_{pb}/I_{pa} values of N_{100}, N_{25} and N_{10} are 0.56, 0.57 and 0.61, respectively. Thus, it can be seen that the β-$Ni(OH)_2$ crystallization obtained at a slower cooling rate (N10) has more complete crystallinity and performs better in terms of electrochemical redox stability.

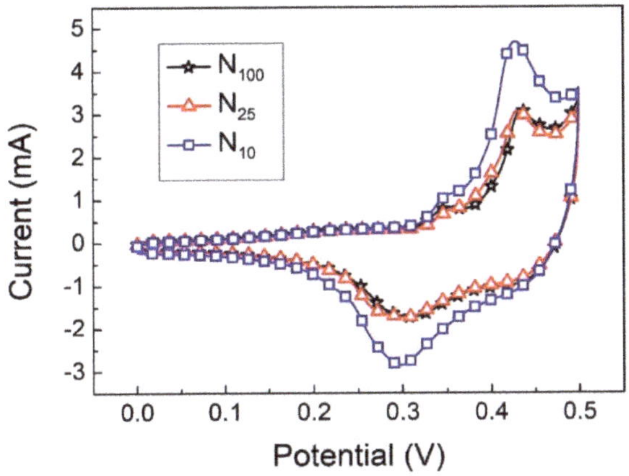

Figure 7. The CV curves of N_{100}, N_{25} and N_{10} at a scan rate of 2 mV/s.

Figure 8a–c presents the GCD curves of N_{100}, N_{25} and N_{10} at different charge–discharge current densities. It can be observed that during the charging process, a plateau interval appears in the GCD curve between 0.4 V and 0.45 V. At this time, nickel hydroxide undergoes an oxidation reaction and absorbs electrons to transform into NiOOH. During the discharge process, there is also an interval with a gentler slope between 0.3 V and 0.2 V. At this time, NiOOH undergoes a reduction reaction, transforms into $Ni(OH)_2$ and emits electrons, which is the standard pseudocapacitance performance.

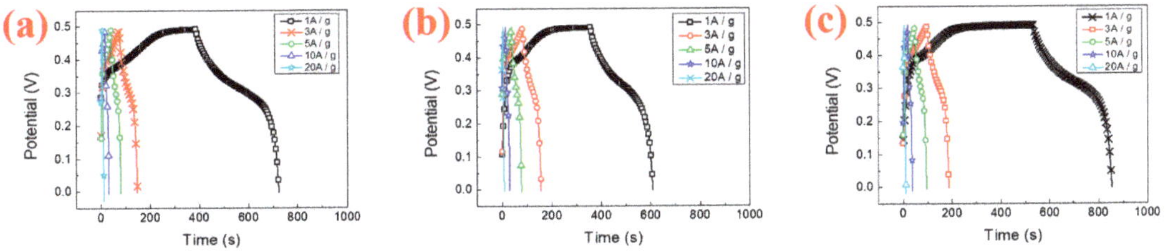

Figure 8. The GCD curves (**a**) N_{100}, (**b**) N_{25} and (**c**) N_{10}.

Figure 9 presents the GCD curves of the three samples under the potential window of 0–0.49 V, with a current density of 3 A/g for the charge and discharge cycles. It can be observed that under the same potential window, the charging and discharging time of N10 is longer than that of the other two. The specific capacitance values are calculated using Equation (2). N_{100}, N_{25} and N_{10} have specific capacitance values of 434.6, 446.3 and

538.7 F/g, respectively. Because CV is measured by changing the potential through the potential difference, and GCD is charged and discharged by passing a current, the values measured with the two measurement methods are different. However, the results measured using CV (Figure 7) and GCD (Figure 9) tend to be consistent; that is, the specific capacitance value increases as the cooling rate becomes slower after the hydrothermal synthesis method is completed. Although the measurement methods are different, both show the same trend results. That is to say, the specific capacitance value of the supercapacitor increases as the cooling rate of the electrode material prepared using the hydrothermal synthesis method slows down.

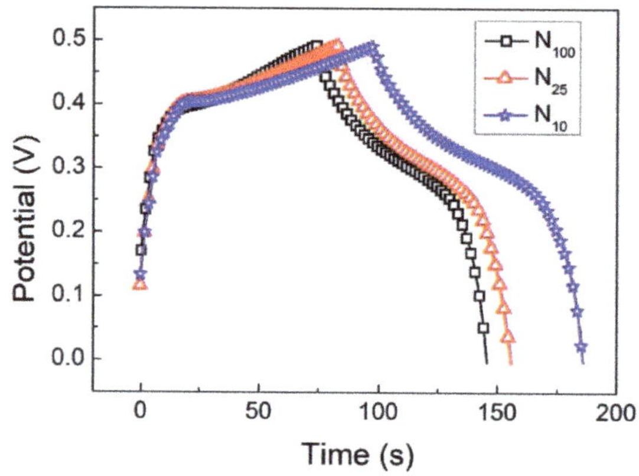

Figure 9. The GCD curves of N_{100}, N_{25} and N_{10} at a current density of 3 A/g.

The energy efficiency can be calculated with the following formula [42]:

$$\text{Energy efficiency} = \frac{energy\ density(discharge)}{energy\ density(charge)} \times 100$$

The calculated results are summarized in Table 2 as shown below.

Table 2. Energy efficiency of the three electrode materials.

Sample Number	N_{100}	N_{25}	N_{10}
Energy efficiency	73.4%	81.2%	90.4%

From the calculated data in Table 2, it is obvious that the N_{10} test piece has the highest energy density. This result is quite consistent with the previous analysis results and inferences.

Figure 10 is the Nyquist diagram of N_{100}, N_{25} and N_{10} tested using EIS. The AC impedance corresponding to different AC frequencies is divided into a real part (Z') and an imaginary part ($-Z''$). The value where the curve intersects the x-axis in the high-frequency region is the equivalent series resistance (ESR), which comes from the solution impedance and the internal resistance of the electrode. In addition, a typical Nyquist diagram has a semicircle in the high-frequency region. The diameter of this semicircle is the charge transfer resistance (R_{ct}). This resistance value can be used as a reference for the charge transfer speed inside the electrode. The smaller the R_{ct} value, the faster the charge transfer rate inside the electrode [43]. It is known from the literature [44] that the EIS Nyquist plot of supercapacitors is between that of typical battery materials and ideal capacitor materials. Observing an obvious semicircle in the plot is not easy. Therefore, in this experiment, we used simulation software and selected multiple points on the bending line in the high-frequency region to estimate ESR and Rct values. The ESR values of N_{100}, N_{25}

and N_{10} measured from the intercept of the x-axis in the plot are 1.052 Ω, 1.012 Ω and 1.010 Ω, respectively. The R_{ct} of these three specimens is calculated by fitting the software simulation to be 0.35 Ω, 0.33 Ω and 0.22 Ω, respectively. Both sets of data indicate that the internal resistance of the electrode of N10 is slightly smaller and that the internal charge transfer rate is faster. This phenomenon is consistent with the results of XRD, CV and GCD. In the low-frequency region, the slope of the straight line of N10 is larger than that of the other two, indicating that the diffusion resistance of OH^- radical ions in the electrolyte is lower in N_{10}. This means that N10 has excellent capacitance material properties, making it easier for OH^- ions in the electrolyte to diffuse into the electrodes.

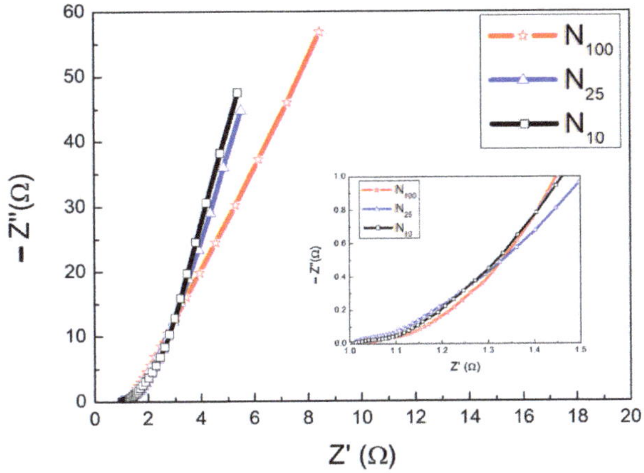

Figure 10. EIS Nyquist plots of N_{100}, N_{25} and N_{10}.

Figure 11 is a graph showing how specific capacitance values of all the samples changed with the charge and discharge current density. It can be found that the specific capacitance of N_{10} with the slowest cooling rate is larger than that of N_{100} and N_{25} at every different current density. At a current density of 3 A/g, the specific capacitance of N_{10} is 539 F/g, and it maintains a specific capacitance of 216 F/g at a high current density of 20 A/g. Compared with N_{100} and N_{25}, N_{10} has a better ability to withstand a high current when charged and discharged. This result is consistent with the XRD deduction results. Because of the more complete crystallization of N10, the carrier transport rate is faster, so it maintains excellent capacitance material characteristics under high current density.

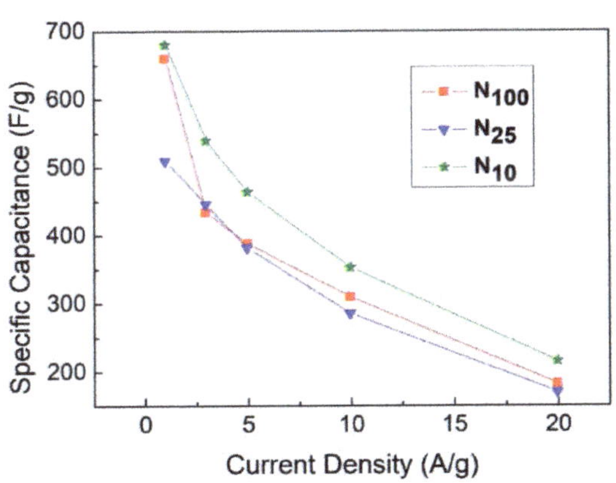

Figure 11. Specific capacitance values of N_{100}, N_{25} and N_{10} as a function of current density.

Table 3 provides a comparison of different preparation methods of nickel hydroxide as a supercapacitor electrode material. It can be seen that using a simple one-step hydrothermal method to directly grow nickel hydroxide on the 3D nickel foam substrate is much simpler. The specific capacitance value is not inferior to that of electrode materials prepared using other processes.

Table 3. Comparison of different preparation methods of nickel hydroxide as supercapacitor electrode material.

Preparation Method	Phase of Crystalline Nickel Hydroxide	GCD Current Density(A/g)	Specific Capacitance Value (F/g)	References
Hydrothermal	α	0.5	516	[45]
AAO-template-supported precipitation	α	5	833	[46]
Ultrasonication and alkalization	β	3	1171	[47]
Electrospun	β	1	763	[48]
Hydrothermal	β	3	539	In this work

Figure 12 is a graph showing changes in the specific capacitance values of N_{100}, N_{25} and N_{10} after 1000 continuous charge and discharge cycles at a charge and discharge current density of 20 A/g. It can be observed from Figure 10 that after 200 to 400 cycles, the specific capacitance value of the capacitor increases from about 50% of the original maximum specific capacitance value to 100%. This phenomenon can be attributed to the increased degree of the wetting effect between the electrode material and the electrolyte. After many charge–discharge cycles, the electrolyte more easily wets the inner surface of the porous electrode, making it easier for the ions in the electrolyte to diffuse into the electrode material. Therefore, the reaction area that can carry out redox increases, so the specific capacitance value quickly rises to its maximum specific capacitance value. After 1000 charge–discharge cycles, the specific capacitance values of N_{100}, N_{25} and N_{10} can still maintain 88.5%, 89.1% and 94.1%, respectively. N10 can maintain 94.1% of the specific capacitance value, which is attributed to the slower cooling rate that makes the crystallization of β-nickel hydroxide more complete and better so that N_{10} still has excellent capacitance characteristics after a large number of charge–discharge cycles. Such an excellent capacitor charge–discharge cycle life can prove that β-nickel hydroxide has high reliability as an electrode material for supercapacitors.

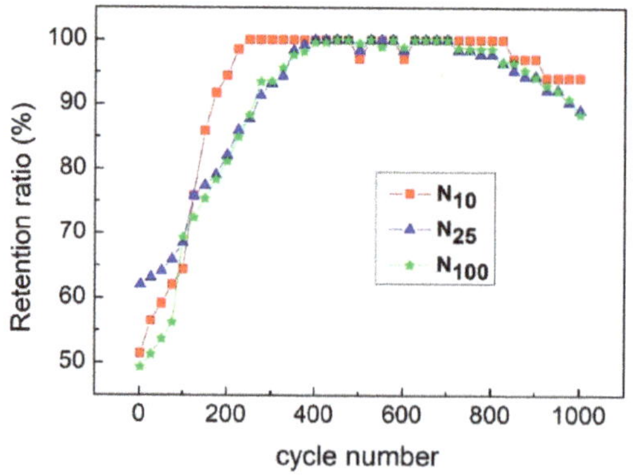

Figure 12. Variation in the specific capacitance values of N_{100}, N_{25} and N_{10} after 1000 continuous charge and discharge cycles.

4. Conclusions

In this experiment, β-nickel hydroxide was successfully prepared as an electrode material for supercapacitors using a simple one-step hydrothermal method. Slowing down the cooling rate of the hydrothermal process can make the crystallization of β-nickel hydroxide more complete. The hydrothermally grown β-nickel hydroxide crystalline state presents the appearance of cone-shaped particles, which effectively increases the surface area of the electrode. According to the research results, the capacitance characteristic of the test piece (N_{10}) with the slowest cooling rate is the best. Under the current density of 3 A/g, the constant current charge and discharge experiment was carried out, and the specific capacitance value was 538.7 F/g. After 1000 continuous charge and discharge cycles, N_{10} still retains 94.1% of the specific capacitance value. These excellent capacitive properties are attributed to the growth of cone-shaped β-nickel hydroxide particles with high specific surface area and more complete crystallinity on the surface of the 3D nickel foam electrode.

5. Future Works

Nickel hydroxide is feasible as an electrode material for supercapacitors, but its conductivity needs to be further improved and its specific surface area increased. Adding conductive substances or conductive layers to the crystalized nickel hydroxide, or doping other elements to form binary or multiple transition-metal composite hydroxides, can effectively improve the capacitance characteristics of nickel hydroxide [49–51]. Growing nanostructures with different morphologies is an effective way to greatly increase the specific surface area. It is known from this study that the cooling rate is directly related to the research results, so it is very important to precisely monitor the cooling rate. In the future, the graph-oriented simulated annealing method proposed by Morteza Mollajafari et al. [52] can be used to make the heat treatment process smoother and more stable for temperature control and optimize the overall cooling process.

Author Contributions: Y.-M.L. is responsible for the experiment planning, directing the research content, and writing this paper. S.-H.H. is responsible for some experiments and data sorting. All authors have read and agreed to the published version of the manuscript.

Funding: This research received funding from the Ministry of Science and Technology, Taiwan, under contract Nos. MOST109-2221-E-024-001.

Institutional Review Board Statement: Not applicable.

Informed Consent Statement: Not applicable.

Data Availability Statement: The data presented in this study are available on request from the corresponding author.

Conflicts of Interest: The authors declare no conflict of interest.

References

1. Liang, C.; Wang, S.; Sha, S.; Lv, S.; Wang, G.; Wang, B.; Li, Q.; Yu, J.; Xua, X.; Zhang, L. Novel semiconductor materials for advanced supercapacitors. *J. Mater. Chem. C* **2023**, *11*, 4288–4317. [CrossRef]
2. Conway, B.E. *Electrochemical Supercapacitors: Scientific Fundamentals and Technological Applications*; Kluwer Academia/Plenum: New York, NY, USA, 1999.
3. Kotz, R.; Carlen, M. Principles and applications of electrochemical capacitors. *Electrochim. Acta* **2000**, *45*, 2483–2498. [CrossRef]
4. Zhao, X.; Sánchez, B.M.; Dobson, P.J.; Grant, P.S. The role of nanomaterials in redox-based supercapacitors for next generation energy storage devices. *Nanoscale* **2011**, *3*, 839. [CrossRef]
5. Frackowiak, E. Carbon materials for supercapacitor application. *Phys. Chem. Chem. Phys.* **2007**, *9*, 1774. [CrossRef]
6. Burke, A. R&D considerations for the performance and application of electrochemical capacitors. *Electrochim. Acta* **2007**, *53*, 1083.
7. Mendoza-Ponce, P.; John, B.; Schroeder, D.; Krautschneider, W.H. Super-capacitors for implantable medical devices with wireless power transmission. In Proceedings of the 14th Conference on Ph. D. Research in Microelectronics and Electronics (PRIME), Prague, Czech Republic, 2–5 July 2018; pp. 241–244. [CrossRef]
8. Horn, M.; MacLeod, J.; Liu, M.; Webb, J.; Motta, N. Supercapacitors: A new source of power for electric cars? *Econ. Anal. Policy* **2019**, *61*, 93–103. [CrossRef]

9. Al-Furjan, M.S.H.; Qi, Z.H.; Shan, L.; Farrokhian, A.; Shen, X.; Kolahchi, R. Nano supercapacitors with practical application in aerospace technology: Vibration and wave propagation analysis. *Aerosp. Sci. Technol.* **2023**, *133*, 108082. [CrossRef]
10. Végvári, Z. Supercapacitors and Their Military Applicability. *Honvédségi Szle.* **2019**, *147*, 38–49. [CrossRef]
11. Shinde, P.A.; Olabi, A.G.; Chodankar, N.R.; Patil, S.J.; Hwang, S.K.; Abdelkareem, M.A. Realizing superior redox kinetics of metal-metal carbides/carbon coordination supported heterointerface for stable solid-state hybrid supercapacitor. *Chem. Eng. J.* **2023**, *454*, 140246. [CrossRef]
12. Lu, J.L.; Li, J.E.; Wan, J.; Han, X.Y.; Ji, P.Y.; Luo, S.; Gu, M.X.; Wei, D.P.; Hu, C.G. A Facile Strategy of In-situ Anchoring of Co_3O_4 on N Doped Carbon Cloth for an Ultrahigh Electrochemical Performance. *Nano Res.* **2020**, *14*, 2410–2417. [CrossRef]
13. Chatterjee, M.; Sain, S.; Roy, A.; Das, S.; Pradhan, S.W. Enhanced Electrochemical Properties of Co_3O_4 with Morphological Hierarchy for Energy Storage Application: A Comparative Study with Different Electrolytes. *J. Phys. Chem. Solids* **2021**, *148*, 109733. [CrossRef]
14. Liang, R.; Du, Y.; Xiao, P.; Cheng, J.; Yuan, S.; Chen, Y.; Yuan, J.; Chen, J. Transition Metal Oxide Electrode Materials for Supercapacitors: A Review of Recent Developments. *Nanomaterials* **2021**, *11*, 1248. [CrossRef]
15. Bello, A.; Makgopa, K.; Fabiane, M.; Dodoo-Ahrin, D.; Ozoemena, K.I.; Manyala, N. Chemical adsorption of NiO nanostructures on nickel foam-graphene for supercapacitor applications. *J. Mater. Sci.* **2013**, *48*, 6707–6712. [CrossRef]
16. Chen, Z.; Ren, W.; Gao, L.; Liu, B.; Pei, S.; Cheng, H.M. Three-dimensional flexible and conductive interconnected graphene networks grown by chemical vapour deposition. *Nat. Mater.* **2011**, *10*, 424–428. [CrossRef] [PubMed]
17. Dong, X.; Wang, X.; Wang, L.; Song, H.; Zhang, H.; Huang, W.; Chen, P. 3D Graphene Foam as a Monolithic and Macroporous Carbon Electrode for Electrochemical Sensing. *ACS Appl. Mater. Interfaces* **2012**, *4*, 3129–3133. [CrossRef] [PubMed]
18. Ni, W.; Wang, T.; Héroguel, F.; Krammer, A.; Lee, S.; Yao, L.; Schuler, A.; Luterbacher, J.S.; Yan, Y.; Hu, X. An efficient nickel hydrogen oxidation catalyst for hydroxide exchange membrane fuel cells. *Nat. Mater.* **2022**, *21*, 804–810. [CrossRef]
19. Pei, P.; Huang, S.; Chen, D.; Li, Y.; Wu, Z.; Ren, P.; Wang, K.; Jia, X. A high-energy-density and long-stable-performance zinc-air fuel cell system. *Appl. Energy* **2019**, *241*, 124–129. [CrossRef]
20. Lee, D.J.; Yu, S.-H.; Lee, H.S.; Jin, A.; Lee, J.; Lee, J.E.; Sung, Y.-E.; Hyeon, T. Facile synthesis of metal hydroxide nanoplates and their application as lithium-ion battery anodes. *J. Mater. Chem. A* **2017**, *5*, 8744–8751. [CrossRef]
21. Bhat, M.Y.; Hashmi, S.A.; Khan, M.; Choi, D.; Qurashi, A. Frontiers and recent developments on supercapacitor's materials, design, and applications: Transport and power system applications. *J. Energy Storage* **2023**, *58*, 106104. [CrossRef]
22. Shi, F.; Li, L.; Wang, X.-L.; Gua, C.-D.; Tu, J.-P. Metal oxide/hydroxide-based materials for supercapacitors. *RCS Adv.* **2014**, *4*, 41910–41921. [CrossRef]
23. Naeem, S.; Patil, A.V.; Shaikh, A.V.; Shinde, U.P.; Husain, D.; Alam, M.T.; Sharma, M.; Tewari, K.; Ahmad, S.; Shah, A.A.; et al. A Review of Cobalt-Based Metal Hydroxide Electrode for Applications in Supercapacitors. *Adv. Mater. Sci. Eng.* **2023**, *2023*, 1133559. [CrossRef]
24. Nilimapriyadarsini, S.; Balasubramaniam, S.; Manab, K.; Lukas, S.M.; Ananthakumar, R. Recent trends in template assisted 3D porous materials for electrochemical supercapacitors. *J. Mater. Chem. A* **2021**, *9*, 25286–25324.
25. Hu, C.C.; Chen, W.C. Effects of substrates on the capacitive performance of $RuOx{\cdot}nH_2O$ and activated carbon–RuO_x electrodes for supercapacitors. *Electrochim. Acta* **2004**, *49*, 3469–3477. [CrossRef]
26. Hu, C.C.; Chen, W.C.; Chang, K.H. How to Achieve Maximum Utilization of Hydrous Ruthenium Oxide for Supercapacitors. *J. Electrochem. Soc.* **2004**, *151*, A281. [CrossRef]
27. Hu, C.C.; Chang, K.H.; Lin, M.C.; Wu, Y.T. Design and Tailoring of the Nanotubular Arrayed Architecture of Hydrous RuO_2 for Next Generation Supercapacitors. *Nano Lett.* **2006**, *6*, 2690–2695. [CrossRef] [PubMed]
28. Sun, H.; Wang, C.; Qi, Z.; Hu, W.; Zhang, Z. Nanostructure Nickel-Based Selenides as Cathode Materials for Hybrid Battery-Supercapacitors. *Front. Chem.* **2021**, *8*, 611032. [CrossRef]
29. Liuqin, L.; Rong, L.; Siyu, S.; Liang, Z.; Yifan, C.; Naili, G.; Wei, S.; Xiaohong, Z. Controllable synthesis of reduced graphene oxide/nickel hydroxide composites with different morphologies for high performance supercapacitors. *J. Alloys Compd.* **2006**, *6*, 2690–2695.
30. Bai, X.L.; Gao, Y.L.; Gao, Z.Y.; Ma, J.Y.; Tong, X.L.; Sun, H.B.; Wang, J.A. Supercapacitor performance of 3D-graphene/MnO_2 foam synthesized via the combination of chemical vapor deposition with hydrothermal method. *Appl. Phys. Lett.* **2020**, *117*, 18. [CrossRef]
31. Qi, J.; Xu, P.; Lv, Z.; Liu, X.; Wen, A. Effect of crystallinity on the electrochemical performance of nanometer Al-stabilized α-nickel hydroxide. *J. Alloys Compd.* **2008**, *462*, 164–169. [CrossRef]
32. Hsiao, Y.-C.; Liao, C.-H.; Hsu, C.-S.; Yougbaré, S.; Lin, L.-Y.; Wu, Y.-F. Novel synthesis of manganese cobalt layered double hydroxide and sulfur-doped nickel cobalt layered double hydroxide composite as efficient active material of battery supercapacitor hybrids. *J. Energy Storage* **2023**, *57*, 106171. [CrossRef]
33. Zhang, M.; Wang, Y.; Guo, X.; Li, R.; Peng, Z.; Zhang, W.; Zheng, Y.; Xie, H.; Zhang, Y.; Zhao, Y. High-Performance Nickel Cobalt Hydroxide Nanosheets/Graphene/Ni foam Composite Electrode for Supercapacitor Applications. *J. Electroanal. Chem.* **2021**, *897*, 115543. [CrossRef]
34. Wang, H.; Xu, X.; Wang, C.; Neville, A.; Hua, Y. Fundamental Insight into the Degradation Mechanism of an rGO-Fe_3O_4 Supercapacitor and Improving Its Capacity Behavior via Adding an Electrolyte Additive. *Energy Fuels* **2021**, *35*, 8406–8416. [CrossRef]

35. Erdemutu, E.; Bai, C.; Ding, L. Electrospun Ni-Ni$(OH)_2$/Carbon Nanofibers as Flexible Binder-Free Supercapacitor Electrode with Enhanced Specific Capacitance. *J. Electron. Mater.* **2020**, *49*, 7211–7218. [CrossRef]
36. Guo, Y.; Chen, C.; Wang, Y.; Hong, Y.; Wu, H.; Wang, K.; Niu, D.; Zhang, C.; Zhang, Q. Cu/Cu_xO@C nanocomposites as efficient electrodes for high-performance supercapacitor devices. *Dalton Trans* **2022**, *51*, 14551–14556. [CrossRef] [PubMed]
37. Kasap, S.; Kaya, I.I.; Repp, S.; Erdem, E. Superbat: Battery-like supercapacitor utilized by graphene foam and zinc oxide (ZnO) electrodes induced by structural defects. *Nanoscale Adv.* **2019**, *1*, 2586–2597. [CrossRef]
38. Samuel, E.; Joshi, B.; Kim, Y.I.; Aldalbahi, A.; Rahaman, M.; Yoon, S.S. ZnO/MnOx Nanoflowers for High-Performance Supercapacitor Electrodes. *ACS Sustain. Chem. Eng.* **2020**, *8*, 3697–3708. [CrossRef]
39. Yi, X.; Sun, H.; Robertson, N.; Kirk, C. Nanoflower Ni$(OH)_2$ grown in situ on Ni foam for high-performance supercapacitor electrodematerials. *Sustain. Energy Fuels* **2021**, *5*, 5236–5246. [CrossRef]
40. Kim, B.K.; Chabot, V.; Yu, A. Carbon nanomaterials supported Ni$(OH)_2$/NiO hybrid flower structure for supercapacitor. *Electrochim. Acta* **2013**, *109*, 370–380. [CrossRef]
41. Lakshmi, K.C.S.; Vedhanarayanan, B. High-Performance Supercapacitors: A Comprehensive Review on Paradigm Shift of Conventional Energy Storage Devices. *Batteries* **2023**, *9*, 202. [CrossRef]
42. Eftekhari, A. Energy efficiency: A critically important but neglected factor in battery research. *Sustain. Energy Fuels* **2017**, *1*, 2053–2060. [CrossRef]
43. Zhao, S.; Wu, F.; Yang, L.; Gao, L.; Burke, A.F. A measurement method for determination of dc internal resistance of batteries and supercapacitors. *Electrochem. Commun.* **2010**, *12*, 242–245. [CrossRef]
44. Negroiu, R.; Svasta, P.; Pirvu, C.; Vasile, A.; Marghescu, C. Electrochemical impedance spectroscopy for different types of supercapacitors. In Proceedings of the 2017 40th International Spring Seminar on Electronics Technology (ISSE), Sofia, Bulgaria, 10–14 May 2017; pp. 1–4. [CrossRef]
45. Wiston, B.R.; Ashok, M. Electrochemical performance of hydrothermally synthesized flower-like α-nickel hydroxide. *Vacuum* **2018**, *160*, 12–17. [CrossRef]
46. Wang, Y.-X.; Hu, Z.-A.; Wu, H.-Y. Preparation and electrochemical performance of alpha-nickel hydroxide nanowire. *Mater. Chem. Phys.* **2011**, *126*, 580–583. [CrossRef]
47. Kuo, T.-R.; Yen, S.-C.; Kubendhiran, S.; Yougbaré, S.; Lin, L.-Y.; Wu, Y.-F. Tailoring morphology of pure terephthalic acid induced beta nickel hydroxide using ultrasonication and alkalization for efficient energy storage. *J. Energy Storage* **2022**, *56*, 106117. [CrossRef]
48. Xie, M.; Xu, Z.; Duan, S.; Tian, Z.; Zhang, Y.; Xiang, K.; Lin, M.; Guo, X.; Ding, W. Facile growth of homogeneous Ni$(OH)_2$ coating on carbon nanosheets for high-performance asymmetric supercapacitor applications. *Nano Res.* **2018**, *11*, 216–224. [CrossRef]
49. Attia, S.Y.; Bedir, A.G.; Barakat, Y.F.; Mohamed, S.G. A two-dimensional nickel-doped bismuth-layered double hydroxide structure as a bifunctional efficient electrode material for symmetric supercapacitors. *Sustain. Mater. Technol.* **2023**, *36*, e00595. [CrossRef]
50. Berrabah, S.E.; Benchettara, A.; Smaili, F.; Benchettara, A.; Mahieddine, A. High performance hybrid supercapacitor based on electrochemical deposed of nickel hydroxide on zinc oxide supported by graphite electrode. *J. Alloys Compd.* **2023**, *942*, 169112. [CrossRef]
51. Emin, A.; Li, J.; Dong, Y.; Fu, Y.; He, D.; Li, Y. Facilely prepared nickel-manganese layered double hydroxide-supported manganese dioxide on nickel foam for aqueous asymmetric supercapacitors with high performance. *J. Energy Storage* **2023**, *65*, 107340. [CrossRef]
52. Mollajafari, M. An efficient lightweight algorithm for scheduling tasks onto dynamically reconfigurable hardware using graph-oriented simulated annealing. *Neural Comput. Appl.* **2023**, *35*, 18035–18057. [CrossRef]

materials

MDPI

Review

Divalent Metal Ion Depletion from Wastewater by RVC Cathodes: A Critical Review

Alessandro Dell'Era [1,*], Carla Lupi [2], Erwin Ciro [3], Francesca A. Scaramuzzo [1] and Mauro Pasquali [1]

1 Department SBAI, Sapienza University of Rome, Via del Castro Laurenziano 7, 00161 Roma, Italy; francesca.scaramuzzo@uniroma1.it (F.A.S.); mauro.pasquali@uniroma1.it (M.P.)
2 Department ICMA, Sapienza University of Rome, Via Eudossiana 18, 00184 Roma, Italy; carla.lupi@uniroma1.it
3 Department of Engineering Sciences, Guglielmo Marconi University, 00193 Rome, Italy; e.ciro@lab.unimarconi.it
* Correspondence: alessandro.dellera@uniroma1.it; Tel.: +39-0649766855

Abstract: In this paper, a critical review of results obtained using a reticulated vitreous carbon (RVC) three-dimensional cathode for the electrochemical depletion of various divalent ions, such as Cu^{+2}, Cd^{+2}, Pb^{+2}, Zn^{+2}, Ni^{+2}, and Co^{+2}, often present in wastewater, has been carried out. By analyzing the kinetics and fluid dynamics of the process found in literature, a general dimensionless equation, $Sh = f(Re)$, has been determined, describing a general trend for all the analyzed systems regardless of the geometry, dimensions, and starting conditions. Thus, a map in the $log(Sh)$ vs. $log(Re)$ plane has been reported by characterizing the whole ion electrochemical depletion process and highlighting the existence of a good correlation among all the results. Moreover, because in recent years, the interest in using this three-dimensional cathode material seems to have slowed, the intent is to revive it as a useful tool for metal recovery, recycling processes, and water treatments.

Keywords: reticulated vitreous carbon (RVC); divalent ion electrodeposition; dimensionless analysis

1. Introduction

In recent years, a lot of research and studies in electrochemistry have been performed to enhance the performance of electrochemical devices by focusing attention on more advanced materials able to deliver or accumulate electrical energy at high and low temperatures. Batteries, fuel cells, or electrolyzers and supercapacitors have been extensively studied and analyzed [1–8]. New catalyst synthesis approaches for achieving high surface areas have been developed to produce nanoparticles, nanowires, or, in general, nanostructures as nanostructured thin film catalysts with extended surface areas, increasing significantly the electrochemical reaction activity [1,2]. Nanowires and nanopowders have also been synthesized having high electron storage densities and high diffusion rates and, thus, can be used to produce high-power sources [2]. Then, research has focused on developing new, cheaper, and more durable catalysts, by improving both the morphology and synthesis protocols, as alternatives to pure platinum, even if, normally, they present lower performances [3–6]. Generally, they are modified platinum-based catalysts containing other metals, such as Co, Cu, and Cr, or nonplatinum-based catalysts, such as non-noble metals, metal oxides, metal nitrides, and covalent organic frameworks, or, finally, organometallic catalysts [3–8]. Moreover, an increase in the performance of membranes or separators at high and low temperatures has been pursued, reaching lower ohmic losses and, thus, enabling higher current densities and higher efficiencies [3–6]. Several kinds of membranes or separators have been studied to work in different environments, allowing the use of less expensive catalysts. Finally, better engineering of the production processes of all the components present in these kinds of devices has enabled their widespread application [1–8].

Moreover, a glimpse has been provided into electrochemical technologies for electrical and/or electronic device waste recycling, for strategic material recovery, and for meeting the necessity to have a circular economy and low-environmental-impact processes, which, today, are becoming more important [9–20]. Herein, in particular, the aim of the authors is to show that an electrochemical technique can be effectively used for the electrochemical depletion of divalent metal ions from wastewater, utilizing a reticulated vitreous carbon (RVC) cathode, and to demonstrate how it is possible to give the general features of the electrochemical reduction process by a critical analysis of literature results obtained either by ourselves or other groups. For this purpose, the authors analyzed the dimensionless approach described in the literature for the depletion process to obtain a more general and exploitable empirical equation describing the electrochemical behavior of an RVC filter and the trend in the data, regardless of the chemical nature of the metallic divalent ions to be removed, geometry, fluid dynamics of the system, and different morphologies of deposits. The use of dimensionless analysis is justified by the reason that normally, the concept of scaling chemical process units can be applied in general to reactor systems, using a dimensionless group correlation [21]. To address scalability issues, the chemical processing industry often refers to the principle of similarity based on setting up the relationships existing between different scales of processing equipment [22–25]. Such relationships are usually determined through the so-called "dimensionless groups", which characterize the process [26]. In this context, the well-known Buckingham theorem, regarding the relationship between the characteristic variables of a system, can be considered as a formal restatement of the requirement of dimensional consistency [27]. Generally speaking, dimensional analysis is considered as a prerequisite in the quantitative study of reacting systems, rate constants, transfer coefficients, and transport properties because it helps both in explaining the mechanisms associated with engineering processes and in developing an effective experimental design [28,29]. An electrochemical reaction can be carried out in the batch or continuous (mixed/plug flow) mode. In principle, reactors can operate with or without recycling, while further classifications are possible based on the flow arrangement (parallel/series flow) or electrical connections. The electrodes may be flat (two-dimensional solid electrodes) or porous (three-dimensional), either with a horizontal or vertical configuration [30]. The electrochemical methodology and reactors, especially when equipped with three-dimensional electrodes, have proved to be very efficient to decrease the amounts of metals in aqueous solutions [31–45] and have allowed the authors, in a previous study [46], to obtain a successful depletion of Ni^{+2} and Co^{+2}. Considering that the hydrodynamic processes of ions transporting and reducing at reticulated vitreous carbon (RVC) three-dimensional cathodes are usually studied under mass-transfer control conditions, the dimensionless analysis is based on (i) the Sherwood number, which takes into account the convective and diffusive mass transports, and (ii) the Reynolds number, which considers the hydrodynamic condition of the flows passing through the RVC three-dimensional electrode.

A third dimensionless number, the Schmidt number, (Sc)μD^{-1}, is also considered. A typical power expression, generally adopted to describe the mass transfer at porous electrodes, is as follows:

$$(Sh) = m(Re)^n(Sc)^{1/3} \tag{1}$$

The analytical equations for the Sherwood and Reynolds numbers are, respectively, as follows:

$Sh = k_m \varepsilon A_e{}^{-1} D^{-1}$ and $Re = v\mu A^{-1} A_e{}^{-1}$, where k_m is the mass transport coefficient; $A_e{}^{-1}$ is the characteristic length; D is the diffusion coefficient, reported in Table 1 for the considered ions; ε is the RVC void volume fraction; μ is the solution kinematic viscosity, equal to about 0.011 cm^2/s at room temperature; and v is the linear velocity.

Table 1. Dimensional cellular features and diffusion coefficients for divalent ions undergoing electrochemical depletion, as described in various reference papers.

Ion	RVC Section (cm^2)	A_e (cm^2/cm^3)	Diffusivity (cm^2/s)	Schmidt Number (Sc)	Reference(s)
Zn^{+2}	6.5	27–40–53–66	1.2×10^{-5} [47–49] Chloride solution	917	[50]
Cu^{+2}	6	7–66	0.49×10^{-5} [31,51–53] Sulfate solution	2200	[31]
	4	27	0.45×10^{-5} [51–53] Sulfate solution	2200	[32]
	2	53	0.5×10^{-5} [51–53] Sulfate solution	2200	[33]
Cd^{+2}	5	40	0.71×10^{-5} [39] Sulfate solution	1550	[34]
	6	40	0.71×10^{-5} [39] Sulfate solution	1550	[35]
Pb^{+2}	6.5	11–27–37–53	2.0×10^{-5} [36] Nitrate solution	550	[36,37]
	6	7–40–66	0.7×10^{-5} [44] Chloride solution	1570	[44]
Co^{+2}	1.5	53	0.65×10^{-5} [51] Sulfate solution	1830	[38]
	10	27	0.65×10^{-5} [51] Sulfate solution	1830	[46]
Ni^{+2}	10	27	0.6×10^{-5} [51] Sulfate solution	1830	[46]

2. Brief Comparison among Wastewater Treatment Techniques

Waste aqueous solutions containing low concentrations of heavy metal ions must be treated before discharging because, from a circular economy perspective, both the recovery/reuse of these metals can be achieved, and damage to the environment can be avoided. Several techniques for wastewater treatment, such as flotation, chemical precipitation, cementation, solvent extraction, ion exchange, and membrane filtration, are processes conventionally used in hydrometallurgy as leach liquor purification techniques [54]. However, the low concentration of metal ions in wastewater, compared to that in leach liquors, does not allow, from an economic point of view, the use of some of these techniques, as in the case of solvent extraction. However, the used organic solvents are flammable, volatile, partially water-soluble, and potentially toxic, with consequent environmental risks [55]. On the other hand, the use of chemical precipitation, although effective, is not selective and produces sludge containing hydroxides or sulfides of heavy metals, which are difficult to dispose, except in special landfills, with increasing operational costs [56,57]. One of the chemical cementation problems is that for presenting a high selectivity, a metal with a reduction potential very close to that of the metal to be recovered must be utilized as a reductive reactant, which involves very low process speeds [58]. The main problems in using membrane filtration are the requirement of a high investment cost and frequent membrane fouling, while the ion exchange method could work better with diluted metal solutions owing to the minor obstruction of the membranes or resins over time [59]. However, common ion exchange resins cannot perform well when utilized for the depletion of wastewater complex contaminants (such as the simultaneous presence of metal ions and organic pollutants), which could induce rapid exhaustion and even severe fouling, with the consequent significant reduction in the removal efficiency and increase in maintenance costs. However, all these recovery/removal methods present several drawbacks, such as the generation of large amounts of sludge, high reagent or energy requirements, incomplete removal of metal ions, difficulty in the treatment of large volumes of wastewater and, last but not least, high operational and maintenance costs.

Lately, metal ion depletion processes, such as adsorption, photocatalysis, and electrochemical techniques, have attracted considerable interest. Adsorption provides high metal recovery capabilities with low operative costs; however, it presents low selectivity, and a large amount of waste production. Furthermore, adsorption needs frequent adsorbent regeneration [60,61]. Although photocatalysis does not produce sludge and instantly removes both metals and any organic pollutants present in wastewater, its industrial application is still limited, while it appears to be effective on a laboratory scale [62]. In this paper, the electrochemical technique, particularly electrodeposition on RVC, has been considered. It could be regarded as one of the most efficient techniques for the depletion of metal ions in wastewater because it does not produce sludge and can be automatically managed.

However, this method is quite expensive, principally for the continuous flow of current that is required, if it is coupled with an analog technique able to directly recover metals in concentrated solutions on suitable support cathodes [20] after the electrochemical stripping of metals from RVC, which is a less costly and more efficient process that can be performed in the context of a circular economy and has a low environmental impact.

3. Materials and Methods

3.1. Theoretical Approach

Before discussing the dimensionless analysis, it is important to briefly describe the equations on which the process is based. The electrolysis rate can be expressed in terms of the reactant concentration, which decreases for a batch reactor containing a three-dimensional electrode as follows:

$$c_t = c_0 \cdot \exp^{\left(-\frac{k_m \cdot A_e \cdot V_r}{V_t}\right) \cdot t} \quad (2)$$

for the global process and is as follows:

$$c_{out} = c_{in} \cdot \exp^{-\frac{k_m \cdot A_e \cdot V_r}{Q_v}} \quad (3)$$

for a single passage inside the reactor according to Equations (2) and (3).

Equations (2) and (3) have been exploited for all the plug flow reactor (PFR) studies cited in this paper. By rearranging Equation (2) and plotting *ln(c/c$_0$)* vs. time, it is possible to obtain lines like those reported in Figure 1, for which slope *m* is as follows:

$$m = \frac{A_e \cdot k_m \cdot V_r}{V_t} \quad (4)$$

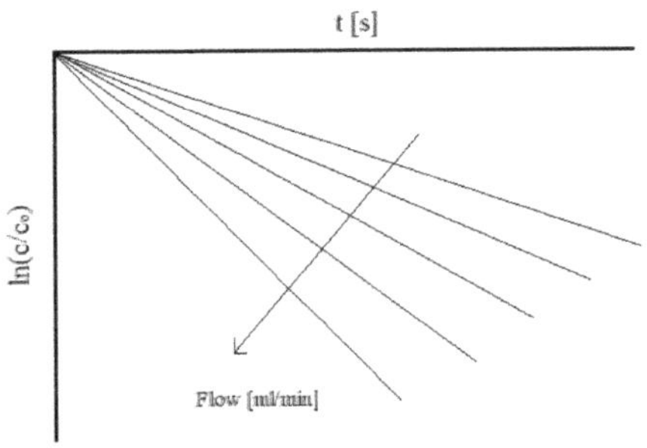

Figure 1. Trend of *ln(c/c$_0$)* vs. time under mass transport conditions and at different flowrates for divalent metal ion depletion.

Under mass transfer control conditions, the value of $A_e k_m$, which increases with the flowrate, can be calculated to finally obtain k_m, which is required for calculating the dimensionless Sherwood number. On the other hand, by knowing the cell geometry and the RVC porosity, the flow passage cross-section can be calculated and therefore, for each flow value, the linear velocity (v) and the dimensionless Reynolds number can be evaluated.

As shown in Figure 2, the mechanism for the ion electrodeposition is similar for all the considered metals: When the mass control condition is established in the solution, the electrochemical reaction at the surface is so fast with respect to the mass transfer flow from the bulk that the ion concentration on the electrode surface becomes zero; and in proximity to the electrode surface, a diffusion layer is formed at the maximum concentration gradient and for which the thickness (δ) is a function of the flowrate. In particular, it decreases as the flow increases, enhancing the mass transfer coefficient and, consequently, the limiting current. To reach this situation, a suitable cathode potential must be imposed, as reported in Table 2. Under this condition, it is possible to use such dimensionless numbers to describe the process. In general, in eletrodeposition, the number of electrons that can be transferred in one or more steps can be evaluated by determining the charge transfer coefficient ($\overline{\alpha}$ using Tafel analysis, but this is outside the scope of this paper; also, the charge transfer in this case does not represent the rate-determining step of the process.

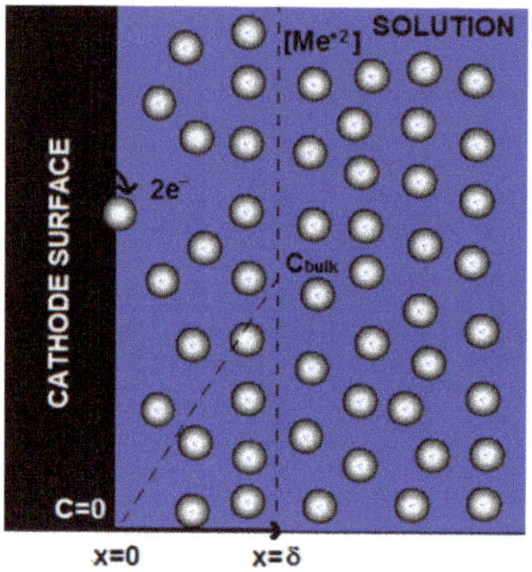

Figure 2. Mechanism for the ion electrodeposition.

Table 2. RVC cathodic potentials (referenced to SCE) required to reach mass transport control conditions, as found in the literature.

Ion		References
Ni^{+2}	−1100 mV vs. SCE	[46]
Co^{+2}	−1200 mV vs. SCE	[46]
Zn^{+2}	−1400 mV vs. SCE	[50]
Cu^{+2}	−500 mV vs. SCE	[31]
Cd^{+2}	−850 mV vs. SCE	[35]
Pb^{+2}	−800 mV vs. SCE	[36,44]

Then, from the electrochemical point of view, to work under the mass transfer control condition means to stay under the limiting current density condition, for which the absolute value is as follows [44]:

$$i_L = \frac{I_L}{A_e V_r} = nFk_m c_t \tag{5}$$

This condition is present when the electrochemical reaction kinetics on electrode surface are faster than the mass flow from the bulk to the electrode and in general, they are experimentally determined. Commonly, the kinetics are quantitatively related to the different ions through c_t, while it is qualitatively bound to them through n and k_m.

3.2. Experimental System Concept

A generic schematic of the experimental apparatus described in all the papers analyzed herein is shown in Figure 3, while Figure 4 shows an example of the porous 3-dimensional RVC cathode on which the depletion process is based. Further details are reported in papers [31–46,50].

Despite the experimental setup being almost the same in all the cases taken into account, it seems that even slight differences, together with the peculiar features of each ionic species, can produce quite different results [31–38,44,46,50]. In principle, such differences can also be ascribed to the limited number of data, different characteristic length definitions, different morphologies of the deposits, and, in general, different systems that are used.

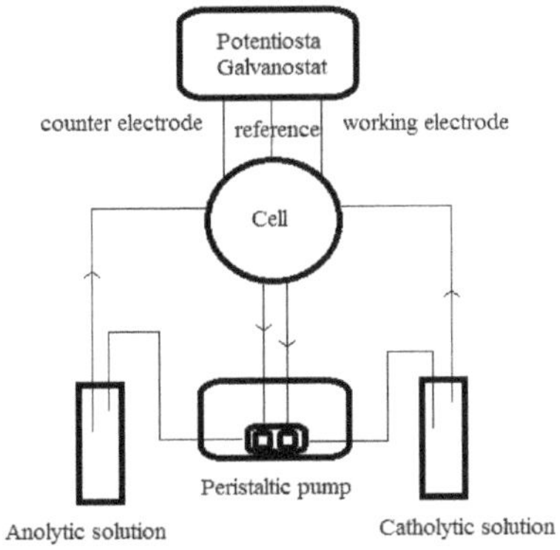

Figure 3. Schematic of the experimental apparatus described in all the papers considered in the present work.

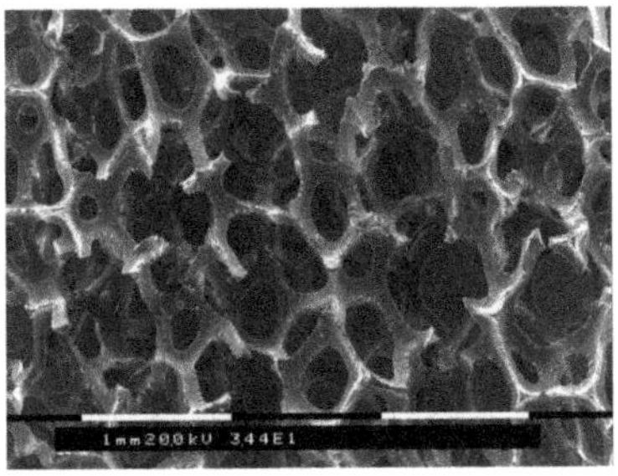

Figure 4. SEM image of an RVC cathode electrode (A_e = 27 cm^2/cm^3).

As far as the difference in the morphology is concerned, in general, it is possible to state that the morphologies of electrodeposited metals depend on several factors, such as the composition of the electrolyte, the current density, the temperature, the nature of the cathode support, fluid dynamics, and the addition of additives to the electrolyte, as well as the characteristics of metals and the interactions associated with the cathode materials [63–65].

However, to verify whether a general trend exists, a dimensionless analysis of the data presented in the literature has been carried out. Therefore, as aforementioned, the intention of this work is to try to overcome the differences found in the literature, thus obtaining a fair comparison of the results. In this sense, the present work can be considered as a critical review in the field of metal ion depletion from wastewater with an RVC cathode. Table 1 shows the geometric and dimensional features of the cells used in the experiments that were taken into account, together with the diffusion coefficient for each cation [35,38,47–49,51,52].

4. Discussion

4.1. Nickel, Cobalt, and Zinc

A previous dimensionless analysis for Ni^{+2} and Co^{+2} ion depletions under the mass control regime for different flowrates [46] was carried out, by the authors, considering a characteristic length ($A_e{}^{-1}$) equal to 27 cm^2/cm^3, a diffusion coefficient (D) of 0.6×10^{-5} cm^2/s [60], a 97% RVC void volume (ε), a solution kinematic viscosity (μ) of 1.1×10^{-2} cm^2/s, and an electrode geometry of 10.0 cm × 10.0 cm × 1.3 cm. The catholyte was a 0.1 M Na_2SO_4 solution, adjusted to pH 6 by the addition of H_2SO_4 and containing an initial concentration (C_i) of Co^{+2} or Ni^{+2} equal to 150 ppm, while the anolyte was a 0.5 M Na_2SO_4 solution. The final Co^{+2} and Ni^{+2} concentrations in the solution were quantified using atomic absorption spectrometry and found to be less than 0.1 ppm in less than one hour. As expected, the $A_e k_m$ values increased with the flowrate [46].

The current efficiencies ranged from 20% to 98% and from 3% to 45% for the nickel and cobalt depletions, respectively, and increased with the flowrate.

In this work, two different equations have been found for Ni^{+2} and Co^{+2} ion depletions on the RVC cathode as follows:

$$Sh = 0.077 \cdot Re^{0.78} Sc^{0.33} \quad (6)$$

$$Sh = 0.207 \cdot Re^{0.293} Sc^{0.33} \quad (7)$$

or considering the Schmidt number value, the relationship is as follows:

$$log(Sh) = 0.78 \cdot log(Re) - 0.036 \quad (8)$$

$$log(Sh) = 0.293 \cdot log(Re) + 0.392 \quad (9)$$

The flow range used for the depletion experiments was from 500 to 1300 mL/min in the case of Ni, while it ranged from 100 to 1650 mL/min in the case of Co. The linear speeds, instead, ranged from 0.66 to 1.70 cm/s and from 0.13 to 2.17 cm/s, respectively. Two different deposit aspects have been obtained. In particular, in the case of the nickel, the deposit appears homogeneous, compact, and uniformly distributed on the electrode, while in the case of the cobalt, it is spongy and dendritic. An anionic membrane was used between the cathode and anode.

Instead, other tests were carried out by Tan et al. [38] for Co^{2+} depletion. In this case, the initial concentration was 100 ppm, and the linear speeds ranged from 0.09 to 0.34 cm/s. In this system, instead of using a galvanostatic/potentiostatic source, a galvanic cell was conceived by coupling the Zn/Zn^{+2} redox couple, with a standard potential (E_0) of −0.763 V, to a Co/Co^{2+} redox couple having a standard potential of E_0 = 0.658 V on an RVC [41]. The catholyte was 0.2 M Na_2SO_4 and 0.1 M H_3BO_4 and contained the specified concentrations of Co^{2+}, and the pH was adjusted to 4 by the addition of sulfuric acid or sodium hydroxide, while the anolyte was 0.1 M Na_2SO_4. The electrode geometry was 5.0 cm × 2.0 cm × 0.7 cm; also, in this case, an anionic membrane was used. Once the mass transfer control conditions were reached, more than 99% of the Co^{+2} was removed in less than two hours, as determined using atomic absorption spectrometry. The current efficiency was equal to about 82%, and the obtained metal deposit appeared particularly rough. The authors [38], in this case, reported different relationships in terms of $A_e k_m$ as a function of the linear velocity (v), as given by Equation (10), where parameters a and b are a function of A_e as follows:

$$A_e k_m = a \cdot v^b \quad (10)$$

As far as zinc ions are concerned, Lanza and Bertazzoli [50] focused on Zn^{+2} depletion either as a function of the flowrate or as a function of the electrode's specific surface area. For this purpose, they started from a $ZnCl_2$ solution with an initial zinc ion concentration of 54 ppm at pH 5.5 and using 0.1 M H_3BO_4 and 0.1 M KCl as the supporting electrolyte. RVC cathode electrodes with specific surface areas of 30, 40, 53, and 66 cm^2/cm^3 and a geometry of 15.0 cm × 5.0 cm × 1.25 cm [37] were used.

The flow range was from 665 to 5300 mL/min, while the linear speed, instead, ranged from 1.7 to 14 cm/s when they used an RVC with a specific surface area of 53 cm^2/cm^3. The flow was kept constant at 2000 mL/mL when they changed the RVC specific surface area from 30 to 66 cm^2/cm^3.

A Nafion membrane was used between the cathodic and anodic compartments and for all the tests, a final concentration of less than 0.1 ppm was reached in less than one hour. The concentrations were measured using atomic absorption spectrometry. In particular, for those experiments, the authors did not report any particular dimensionless relationship between the Reynolds and Sherwood numbers, but they stated that good performance was obtained. Finally, the deposit quality was not mentioned.

4.2. Copper, Cadmium, and Lead

On the other hand, Pletcher et al. [31] performed Cu^{+2} removal by working at different flowrates, with electrodes having specific surface areas equal to either 7 cm^2/cm^3 or 66 cm^2/cm^3, using the catholyte $CuSO_4$ at an initial concentration of 10 ppm as a cation source. In this case, both the catholyte and anolyte were 0.5 M Na_2SO_4 solutions, kept at pH 2 by the addition of H_2SO_4, while the geometry of the electrode was 5.0 cm × 5.0 cm × 1.2 cm. The final concentration reached less than 0.1 ppm in a range of times from a few minutes to a few hours, depending on the RVC porosity. The authors did not describe the quality of the copper deposits, but they stated that good efficiency was achieved. In this case, a different definition of the characteristic length was taken into account. The authors defined d_e as follows:

$$d_e = \frac{4A}{P} = \frac{2BS}{(B+S)} \tag{11}$$

where A is the cross-sectional area of the flow, P is the wetting perimeter of the cross-section, while B and S are the width and thickness of the RVC, respectively.

Using the above-mentioned d_e definition, the authors found the following dimensionless relationship among the Sherwood, Reynolds, and Schmidt numbers:

$$Sh = 2.7 \cdot Re^{0.48} Sc^{0.33} \tag{12}$$

or considering the Schmidt number value, the relationship is as follows:

$$log(Sh) = 0.48 \cdot log(Re) + 1.53 \tag{13}$$

However, it is worth stressing that the values of $log(Re)$ and $log(Sh)$, for our purpose, were recalculated using the definition of d_e used in this work, namely A_e.

The flow range used in the experiment was from 420 to 7000 mL/min, while the linear speed ranged from 1.2 to 20 cm/s for the considered cases.

On the other hand, Podlaha and Fenton [32] performed Cu^{2+} reduction using an RVC electrode with a specific surface area of 27 cm^2/cm^3 and a geometry of 10.2 cm × 4.5 cm × 0.87 cm. The concentrations of the Cu^{2+} copper ion ranged from 10^{-4} M to 10^{-2} M in a 0.8 M $NaSO_4$ supporting electrolyte solution kept at pH 3 upon the addition of H_2SO_4. A Nafion membrane was used between the cathodic and anodic compartments. The electrolyte flowrate through the RVC electrode was varied between 0.012 and 2.6 cm/s (2.8–620 mL/min). In this case, the obtained deposit thus had a granular morphology.

The copper concentration was determined spectrophotometrically, using a wavelength of 560 nm. In this work, the authors found the following dimensionless relationship for Cu^{+2} ion depletion on the RVC cathode, using A_e as the characteristic length:

$$Sh = 0.44 \cdot Re^{0.69} Sc^{0.33} \tag{14}$$

or considering the Schmidt number value, the relationship is as follows:

$$log(Sh) = 0.69 \cdot log(Re) + 0.74 \tag{15}$$

Moreover, by always considering Cu^{+2} ions, Y.P. Hor and N. Mohamed [33] instead operated a galvanic cementation system. In this system, instead of using a galvanostatic/potentiostatic source, a galvanic cell was conceived by coupling the Fe/Fe^{+2} redox couple, with a standard potential (E_0) of −0.44 V, to a Cu/Cu^{2+} redox couple having a standard potential of E_0 = 0.34 V. Therefore, in this case, the copper reduction occurs spontaneously on an RVC cathodic support. The authors used an RVC electrode with a specific surface area of 53 cm^2/cm^3 and a geometry of 5 cm × 2 cm × 1 cm and an anion exchange membrane. The electrolyte solutions were pumped by a peristaltic pump into the cell flowing from the bottom to the top of the cell. The initial concentration of the Cu^{2+} copper ion was 100 ppm in 0.5 M H_2SO_4. The flow range used in the experiment was from 2 to 20.1 mL/min, while the linear speed ranged from 0.016 to 0.160 cm/s, values which were much lower with respect to those obtained by other authors.

The concentrations of the effluent were determined using an absorption spectrophotometer. During operation in the recycling mode, they reached almost 100% Cu^{+2} removal, showing that reticulated vitreous carbon is a very good electrode material for copper ion removal. In this work, the authors found the following dimensionless relationship for Cu^{+2} ion depletion on the RVC cathode, using A_e as the characteristic length for defining the Reynolds and Sherwood dimensionless numbers:

$$Sh = 0.94 \cdot Re^{0.63} Sc^{0.33} \quad (16)$$

or considering the Schmidt number value, the relationship is as follows:

$$log(Sh) = 0.63 \cdot log(Re) + 1.07 \quad (17)$$

Considering Cd^{+2} ions, Llovera-Hernandez et al. [34] performed depletion by working in the galvanostatic mode with a de-aerated electrolyte and an electrode with 5.0 cm × 5.0 cm × 1.0 cm size and an RVC specific surface area of 40 cm^2/cm^3. The electrolyte composition was different for different experiments. Only the results for the 0.5 molar concentrations of NaCl and Na_2SO_4 were considered, with an initial pH value equal to 6 and an initial Cd^{+2} ion concentration of 225 ppm. The flowrate and the linear speed were 4600 mL/min and 16 cm/s, respectively. The final Cd^{+2} concentration was in the range of about 2–10 ppm, which was reached in about 2 h, and an electrochemical method (differential pulse-stripping voltammetry) was used to measure the Cd^{+2} concentration in the solution. A Nafion membrane was used between the anodic and cathodic compartments. A constant current density equal to 3.5 A/m^2 was used under mass transport control conditions. The authors did not report any relationship between the Reynolds and Sherwood dimensionless numbers. So, in this case, the data were used to calculate them.

Furthermore, Dutra et al. [35] performed tests, in the potentiostatic mode, for Cd^{2+} removal, using a de-aerated electrolyte and an electrode having a specific surface area of 40 cm^2/cm^3 and a geometry of 12.0 cm × 5.0 cm × 1.2 cm. The anolyte and catholyte were 0.5 M Na_2SO_4, adjusted to pH 2 upon the addition of H_2SO_4. Starting from 210 ppm of Cd^{2+} dissolved in the catholyte, a final concentration of 0.1 ppm, as measured using atomic absorption, was obtained in about one hour and a half. A protonic membrane (Nafion) was used to separate the anodic and cathodic compartments. The metallic deposit thus obtained was also, in this case, nodular [35,39]. The authors did not report any dimensionless relationship. In any case, the best test results were obtained using a flowrate and a linear speed of 4200 mL/min and 12 cm/s, respectively.

Finally, considering Pb^{+2} ions, depletion tests were carried out by Bertazzoli et al. [36] and Widner et al. [37] by operating at three different flowrates ranging from 1000 to 4000 mL/min, with electrodes having specific surface areas equal to 53, 37, 27, and 11 cm^2/cm^3 and a geometry equal to 15.0 cm × 5.0 cm × 1.25 cm.

A protonic membrane (Nafion) was used in the experimental setup, and a mixed solution of 0.5 M H_3BO_4 and 0.05 M KNO_3 at pH 4.8 was used as the supporting electrolyte. Starting from a 50 ppm initial Pb^{2+} concentration, obtained using $Pb(NO_3)_2$, a final concentration of less than 0.1 ppm was obtained in a time range from about 20 min to 2 h, depending on the RVC porosity and flowrate. In this case, the current efficiencies ranged from 4.6% to 22.3% and from 1.8% to 14.1% for 90% and 99% Pb^{+2} removals, respectively, and it increased with increasing specific electrode surface (A_e) of the RVC.

As for Cu and Co, a rough lead deposit was obtained. Also, in this case, like Pletcher, the authors used Equation (11) as the characteristic length and found the following dimensionless relationship [36]:

$$Sh = (24 \pm 7)Re^{(0.32\pm0.05)}Sc^{0.33} \tag{18}$$

or considering the Schmidt number value, the relationship is as follows:

$$log(Sh) = (0.32 \pm 0.05)\cdot log(Re) + (2.28 \pm 0.12) \tag{19}$$

Nevertheless, as before, $log(Re)$ and $log(Sh)$, for our intent, were recalculated using A_e as the characteristic length.

In addition, other tests have been also carried out by C. Ponce de Leon and D. Pletcher [44] for Pb^{2+} depletion at six different flowrates in a chloride solution, using RVC cathodes with different specific surface areas and having a geometry equal to 5.0 cm × 5.0 cm × 1.2 cm. The flow range used for the tests was from 60 to 6050 mL/min, while the linear speed ranged from 1.6 to 16 cm/s. A 0.5 M NaCl solution at pH 2 was used as the supporting electrolyte. Starting from a 7 ppm initial Pb^{2+} concentration, a final concentration of less than about 0.5 ppm was obtained in less than 60 min, working at a current efficiency of about 80% when 90% of the Pb^{2+} had been removed. The concentrations were determined using atomic absorption spectroscopy. In this case, the authors found k_m values, using the voltammetric technique to determine the limiting current for different flows and different RVC cathodes, and using Equation (5).

These results are in agreement with those obtained by other authors in other experiments. Also, in this case, the authors reported several relationships similar to Equation (10), with different *a* and *b* coefficient values, depending on the specific RVC electrode surface (A_e) that was used.

Moreover, for lead and copper ions, respectively, the authors found the following dimensionless relationships among the Sherwood, Reynolds, and Schmidt numbers, using Equation (11) as the characteristic length:

$$Sh = 5.6\cdot Re^{0.44}Sc^{0.33} \tag{20}$$

for Pb^{+2} ion depletion in a solution of 0.5 M NaCl at pH 2 and a 60 ppi (40 cm^2/cm^3) RVC and

$$Sh = 2.4\cdot Re^{0.48}Sc^{0.33} \tag{21}$$

for Cu^{+2} ion depletion in a solution of 0.5 M Na_2SO_4 at pH 2 and a 10 ppi (7 cm^2/cm^3) RVC, or considering the Schmidt number values, the respective relationships are as follows:

$$log(Sh) = 0.44\cdot log(Re) + 1.48 \tag{22}$$

and

$$log(Sh) = 0.48\cdot log(Re) + 1.8 \tag{23}$$

Finally, the authors concluded that electrolytic treatment using a reticulated vitreous carbon cathode could be usefully applied to the waste stream to decrease the ion level to a concentration suitable for direct discharge.

To maximize the process results, it is necessary to work under mass transport control conditions, i.e., at the limiting current value, which depend on the ion concentrations in the bulk, flowrates, and temperatures. In every case, it is necessary to determine the cathodic potentials beyond which the mass control condition is present, such as those found in the literature and reported in Table 2. With some exceptions, most authors worked in the potentiostatic mode.

It is worth stressing that they could be different because these cathodic potential values depend on different parameters and always have to be experimentally determined. This means that when working under galvanostatic conditions (as in the case of Llovera-Hernandez et al. [34]), a current that implies those cathodic potential values must be set up. Straying too much toward more cathodic potentials, for example, by increasing the value of the current too much, means decreasing the process efficiency because other reactions, such as hydrogen evolution, may occur. In all the cases, either a protonic membrane (Nafion) [31,32,34,35,37,50] or an anionic membrane [29,33,46] was used between the cathodic and anodic compartments.

5. Dimensionless Analysis

For comparing the literature results, it is necessary to point out that the definition of the characteristic length is crucial because it has a drastic effect on the coherence of the results. Actually, even though some authors, like Pletcher and Bertazzoli, used different definitions [31,36,37,50], the use of the A_e^{-1} term seems to be the most appropriate because it takes into account the RVC porosity (in terms of the number of pores per inch, ppi), which can also be different for cathodes with the same geometry. For a fair comparison, using the literature results and A_e^{-1} as the characteristic length, pairs of dimensionless numbers were calculated and evaluated for all the examined and cited works, and Figure 5 shows where the calculated pairs of dimensionless numbers fall in the *log*(*Sh*) vs. *log*(*Re*) plane more frequently for each different divalent ion if this kind of experiment is carried out. This clearly shows a general trend in a well-defined region of the graph.

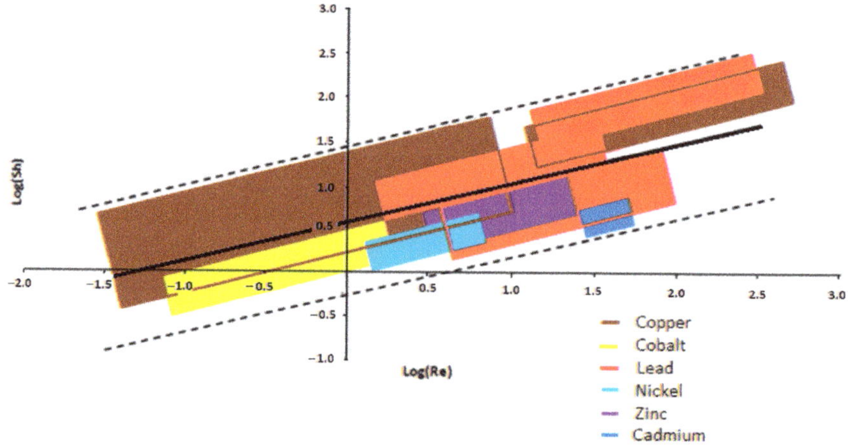

Figure 5. Recalculated trend of log(Sh) vs. log(Re) equations for the depletions of Ni^{+2} [46], Co^{+2} [38,46], Cu^{+2} [31–33,44], Cd^{+2} [34,35], Pb^{+2} [36,37,44], and Zn^{+2} [50].

The statistical analysis highlights a strong correlation among the experimental results as well. Indeed, from Table 3, where some statistics, such as the standard deviations (σ_x and σ_y), covariance (Cov_{xy}), Pearson correlation coefficient (r), and degrees of freedom (number of pairs minus two), are reported, it is possible to notice that the r value (0.71) exceeds the critical one, r_c (0.174), highlighting a strong correlation.

Table 3. Statistical data: standard deviations (σ_x and σ_y), covariance (Cov_{xy}), Pearson correlation coefficient (r), critical Pearson coefficient (r_c), alpha level (α), degrees of freedom (fd).

σ_x	σ_y	Cov_{xy}	r	r_c	α	fd
0.88	0.58	0.37	0.71	0.174	0.05	125

This behavior can be best described by the equation of the straight black line shown in Figure 5, for which the coefficients were calculated by minimizing the error, using the least square method, of the difference between the experimentally measured and theoretically calculated points, using Equation (24) as follows:

$$log(Sh) = 0.47 \cdot log(Re) + 0.57 \pm 0.85 \quad (24)$$

In Equation (24), ± 0.85 represents the region of the graph, in Figure 5, delimited by the dotted lines, where the totality of the points falls. In summary, if Equation (11) is used as the equivalent diameter for the dimensionless numbers, the constant term of the resulting *Sh* vs. *Re* equation is too high, and the curve, in general, does not fall in this region; however, if the specific area of RVC (A^{-1}) is used instead, even presenting, in some cases, quite different slopes and constant terms, the curves lie between the two dotted lines in all the cases. This does not mean that Equation (11) cannot be used but only that to compare the data, it is very important to use the same equivalent diameter definition, whatever it is, but in our opinion, A^{-1} is more suitable. Moreover, it is worth stressing that in our opinion, if further experiments will be considered, the sizes of the colored regions could be extended and overlap with each other but very probably would remain between the dotted lines all the same. Finally, all these considerations obviously are not valid, for example, if different kinds of electrodes (other than RVC), different types of solvents (other than water), or other additives in the electrolyte solutions (besides those mentioned in this work) are considered.

6. Conclusions

A suitable analysis of the literature results related to the kinetics and fluid dynamics of the depletion processes of different divalent metal ions from wastewater allowed a fair comparison among the different outcomes obtained by the authors considered in this work. This, in turn, allowed the identification of a particular region in the *log*(*Re*) vs. *log*(*Sh*) plane, where it is more probable to find these outcomes in terms of such dimensionless numbers. The discussion and description of such results highlighted that the electrochemical technique using the RVC cathode material should have more consideration as it can be taken into account as an efficient tool for wastewater treatment in the removal of divalent metal ions. The outcomes suggest that if the same dimensionless number definition is used, general and similar behaviors for all the described systems could be found regardless of the differences in geometry, dimensions, and initial conditions. Thus, a general dimensionless trend equation has been proposed to describe the electrochemical depletion of such divalent ions from wastewater, using an RVC cathode. The statistical analysis highlights a good correlation. The obtained empirical equation can be exploited, first and foremost, for designing an RVC electrochemical filter for the divalent ion depletion process.

Author Contributions: Formal analysis, A.D. and M.P.; data curation, A.D.; writing—original draft preparation, A.D., C.L., E.C., F.A.S. and M.P.; writing—review and editing A.D. and C.L.; supervision A.D., C.L., E.C., F.A.S. and M.P. All authors have read and agreed to the published version of the manuscript.

Funding: This research received no external funding.

Institutional Review Board Statement: Not applicable.

Informed Consent Statement: Not applicable.

Data Availability Statement: Data are contained within the article.

Conflicts of Interest: The authors declare no conflicts of interest.

Nomenclature

A_e is the three-dimensional electrode's specific surface (cm^{-1})
${A_e}^{-1}$ is the characteristic length (cm)
c_t is the concentration at time t (mol/cm^3)
D is the diffusion coefficient (cm^2/s)

F is the Faraday constant
k_m is the mass transport coefficient (cm/s)
L is the RVC cathode length (cm)
Q_v is the volumetric flowrate (cm^3/s)
n is the ion valence
N is the number of recirculations
N_{cells} is the number of cells
S is the RVC cathode section (cm^2)
t_r is the reactor residence time (s)
t_f is the total depletion time (s)
V_r is the three-dimensional electrode's volume (cm^3)
V_t is the solution volume (cm^3)
v is the linear velocity (cm/s)
ε is the RVC void's volume
μ is the solution's kinematic viscosity (cm^2/s)
α is the alpha level
σ_x and σ_y are the standard deviations
Cov_{xy} is the covariance
r is the Pearson correlation coefficient
r_c is the critical Pearson coefficient
fd is the degrees of freedom

References

1. Hu, M.; Wang, Y.; Ye, D. A Timely Review of Lithium-Ion Batteries in Electric Vehicles: Progress, Future Opportunities and Challenges. In Proceedings of the E3S Web of Conferences 6th International Conference on Materials Science, Energy Technology and Environmental Engineering (MSETEE), Hangzhou, China, 13–15 August 2021; Volume 308, p. 01015.
2. Pruthvija, B.; Lakshmi, K.P. Review on Battery Technology and its Challenges. *Int. J. Sci. Eng. Res.* **2020**, *11*, 1706–1713.
3. Babic, U.; Suermann, M.; Buchi, F.N.; Gubler, L.; Schmidt, T.J. Critical Review—Identifying Critical Gaps for Polymer Electrolyte Water Electrolysis Development. *J. Electrochem. Soc.* **2017**, *164*, F387–F399. [CrossRef]
4. Santos, A.L.; Cebola, M.-J.; Santos, D.M.F. Towards the Hydrogen Economy-A Review of the Parameters That Influence the Efficiency of Alkaline Water Electrolyzers. *Energies* **2021**, *14*, 3193. [CrossRef]
5. Valente, A.; Iribarren, D.; Dufour, J. End of life of fuel cells and hydrogen products: From technologies to strategies. *Int. J. Hydrogen Energy* **2019**, *44*, 20965–20977. [CrossRef]
6. Fan, L.; Tu, Z.; Chan, S.H. Recent development of hydrogen and fuel cell technologies: A review. *Energy Rep.* **2021**, *7*, 8421–8446.
7. Poonam, K.; Sharma, A.; Tripathi, S.K. Review of supercapacitors: Materials and devices. *J. Energy Storage* **2019**, *21*, 801–825. [CrossRef]
8. Chakraborty, S.; Mary, N.L. Review—An Overview on Supercapacitors and Its Applications. *J. Electrochem. Soc.* **2022**, *169*, 020552.
9. Jena, K.K.; Mayyas, A.T.; Mohanty, B.; Jena, B.K.; Jos, J.R.; Al Fantazi, A.; Chakraborty, B.; Almarzooqi, A.A. Recycling of Electrode Materials from Spent Lithium-Ion Batteries to Develop Graphene Nanosheets and Graphene−Molybdenum Disulfide Nanohybrid: Environmental Benefits, Analysis of Supercapacitor Performance, and Influence of Density Functional Theory Calculations. *Energy Fuels* **2022**, *36*, 2159–2170.
10. Pražanová, A.; Knap, V.; Stroe, D.-I. Literature Review, Recycling of Lithium-Ion Batteries from Electric Vehicles, Part II: Environmental and Economic Perspective. *Energies* **2022**, *15*, 7356. [CrossRef]
11. Steward, D.; Mayyas, A.; Mann, M. Economics and Challenges of Li-Ion Battery Recycling from End of Life Vehicles. *Procedia Manuf.* **2019**, *33*, 272–279. [CrossRef]
12. Carmo, M.; Keeley, G.P.; Holtz, D.; Grube, T.; Robinius, M.; Muller, M.; Stolten, D. PEM water electrolysis: Innovative approaches towards catalyst separation, recovery and recycling. *Int. J. Hydrogen Energy* **2019**, *44*, 3450–3455. [CrossRef]
13. Wittstock, R.; Pehlken, A.; Wark, M. Challenges in Automotive Fuel Cells Recycling. *Recycling* **2016**, *1*, 343–364. [CrossRef]
14. Eh, C.L.M.; Tiong, A.N.T.; Kansedo, J.; Lim, C.H.; How, B.S.; Ng, W.P.Q. Circular Hydrogen Economy and Its Challenges. *Chem. Eng. Trans.* **2022**, *94*, 1273–1278.
15. Yanshyna, O.; Weissman, H.; Rybtchinski, B. Recyclable electrochemical supercapacitors based on carbon nanotubes and organic nanocrystals. *Nanoscale* **2020**, *12*, 8909–8914. [CrossRef]
16. Jiang, G.; Pickering, S.J. Recycling supercapacitors based on shredding and mild thermal treatment. *Waste Manag.* **2016**, *48*, 465–470. [CrossRef]
17. Lupi, C.; Pasquali, M.; Dell'Era, A. Nickel and cobalt recycling from lithium-ion batteries by electrochemical processes. *Waste Manag.* **2005**, *25*, 215–220. [CrossRef] [PubMed]
18. Lupi, C.; Pasquali, M.; Dell'Era, A. Studies concerning nickel electrowinning from acidic and alkaline electrolytes. *Miner. Eng.* **2006**, *19*, 1246–1250.

19. Dell'Era, A.; Lupi, C.; Pasquali, M. Electrochemical Purification of Cobalt Polluted Effluents. In Proceedings of the REWAS 2008: Global Symposium on Recycling, Waste Treatment and Clean Technology, Cancun, Mexico, 12–15 October 2008; pp. 1413–1418.
20. Ciro, E.; Dell'Era, A.; Pasquali, M.; Lupi, C. Indium electrowinning study from sulfate aqueous solution using different metal cathodes. *J. Environ. Chem. Eng.* **2020**, *8*, 103688.
21. Perry, R.H.; Chilton, C.H. *Chemical Engineers Handbook*, 5th ed.; McGraw Hill: New York, NY, USA, 1973.
22. Sulaymon, A.H.; Abbar, A.H. Scale-Up of Electrochemical Reactors. In *Electrolysis*; InTech: London, UK, 2012; Chapter 9; pp. 189–202, licensee InTech. [CrossRef]
23. Levin, M. How to scale up scientifically: Scaling up manufacturing process. *Pharm. Technol.* **2005**, *3*, 4s–12s.
24. Wichterle, K.; Ruzicka, M.C. Chapter: Scale-up fundamentals. In *Scale-Up in Metallurgy*; Lackner, M., Ed.; Verlag ProcessEng Engineering GmbH: Wien, Austria, 2010; pp. 1–33, ISBN 978-3-902655-10-3.
25. Zlokarnik, M. Scale-up of Processes Using Material Systems with Variable Physical Properties. *Chem. Biochem. Eng. Q.* **2001**, *15*, 43–47. [CrossRef]
26. Mithani, R.; Shenoi, S.; Fan, L.T.; Walawender, W.P. Ranking Dimensionless Groups in Fluidized-Bed Reactor Scale-Up. *Int. J. Approx. Reason.* **1990**, *4*, 69–85. [CrossRef]
27. Buckingham, E. On Physically Similar Systems; Illustrations of the Use of Dimensional Equations. *Phys. Rev.* **1914**, *4*, 345–376.
28. Rase, H. *Chemical Reactor Design for Process Plants*; Wiley-Interscience: New York, NY, USA, 1977.
29. Kulkarni, M.M.; Sharma, A.P.; Rao, D.N. Use of dimensional analysis for scaling immiscible gas assisted gravity drainage (GAGD) experiments. In Proceedings of the International Symposium of the Society of Core Analysts, Toronto, ON, Canada, 21–25 August 2005; pp. 21–25.
30. Mantall, C.L. *Electrochemical Engineering*, 4th ed.; Mc-Graw Hill: New York, NY, USA, 1960.
31. Pletcher, D.; Whyte, I.; Walsh, F.C.; Millington, J.P. Reticulated vitreous carbon cathodes for metal ion removal from process streams Part I: Mass transport studies. *J. Appl. Electrochem.* **1991**, *21*, 659–666. [CrossRef]
32. Podlaha, E.J.; Fenton, J.M. Characterization of a flow-by RVC electrode reactor for the removal of heavy metals from dilute solutions. *J. Appl. Electrochem.* **1995**, *25*, 299–306. [CrossRef]
33. Hor, Y.P.; Mohamed, N. Removal and recovery of copper via a galvanic cementation system Part I: Single-pass reactor. *J. Appl. Electrochem.* **2003**, *33*, 279–285. [CrossRef]
34. de Lourdes Llovera-Hernandez, M.; Alvarez-Gallegos, A.; Hernandez, J.A. Cadmium removal from dilute aqueous solutions under galvanostatic mode in a flow-through cell. *Desalination Water Treat.* **2016**, *57*, 22809–22817. [CrossRef]
35. Dutra, A.J.B.; Espinola, A.; Borges, P.P. Cadmium removal from diluitedacqueous solutions by electrowinning in a flow by cell. *Miner. Eng.* **2000**, *13*, 1139–1148. [CrossRef]
36. Bertazzoli, R.; Rodrigues, C.A.; Dallan, E.J.; Fukunaga, M.T.; Lanza, M.R.V.; Leme, R.R.; Widner, R.C. Mass transport properties of a flow-through electrolytic reactor using a porous electrode: Performance and figures of merit for Pb+2 removal. *Braz. J. Chem. Eng.* **1998**, *15*, 4.
37. Widner, R.C.; Sousa, M.F.B.; Bertazzoli, R. Electrolytic removal of lead using a flow-through cell with a reticulated vitreous carbon cathode. *J. Appl. Electrochem.* **1998**, *28*, 201–207. [CrossRef]
38. Tan, W.X.; Hasnat, M.A.; Ramalan, N.H.M.; Soh, W.M.; Mohamed, N. Influence of flow rates on the electrogenerative Co^{2+} recovery at a reticulated vitreous carbon cathode. *Chem. Eng. J.* **2012**, *189–190*, 182–187. [CrossRef]
39. Tramontina, J.; Azambuja, D.S.; Piatnicki, C.M.S. Removal of Cd2+ Ion from Diluted Aqueous Solutions by Electrodeposition on Reticulated Vitreous Carbon Electrodes. *J. Braz. Chem. Soc.* **2002**, *13*, 4. [CrossRef]
40. Ramalan, N.H.M.; Karoonian, F.S.; Etesami, M.; Wen-Min, A.; Hasnat, S.M.; Mohamed, N. Impulsive removal of Pb(II) at a 3-D reticulated vitreous carbon cathode. *Chem. Eng. J.* **2012**, *203*, 123–129. [CrossRef]
41. Friedrich, J.M.; Ponce-de-Leon, C.; Reade, G.W.; Walsh, F.C. Reticulated vitreous carbon as an electrode material. *J. Electroanal. Chem.* **2004**, *561*, 203–217. [CrossRef]
42. Czerwinski, A.; Obrebowski, S.; Kotowski, J.; Rogulski, Z.; Skowronskic, J.M.; Krawczyk, P.; Rozmanowski, T.; Bajsert, M.; Przystałowski, M.; Buczkowska-Biniecka, M.; et al. Electrochemical behavior of negative electrode of lead-acid cells based on reticulated vitreous carbon carrier. *J. Power Sources* **2010**, *195*, 7524–7529. [CrossRef]
43. Tangirala, R.C.; Low, C.T.J.; Ponce-de-Leon, C.; Campbell, S.A.; Walsh, F.C. Copper deposition at segmented, reticulated vitreous carbon cathode in Hull cell. *Trans. Inst. Met. Finish.* **2010**, *88*, 84–92. [CrossRef]
44. Ponce de Leon, C.; Pletcher, D. The removal of Pb from aqueous solutions using a reticulated vitreous carbon cathode cell-the influence of the electrolyte medium. *Electrochim. Acta* **1996**, *41*, 533–541. [CrossRef]
45. Polcaro, A.M.; Palmas, S.; Mascia, M.; Renoldi, F.; Correa dos Santos, R.L. Reticulated Vitreous Carbon Electrode for Copper(II) Removal from Dilute Acid Solutions: Effect of the Dissolved Oxygen. *Ind. Eng. Chem. Res.* **1999**, *38*, 1400–1405. [CrossRef]
46. Dell'Era, A.; Pasquali, M.; Lupi, C.; Zaza, F. Purification of nickel or cobalt ion containing effluents by electrolysis on reticulated vitreous carbon cathode. *Hydrometallurgy* **2014**, *150*, 1–8. [CrossRef]
47. Dimaculangan, D.A.A.; Bungay, V.C.; Sorian, A.N. Determination of Diffusion Coefficients of Heavy Metal Ions (Ni^{3+}, Zn^{2+}, Ba^{2+}, and Mn^{2+}) at Infinite Dilution through Electrolytic Conductivity Measurements. *ASEAN J. Chem. Eng.* **2022**, *22*, 58–71.
48. Patil, S.F.; Adhyapak, N.G.; Patil, P.R. Diffusion of Zinc Chloride in Aqueous Solution at 25 °C. *Int. J. Appl. Radiat. Isot.* **1981**, *35*, 761–766.
49. Patil, S.F.; Borhade, A.V.; Nath, M. Diffusivity of some zinc and cobalt salt in water. *J. Chem. Eng. Data* **1993**, *38*, 574–576.

50. Lanza, M.R.V.; Bertazzoli, R. Removal of Zn(II) from chloride medium using a porous electrode: Current penetration within the cathode. *J. Appl. Electrochem.* **2000**, *30*, 61–70.
51. Leaist, D.G. Tracer diffusion of cobalt(II), nickel(II), copper(II) and hydrogen ions in aqueous sodium sulphate solutions. *Electrochim. Acta* **1989**, *34*, 651–655.
52. Moats, M.S.; Hiskey, J.B.; Collins, D.W. The effect of copper acid and temperature on the diffusion coefficient of cupric ions in simulated electrorefining electrolytes. *Hydrometallurgy* **2000**, *56*, 255–268.
53. Noulty, R.A.; Leaist, D.G. Diffusion in Aqueous Copper Sulfate and Copper Sulfate-Sulfuric Acid Solutions. *J. Sol. Chem.* **1987**, *16*, 813–825. [CrossRef]
54. Hayes, P.C. Chapter 6, Hydrometallurgy. In *Process Selection in Extractive Metallurgy*, 1st ed.; Hayes Publishing Co.: Brisbane, Australia, 1985.
55. Leyma, R.; Platzer, S.; Jirsa, F.; Kandioller, W.; Krachler, R.; Keppler, B.K. Novel thiosalicylate-based ionic liquids for heavy metal extractions. *J. Hazard. Mater.* **2016**, *314*, 164–171. [CrossRef] [PubMed]
56. Saleh, T.A.; Mustaqeem, M.; Khaled, M. Water treatment technologies in removing heavy metal ions from wastewater: A review. *Environ. Nanotechnol. Monit. Manag.* **2022**, *17*, 100617. [CrossRef]
57. Chen, Q.; Yao, Y.; Li, X.; Lu, J.; Zhou, J.; Huang, Z. Comparison of heavy metal removals from aqueous solutions by chemical precipitation and characteristics of precipitates. *J. Water Process Eng.* **2018**, *26*, 289–300.
58. Shamsuddin, M. *Physical Chemistry of Metallurgical Processes*; Wiley: Hoboken, NJ, USA, 2016.
59. Bolisetty, S.; Peydayesh, M.; Mezzenga, R. Sustainable technologies for water purification from heavy metals: Review and analysis. *Chem. Soc. Rev.* **2019**, *48*, 463–487. [CrossRef]
60. Sitko, R.; Musielak, M.; Serda, M.; Talik, E.; Gagor, A.; Zawisza, B.; Malecka, M. Graphene oxide decorated with fullerenol nanoparticles for highly efficient removal of Pb(II) ions and ultrasensitive detection by total- reflection X-ray fluorescence spectrometry. *Sep. Purif. Technol.* **2021**, *277*, 119450–119460. [CrossRef]
61. Ghassabzadeh, H.; Torab-Mostaedi, M.; Mohaddespour, A.; Maragheh, M.G.; Ahmadi, S.J.; Zaheri, P. Characterizations of Co(II) and Pb(II) removal process from aqueous solutions using expanded perlite. *Desalination* **2010**, *261*, 73–79. [CrossRef]
62. Daghrir, R.; Drogui, P.; Robert, D. Modified TiO_2 for environmental photocatalytic applications: A review. *Ind. Eng. Chem. Res.* **2013**, *52*, 3581–3599. [CrossRef]
63. Niu, J.; Liu, X.; Xia, K.; Xu, L.; Xu, Y.; Fang, X.; Lu, W. Effect of Electrodeposition Parameters on the Morphology of Three-Dimensional Porous Copper Foams. *Int. J. Electrochem. Sci.* **2015**, *10*, 7331–7340. [CrossRef]
64. Sengupta, S.; Patra, A.; Jena, S.; Das, K.; Das, S. A Study on the Effect of Electrodeposition Parameters on the Morphology of Porous Nickel Electrodeposits. *Metall. Mater. Trans. A* **2018**, *49*, 920–937.
65. Nikolic, N.D.; Maksimovic, V.M.; Avramovic, L. Correlation of Morphology and Crystal Structure of Metal Powders Produced by Electrolysis Processes. *Metals* **2021**, *11*, 859.

materials

MDPI

Article

Recovery of Metal Ions (Cd^{2+}, Co^{2+}, and Ni^{2+}) from Nitrate and Sulfate on Laser-Induced Graphene Film Using Applied Voltage and Its Application

Xiu-man Wang [1,2,*], Tong Su [3] and Yujun Chai [3]

1 State Key Laboratory of Mountain Bridge and Tunnel Engineering, Chongqing Jiaotong University, Chongqing 400074, China
2 School of Civil Engineering, Chongqing Jiaotong University, Chongqing 400074, China
3 College of Chemistry and Materials Science, Hebei Normal University, Shijiazhuang 050024, China; 20190801199@cqu.edu.cn (T.S.); yjchaiwn@gmail.com (Y.C.)
* Correspondence: wangxiuman@cqjtu.edu.cn

Abstract: The urgent removal of Cd, Co, and Ni from nitrate and sulfate is essential to mitigate the potential risk of chemical pollution from large volumes of industrial wastewater. In this study, these metal ions were rapidly recovered through applying voltage on nitrate and sulfate, utilizing laser-induced graphene/polyimide (LIG/PI) film as the electrode. Following the application of external voltage, both the pH value and conductivity of the solution undergo changes. Compared to Co^{2+} and Ni^{2+}, Cd^{2+} exhibits a lower standard electrode potential and stronger reducibility. Consequently, in both nitrate and sulfate solutions, the reaction sequence follows the order of $Cd^{2+} > Co^{2+} > Ni^{2+}$, with the corresponding electrode adsorption quantities in the order of $Cd^{2+} > Co^{2+} \sim Ni^{2+}$. Additionally, using the recovered $Co(OH)_2$ as the raw material, a $LiCoO_2$ composite was prepared. The assembled battery with this composite exhibited a specific capacity of 122.8 mAh g^{-1}, meeting practical application requirements. This research has significance for fostering green development.

Keywords: recovery; transition metal; LIG/PI film electrode

1. Introduction

Large amounts of residual metal ions remain in industrial wastewater. Once metal ions contaminate water or soil, they pollute the environment and affect people's health [1]. Therefore, an efficient and environmentally friendly disposal method for treating polluted waterbodies is an urgent issue.

Researchers have employed various methods to recover valuable metal ions from waste batteries, including leaching, chemical reduction, metallurgy, and other techniques [2–5]. Meshram et al. [6] reviewed the recovery of metals using various acids, which unfortunately led to secondary pollution caused by the chemical agents. In recent decades, electrosorption or capacitive deionization (CDI), which rapidly pushes the migration of ionics using electrical force, has been widely utilized for the recovery of metal ions in wastewater due to its low cost, ease of control, and absence of secondary pollution from the chemical agents [7–9]. The adsorption electrode serves as the central component of the electric deionization device, playing a pivotal role in determining the device's adsorption capacity [8,10]. Electrode materials featuring a carbon network structure have been extensively researched due to their high specific surface area, well-balanced pore size distribution, exceptional electrical conductivity, and favorable adsorption properties [9,11–13]. Li et al. [14] reported on the removal of Ni^{2+} ions using an α-MnO_2/carbon fiber paper composite, which exhibited pseudocapacitor properties, facilitating electron/charge transfer. Xue et al. developed a nitric acid-modified activated carbon electrode for the removal of Co^{2+}, Mn^{2+}, and Ni^{2+}

ions [15]. Coconut-shell activated carbon, activated by CO_2, served as the electrode, demonstrating a Ni removal capacity of 5.32 mg g^{-1} [16]. This implies that modifying carbon materials with various functional groups can effectively be used to adsorb metal ions.

Laser-induced graphene (LIG), as an easily obtainable carbon material, has been utilized across various fields due to its high thermal stability, excellent conductivity, rich 3D porous network, and abundant functional groups [17]. In our previous research, employing LIG as an electrode material exhibited outstanding capability in recovering rare earth ions [18]. Li et al. reported that a $Co_4S_3^-$ modified LIG electrode exhibited a high adsorption capacity of 2702.79 mg g^{-1} for UO_2^{2+} [19], indicating its potential as an absorbent electrode.

To our knowledge, the use of LIG films as electrodes for recovering Cd^{2+}, Co^{2+}, and Ni^{2+} ions has not been reported previously. Therefore, the recovery of Cd^{2+}, Co^{2+}, and Ni^{2+} ions in nitrate and sulfate forms was investigated in this work. The standard potential of Cd ($E^{\theta} = -0.403$ V vs. SHE) is lower than that of Co ($E^{\theta} = -0.277$ V vs. SHE) and Ni ($E^{\theta} = -0.257$ V vs. SHE), indicating the easy recovery of cadmium ions under identical solution concentrations. Due to the superior oxidation capabilities of nitrate ions, the ions recovered from nitrate solutions are mainly hydroxides, whereas the products recovered from sulphate solutions are metals. Furthermore, the cobalt hydroxide recovered could serve as a raw material for the preparation of lithium-ion electrode materials. This has significance for fostering green development.

2. Materials and Methods

2.1. Materials and Characteristics

$Co(NO_3)_2$, $Ni(NO_3)_2$, $Cd(NO_3)_2$, $CoSO_4$, $NiSO_4$, and $CdSO_4$ were procured from Sinopharm Group Chemical Reagent Co., Ltd, Shanghai, China. A 100 μm polyimide (PI) film was supplied by Texiang Electrical Insulation Material Company, PRC, Tianjin. Ethanol was purchased from Chongqing Colon Chemical Co., Ltd., Chongqing, China. Deionized water specifically prepared for laboratory use was employed. The aforementioned chemicals are of analytical purity and were used without any further treatment.

The structure and morphology of the product were characterized using X-ray diffractometry (XRD, SmartLab, Shanghai, China) and field-emission scanning electron microscopy (FESEM, Hitachi S-4800, Tokyo, Japan). The bonds in the sample were analyzed using Fourier-transform infrared spectroscopy (FTIR, Thermo Scientific, Suzhou, China) and X-ray photoelectron spectroscopy (XPS, Thermo Fisher Scientific Escalab QXi, Waltham, America). The pH of the solution was determined using a PHS-3C pH acidity meter. The concentration of ions in solution during the application of external voltage was measured using atomic absorption spectrometry (AAS, TAS-990 SUPER, Weizhou, China). The electrochemical performance of the electrode was carried out using VMP3 multi-channel electrochemical impedance system.

2.2. Preparation of LIG/PI Film Electrodes via Laser Scribing

Firstly, commercial PI films were cleaned with ethanol and deionized water. Subsequently, a 100 μm thick PI film was affixed to a glass plate, and a CO_2 laser with a wavelength of 10.6 μm was utilized to carve the PI film, resulting in the production of an LIG/PI film electrode. The scanning speed and scanning power of the laser engraving machine were 100 mm/s and 6.6 W, respectively. Finally, an LIG/PI thin film electrode with a size of 2.5 × 1 cm^2 was obtained. Figure 1a shows the preparation process of the LIG/PI thin film electrode.

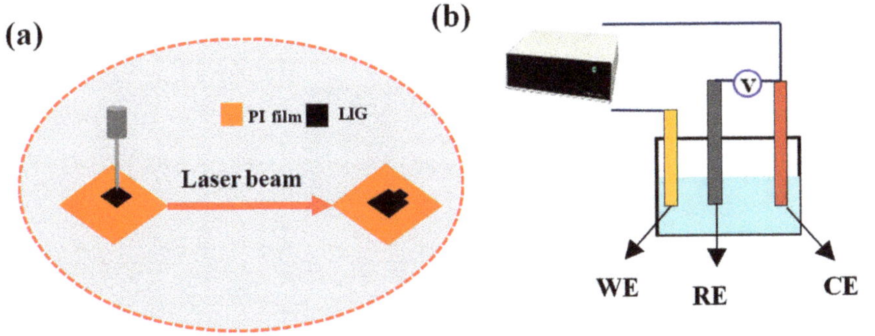

Figure 1. (**a**) Schematic diagram of the fabrication of LIG/PI film electrodes. (**b**) Schematic diagram of the three-electrode system.

In this experiment, a traditional three-electrode system (Figure 1b) was employed to the simulated solutions containing 0.005–0.02 M Co^{2+}, Ni^{2+}, and Cd^{2+}. The LIG/PI film served as the working electrode, while a Pt electrode was utilized as the counter electrode. An Ag-AgCl electrode was employed as the reference electrode. Upon applying a voltage to the reference electrode, the ions were recovered. The formulas for calculating the adsorption capacity (Q) and removal rate (W) are provided in detail in the Supplementary Materials.

2.3. Application of Recovered $Co(OH)_2$

$LiCoO_2$ was prepared using the sol-gel method. First, the recovered $Co(OH)_2$ and citric acid in a stoichiometric ratio of 1:2 were dissolved in 50 mL distilled water; then, the mixture was heated in an oil bath for 1 h at 80 °C with stirring. A certain amount of LiOH was added; meanwhile, an ammonia solution was used to adjust the pH of solution. The mixture was dried at 120 °C to evaporate the water. Finally, it was annealed at 800 °C for 8 h to obtain $LiCoO_2$.

A slurry was prepared by mixing $LiCoO_2$ (80 wt%), Super P (10 wt%), and polyvinylidene fluoride (PVDF, 10 wt%) in N-methyl-pyrrolidone (NMP). Subsequently, the half-cell was assembled and measured following the procedures outlined in previous studies [1].

3. Results

3.1. Change in pH and Conductivity after Applying the External Potential

Figure 2 illustrates the changes in pH and conductivity following the application of an external potential to a 0.02 M solution of $Cd(NO_3)_2$, $Co(NO_3)_2$, and $Ni(NO_3)_2$. When applying a potential of −1.2 V (vs. Ag/AgCl) to a 0.02 M solution of $Cd(NO_3)_2$, a silver-white product was obtained (inset of Figure 2a). In the $Co(NO_3)_2$ solution, both a gray product and a light blue product were obtained (inset of Figure 2b). And a green product was observed in the $Ni(NO_3)_2$ solution (inset of Figure 2c). Correspondingly, the pH decreased from ~6.0 to ~2.5, and the conductivity increased from ~2.0 to ~2.9 in these solutions. After the potential became −1.5 V, different color products were obtained, as shown in the inset of Figure S1. This illustrated a difference in the composition of products. Herein, a more obvious change in the pH and conductivity was detected. A higher applied voltage implies a strong electric driving force, which effectively influences the mobility of Co^{2+}, Ni^{2+}, and Cd^{2+} ions.

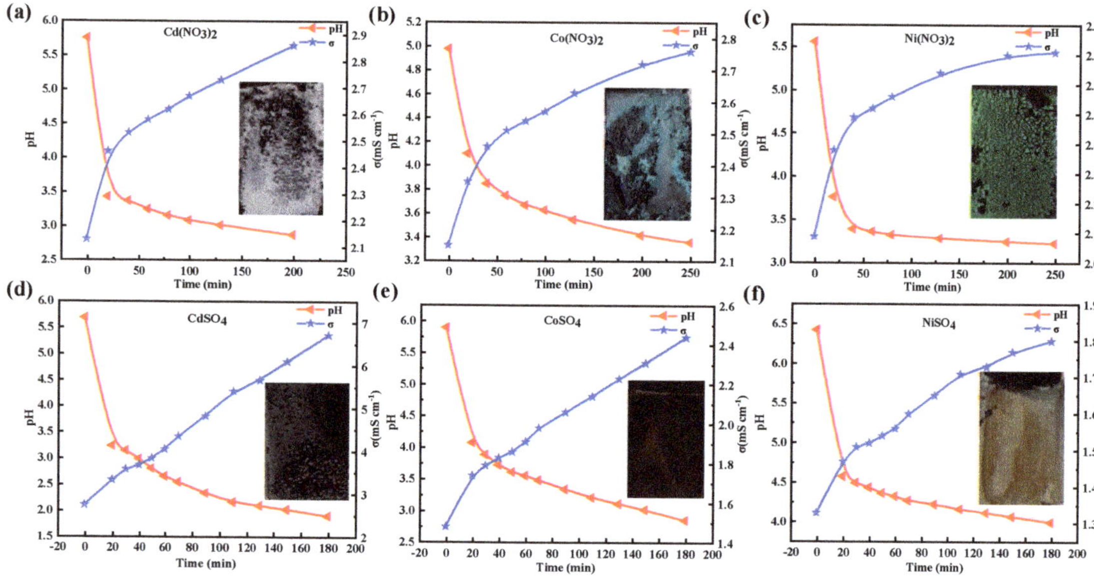

Figure 2. Changes in pH and conductivity after applying the potential (−1.2 V). (**a**) $Cd(NO_3)_2$, (**b**) $Co(NO_3)_2$, (**c**) $Ni(NO_3)_2$, (**d**) $CdSO_4$, (**e**) $CoSO_4$, (**f**) $NiSO_4$.

When applying a potential of −1.2 V to the sulfate solutions (Figure 2d–f), the variation in the pH and conductivity was notably significant in the $CdSO_4$ solution ($\Delta pH = 3.78$, $\Delta\sigma = 3.97$ mS cm^{-1}). However, it was relatively small in $CoSO_4$ ($\Delta pH = 3.04$, $\Delta\sigma = 0.96$ mS cm^{-1}) and $NiSO_4$($\Delta pH = 2.43$, $\Delta\sigma = 0.47$ mS cm^{-1}). Changing the potential to −1.5 V (Figure S1d–f) resulted in a dramatic change in pH ($\Delta pH = 4$) and conductivity ($\Delta\sigma = 8$ mS cm^{-1}) observed in the $CdSO_4$ solution. It was smallest in the $NiSO_4$ solution ($\Delta pH = 3$, $\Delta\sigma = 2$ mS cm^{-1}). These results illustrate that the reaction kinetics in sulfate and nitrate are different. The reaction ability in this solution follows the order of $Cd^{2+} > Co^{2+} > Ni^{2+}$ under these conditions, disregarding the effect of anions.

3.2. Analysis of the Adsorption Product

The product compositions obtained on electrode surfaces after application of −1.2 V voltage in nitrate and sulfate solutions was determined using X-ray diffraction (XRD) analysis, as depicted in Figure 3. It can be observed that the sharp and strong peaks at $2\theta = 18.7$, 29.5, 35.1, and 59.7° belonged to $Cd(OH)_2$ (JCPDS 31-0228) and the peaks at $2\theta = 19.1$, 32.5, 37.9, and 51.4° were identified as well-crystallized $Co(OH)_2$ (JCPDS 30-0443) in nitrates [20,21]. However, the peaks around $2\theta = 33.5$, 38.5, and 59.7° were broad and poor, indicating the formation of amorphous $Ni(OH)_2$/NiOOH (JCPDS 38-0715). The SEM further revealed that irregular blocks of $Cd(OH)_2$ formed due to the rapid reaction (Figure 4a). A sheet-like structure of $Co(OH)_2$ (Figure 4b) and an irregular rod-like shape of $Ni(OH)_2$/NiOOH (Figure 4c) were generated, with the particles aggregating to form blocks. However, the adsorption products in the sulfate solutions consisted of metallic Cd (JCPDS 85-1328), Co (JCPDS 05-0727), and Ni (JCPDS 87-0712). Among them, metallic Cd and Co exhibited dendritic morphology, as depicted in Figure 4d,e. The morphology of Ni appeared as irregular spherical shapes (Figure 4f).

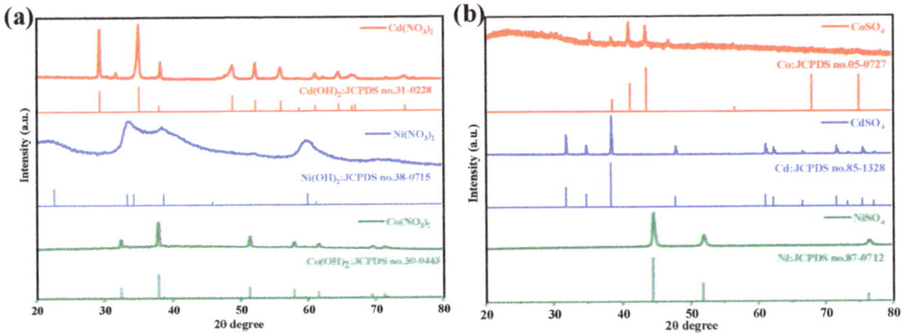

Figure 3. XRD pattern of electrode surface products in (**a**) nitric acid and (**b**) sulfate solutions with a voltage of −1.2 V.

Figure 4. SEM images of the products in (**a**) $Cd(NO_3)_2$, (**b**) $Co(NO_3)_2$, (**c**) $Ni(NO_3)_2$, (**d**) $CdSO_4$, (**e**) $CoSO_4$ and (**f**) $NiSO_4$.

When the voltage was −1.5 V, hydroxides could be obtained in these nitrates (Figure S2). However, the morphology of the $Cd(OH)_2$, $Ni(OH)_2$/NiOOH and $Co(OH)_2$ obviously changed (Figure S3). Although no change in the adsorbed products was detected from the XRD profiles in sulfate solution, the dendritic crystal became more obvious (Figure S4).

The chemical state of the product on the surface of the LIG film was analyzed using XPS, as shown in Figure 5. All XPS data were calibrated and fitted using PeakFit software 4.12. In the fitted spectra of products in $Cd(NO_3)_2$ solution (Figure 4a), two characteristic peaks were observed at 405.3 eV (Cd $3d_{5/2}$) and 412.11 eV (Cd $3d_{3/2}$), indicative of Cd^{2+} in $Cd(OH)_2$ [22]. The position of the fitted peak in the Cd 3d spectrum of $CdSO_4$ was identical to that of $Cd(NO_3)_2$, as zero-valent cadmium exhibits no characteristic peak. The results of XRD show that the adsorbed product of $CdSO_4$ is Cd. This illustrates that besides the metallic Cd, trace of hydroxide existed. In the $Co(NO_3)_2$ solution, the peaks observed at 781.44 and 797.29 eV corresponded to Co $2p_{3/2}$ and Co $2p_{1/2}$ in $Co(OH)_2$, respectively [23]. The satellite peaks at 787.18 and 803.54 eV, respectively, corresponded to the high spin state of Co^{2+} [24]. The fitting of the Co 2p spectra from $CoSO_4$ (Figure 5e) exhibited significant differences compared to that from $Co(NO_3)_2$. Herein, characteristic peaks emerge at 778.1, 781.1, and 783.1 eV, respectively, corresponding to metallic Co [25]. The Ni 2p spectrum in

$Ni(NO_3)_2$ was fitted with six characteristic peaks located at 854.8, 855.7, 857.7, 860.1, 861.5 and 866.5 eV, respectively. This indicates that the product is $Ni(OH)_2$ or NiOOH, aligning with the reports in the literature [26]. The characteristic peaks of the Ni 2p spectra from $NiSO_4$ (Figure 5f) were observed at 852.9, 856.4 and 859.2 eV, respectively, corresponding to metallic Ni.

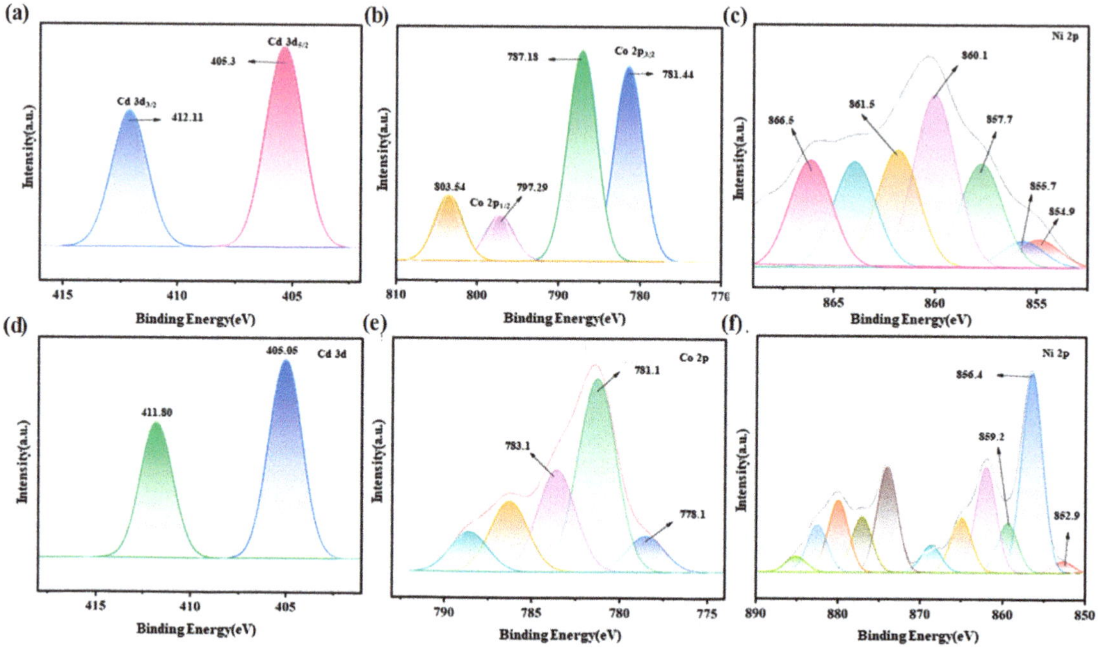

Figure 5. XPS the of product in (**a**) $Cd(NO_3)_2$, (**b**) $Co(NO_3)_2$, (**c**) $Ni(NO_3)_2$, (**d**) $CdSO_4$, (**e**) $CoSO_4$, and (**f**) $NiSO_4$.

3.3. Electrochemical Properties

Figure 6 illustrates the cyclic voltammetry (CV) curves of the LIG/PI thin film electrode under different scan rates in various nitrate and sulfate solutions. At scan rates of 5 to 20 mV s^{-1} in the $Cd(NO_3)_2$ solution, a quasi-rectangular shape was observed, illustrating stable capacitance behavior without a redox reaction. However, an oxidation peak appeared in $CdSO_4$, indicating the possible decomposition of water. No obvious reduction peak from Cd^{2+} to Cd was detected due to the rapid scan. An obvious difference in $Co(NO_3)_2$ and $CoSO_4$ was detected; however, a slight difference was observed in $Ni(NO_3)_2$ and $NiSO_4$. Because the applied potential was higher (−1.2 V), both H^+ and H_2O participated in the reaction, reducing the pH of the solutions and increasing the conductivity.

After applying a potential of −1.2 V, the current density remained stable in the nitrate solution (Figure 7a). At the moment the voltage was applied, the current density surged in the nitrate solutions. Afterward, the current density decreased and stabilized. The stable current density values were approximately 1.5 mA cm^{-2} in the $Co(NO_3)_2$ solution, 2 mA cm^{-2} in the $Ni(NO_3)_2$ solution, and 4 mA cm^{-2} in the $Cd(NO_3)_2$ solution, respectively. Combining the changes in pH and conductivity, water decomposition occurred, and the generated H^+ ions compensated for the adsorption of the Cd^{2+}, Co^{2+}, and Ni^{2+} ions, thereby ensuring that the mobility of the ions in the nitrate salt solutions remained unchanged. The current density was relatively stable in the $NiSO_4$ (approximately 1.3 mA cm^{-2}) and $CoSO_4$ solution (approximately 2.3 mA cm^{-2}). However, the current density in $CdSO_4$ rapidly and linearly increased over time. This further illustrates that the stable reduction reaction in $NiSO_4$ and $CoSO_4$ was supported by the generated H^+ ions. In addition, the reduction of Cd^{2+} ions was dramatically affected by the polarization.

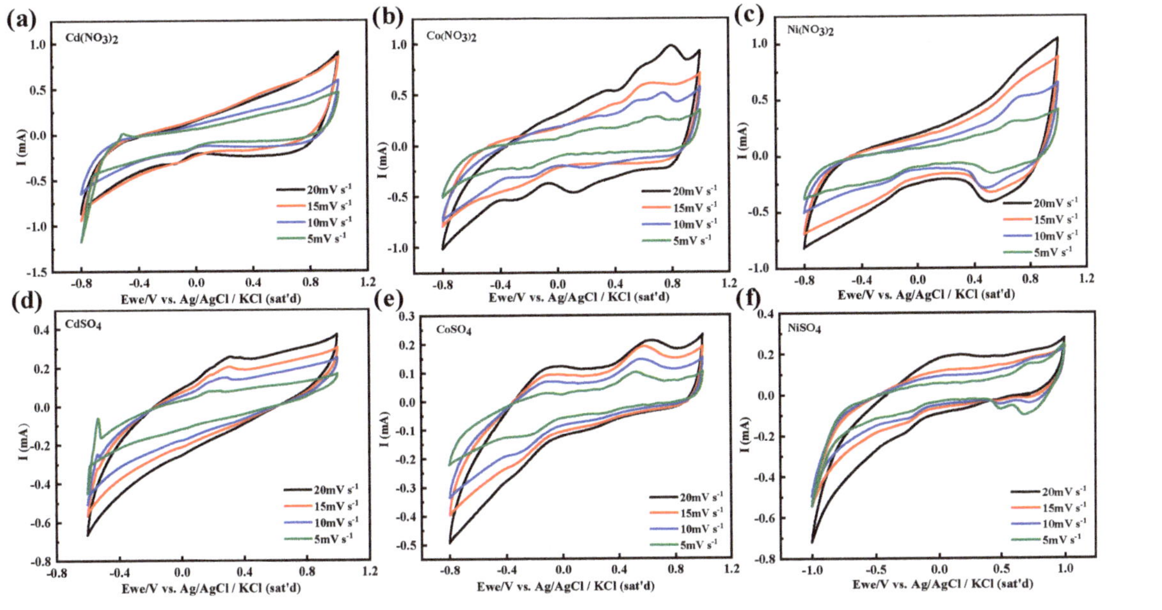

Figure 6. The CV curves at different scan rates in nitrate and sulfate solutions of LIG/PI film electrodes. (**a**) $Cd(NO_3)_2$, (**b**) $Co(NO_3)_2$, (**c**) $Ni(NO_3)_2$, (**d**) $CdSO_4$, (**e**) $CoSO_4$ and (**f**) $NiSO_4$.

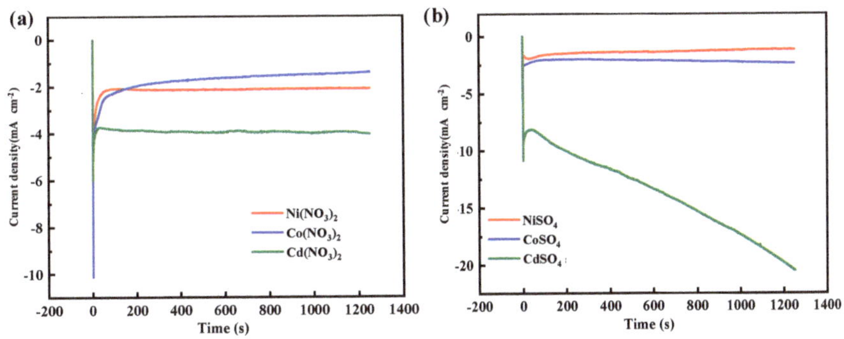

Figure 7. Current changes in (**a**) nitrate solution and (**b**) sulfate solution (−1.2 V).

When the voltage was −1.5 V, the current density exhibited changes as depicted in Figure S5. Except for $CdSO_4$ and $CoSO_4$, the current density changes in other solutions were consistent with the changes observed at −1.2 V, wherein the current density initially increased sharply before decreasing and ultimately stabilizing. The change was more pronounced in the $CdSO_4$ solution, with a current density of 20 mA cm^{-2} at −1.2 V and 45 mA cm^{-2} at −1.5 V at 1200 s, respectively. In the $CoSO_4$ solution, the current density did not remain stable but instead increased with time.

3.4. Adsorption Capacity

The adsorption amount of metal ions was measured using an atomic absorption spectrometer to evaluate the adsorption capability of the LIG/PI membrane electrode towards metal ions. We calculated Q and W using Formulas (1) and (2) (Supplementary Materials). The standard curves for different metal ions are depicted in Figure S6, while Figure 8 illustrates the temporal changes in the adsorption capacity. It was evident that with the increasing concentration, the adsorption capacity also increased. The adsorption capacity in $Cd(NO_3)_2$ increased from 913.5 mg g^{-1} at 0.005 M to 3479.8 mg g^{-1} at 0.02 M. Approximately 39% of Cd^{2+} ions were adsorbed on the electrode surface in 0.02M $Cd(NO_3)_2$ over 300 min. The adsorption capacity of 1381.5 mg g^{-1} in 0.02 M $Co(NO_3)_2$ solution was slightly lower than that of 1448.7 mg g^{-1} in $Ni(NO_3)_2$ solution. The adsorption capacity in the 0.005 M $CdSO_4$ solution reached 2634.27 mg g^{-1}, indicating the easy migration of Cd^{2+} and its reduction in the sulfate solution. Moreover, the adsorption capacity of 0.02 M $CoSO_4$ and $NiSO_4$ was also higher than that in the nitric acid solution. However, the recovery capability at a low concentration of $CoSO_4$ and $NiSO_4$ was similar. In sulfuric acid solution, the adsorption capacity followed the order of Cd $^{2+}$ > Co $^{2+}$ > Ni^{2+}.

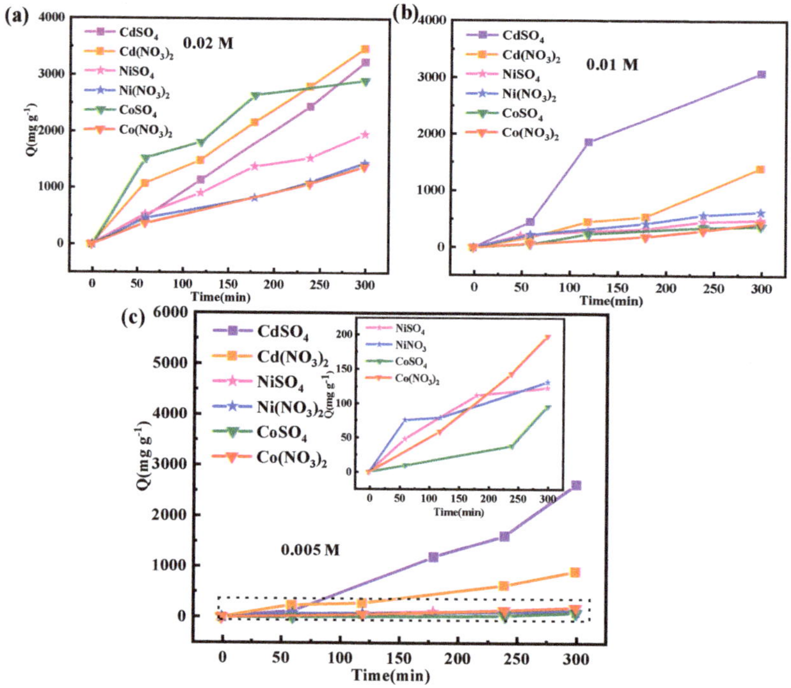

Figure 8. Adsorption of metal ions at different concentrations of nitrate and sulfate (−1.2 V). **(a)** 0.02 M, **(b)** 0.01 M, and **(c)** 0.005 M.

As the voltage was −1.5 V, the adsorption capacity also changed, as depicted in Figure S7. Approximately 40% Ni^{2+} in $Ni(NO_3)_2$ was covered in 100 min; then, the adsorption showed no obvious change due to the occupation of the active sites. It showed a faster adsorption with a longer adsorption time in $Cd(NO_3)_2$ than in $Co(NO_3)_2$. The recovery of Cd^{2+}, Ni^{2+}, and Co^{2+} was lower than approximately 20% when the concentration was 0.002 M. The recovery efficiency in the 0.002 M $CoSO_4$ solution remained approximately 10%, while it increased in the $NiSO_4$ and $CdSO_4$ solution.

The recycling stability of the LIG electrode was one of the important factors for application. After the product was collected, which involved soaking in 5% nitric acid for 1 h, the electrode was washed with deionized water and then dried at 30 °C. The conductivities in this solution were stable in five adsorptions–desorption cycles, indicating that the LIG film possesses regenerative properties, as shown in Figure 9.

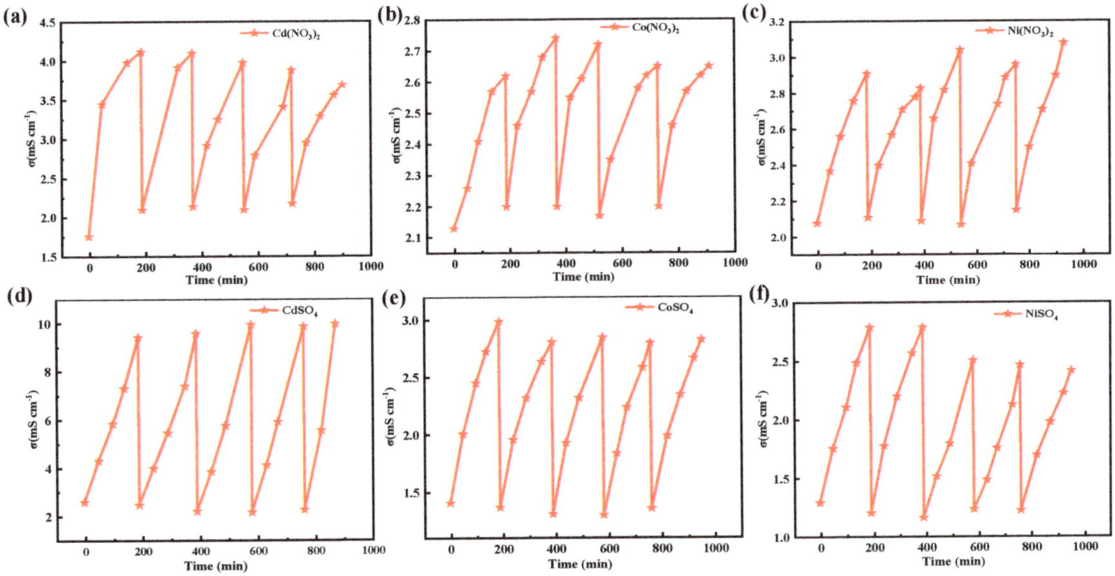

Figure 9. Cyclic performance of the electrodes. (**a**) $Cd(NO_3)_2$, (**b**) $Co(NO_3)_2$, (**c**) $Ni(NO_3)_2$, (**d**) $CdSO_4$, (**e**) $CoSO_4$, (**f**) $NiSO_4$.

3.5. Reaction Mechanism

In nitrate, NO^{3-} ions migrate towards the counter electrode after the application of voltage, where they may acquire electrons and undergo reduction to form NO^{2-} ions (Equation (1)). Meanwhile, reaction Equation (2) occurs, generating H^+ ions and reducing the pH of these solutions. Correspondingly, metallic ions swiftly migrate towards and combine with hydroxyl and carboxyl groups on the surface of the LIG film. Moreover, the generated OH^- ions further combine with metal ions to facilitate the formation of hydroxide precipitates (Equation (3)). However, water is decomposed in sulfates, and the acidic condition accelerates the reduction to form the corresponding metals (Equation (4)).

$$NO^{3-} + H_2O + 2e^- \rightarrow NO_2^- + 2OH^- \; E = 0.93 \text{ V vs. SHE} \tag{1}$$

$$2H_2O + 4e \rightarrow O_2 + 4H^+ \tag{2}$$

$$M^{2+} + 2OH^- \rightarrow M(OH)_2 \tag{3}$$

$$M^{2+} + 2e \rightarrow M \tag{4}$$

3.6. Application of the Product

Taking recycled cobalt as an example, the investigation focused on the reutilization of the recovered product. The obtained $Co(OH)_2$ served as the raw material for preparing $LiCoO_2$. The CV curves (Figure 10) displayed a distinct redox peak at 4.1/3.7 V, corresponding to the insertion/extraction of Li. The specific capacity reached 122.8 mAh g^{-1} at a current density of 0.5 C, highly consistent with the reported data from commercially prepared materials [27]. The initial discharge capacity exhibited no significant change at a current density of 1 C. However, it rapidly decreased with increasing cycles, a phenomenon attributed to the particle size of the $LiCoO_2$. In summary, the products recovered through this method also hold potential application value in other fields.

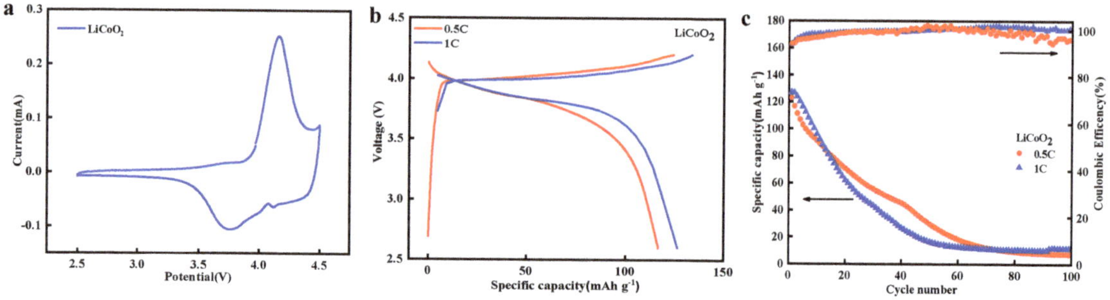

Figure 10. (**a**) CV curve, (**b**) charging/discharging curves, and (**c**) cyclic stability at 0.5 C and 1 C.

4. Conclusions

In this study, the transition metal ions, Co^{2+}, Cd^{2+}, and Ni^{2+}, were recovered using LIG/PI thin films as the working electrodes through electrosorption and electrodeposition methods in nitrate and sulfate solutions, respectively. In nitrate and sulfate solutions, the LIG/PI electrode surface was found to accumulate hydroxides and dendritic metal deposits, respectively. After applying an external voltage, the pH of the solution decreased, while the conductivity increased. Compared to Co^{2+} and Ni^{2+}, Cd^{2+} exhibited a lower standard electrode potential and stronger reducibility. Consequently, in both nitrate and sulfate solutions, the reaction sequence followed the order of $Cd^{2+} > Co^{2+} > Ni^{2+}$, with the corresponding electrode adsorption quantities in the order of $Cd^{2+} > Co^{2+} \sim Ni^{2+}$ in 0.005 to 0.02 M. Additionally, using the recovered $Co(OH)_2$ as raw material, a $LiCoO_2$ composite was prepared. The assembled battery with this composite exhibited a specific capacity of 122.8 mAh g^{-1}, meeting practical application requirements. This has significance for fostering green development.

Supplementary Materials: The following supporting information can be downloaded at: https://www.mdpi.com/article/10.3390/ma17122965/s1, Figure S1. Changes in pH and conductivity during the electro adsorption process (−1.5 V). (a) $Cd(NO_3)_2$, (b) $Ni(NO_3)_2$, (c) $Co(NO_3)_2$, (d) $CdSO_4$, (e) $NiSO_4$, and (f) $CoSO_4$. Figure S2. XRD pattern of electrode surface products in (a) nitric acid and (b) sulfate solutions with a voltage of −1.5 V. Figure S3. SEM of the product in (a,b) $Co(NO_3)_2$, (c,d) $Ni(NO_3)_2$, (e,f) $Cd(NO_3)_2$ solution (−1.5 V). Figure S4. SEM of the product in (a–c) $CoSO_4$, (d–f) $CdSO_4$, and (g–i) $NiSO_4$ solution (−1.5 V). Figure S5. Current changes in nitrate solution and sulfate solution (−1.5 V). Figure S6. Standard Curve of (a) $Cd(NO_3)_2$, (b) $Co(NO_3)_2$, (c) $Ni(NO_3)_2$ solution (−1.2 V). Figure S7. Standard Curves of (a) $Co(NO_3)_2$, (b) $Cd(NO_3)_2$, (c) $Ni(NO_3)_2$ solution (−1.5 V), (d,e) Change in adsorption capacity over time at different concentrations, (f,g) Change in removal rate over time.

Author Contributions: X.-m.W. and T.S. are responsible for the experiment planning, directing the research content, and writing this paper. Y.C. is responsible for some experiments and data sorting. All authors have read and agreed to the published version of the manuscript.

Funding: This research received no external funding.

Institutional Review Board Statement: Not applicable.

Informed Consent Statement: Not applicable.

Data Availability Statement: The data presented in this study are available on request from the corresponding author.

Conflicts of Interest: The authors declare no conflict of interest.

References

1. Zhao, Y.; Kang, Y.; Fan, M.; Li, T.; Wozny, J.; Zhou, Y.; Wang, X.; Chueh, Y.-L.; Liang, Z.; Zhou, G.; et al. Precise separation of spent lithium ion cells in water without discharging for recycling. *Energy Storage Mater.* **2022**, *45*, 1092–1099. [CrossRef]
2. Zhou, L.; Yang, D.; Du, T.; Gong, H.; Luo, W. The current process for the recycling of spent lithium ion batteries. *Front. Chem.* **2021**, *8*, 578044. [CrossRef] [PubMed]
3. Windisch-Kern, S.; Li, B.; Li, J.; Xu, Z. Pyrometallurgical technology in the recycling of a spent lithium ion battery: Evolution and the challenge. *Processes* **2021**, *1*, 1369–1382.
4. Du, K.D.; Ang, E.H.; Wu, X.L.; Liu, Y.C. Progresses in sustainable recycling technology of spent lithium-ion batteries. *Energy Environ. Mater.* **2022**, *5*, 1012–1036. [CrossRef]
5. Gong, R.; Li, C.C.; Meng, Q.; Dong, P.; Zhang, Y.J.; Zhang, B.; Yan, J.; Li, Y. A sustainable closed-loop method of selective oxidation leaching and regeneration for lithium iron phosphate cathode materials from spent batteries. *J. Environ. Manage.* **2022**, *319*, 115740. [CrossRef] [PubMed]
6. Sankar, T.K.; Abhilash; Meshram, P. Environmental impact assessment in the entire life cycle of lithium-ioin batteries. *Rev. Environ. Contam. Toxicol.* **2024**, *262*, 5.
7. Liu, Y.; Deng, Y.Y.; Zhang, Q.; Liu, H. Overview of recent developments of resource recovery from wastewater via electrochemistry-based technologies. *Sci. Total Environ.* **2021**, *757*, 143901. [CrossRef] [PubMed]
8. Yang, L.M.; Hu, W.B.; Chang, Z.; Liu, T.; Fang, D.F.; Shao, P.H.; Shi, H.; Luo, X.B. Electrochemical recovery and high value-added reutilization of heavy metal ions from wastewater: Recent advances and future trends. *Environ. Int.* **2021**, *152*, 106512. [CrossRef]
9. Mousset, E.; Fournier, M.; Su, X. Recent advances of reactive electroseparation systems for water treatment and selective resource recovery. *Curr. Opin. Electrochem.* **2023**, *42*, 101384. [CrossRef]
10. Huang, L.; Sheng, L.; Wan, K.L.; Wang, M.C.; Zhang, H.G.; Yan, J.; Liu, Y.H.; Alhassan, S.I.; Chen, Y.S.; Arulmani, S.R.B. The diversified mechanism of adsorption and electro-adsorption technologies by using Ti_3AlC_2 for removing fluoride. *Desalination* **2024**, *578*, 117449. [CrossRef]
11. Shin, J.; Lee, G.; Choi, M.; Jang, H.; Lim, Y.; Kim, G.S.; Nam, S.H.; Baek, S.H.; Song, H.C.; Kim, J.; et al. Atomically mixed catalysts on a 3D thin-shell TiO_2 for dual-modal chemical detection and neutralization. *J. Mater. Chem. A* **2023**, *11*, 18195–18206. [CrossRef]
12. Wang, X.M.; Chai, Y.J.; Wang, Z.P.; Yu, J.B.; Chen, X.P. A linear and large-range pressure sensor based on hierarchical structural SnO_2@carbon nanotubes/polyurethane sponge. *Ceram. Int.* **2023**, *49*, 30579–30585. [CrossRef]
13. Kwak, D.; Kim, H.; Jang, S.; Kim, B.G.; Cho, D.; Chang, H.; Lee, J.O. Investigation of laser-induced graphene (LIG) on a flexible substrate and its functionalization by metal doping for gas-sensing applications. *Int. J. Mol. Sci.* **2024**, *25*, 1172. [CrossRef]
14. Li, P.J.; Gui, Y.; Blackwood, D.J. Development of a nanostructured α-MnO_2/carbon paper composite for removal of Ni^{2+}/Mn^{2+} ions by electrosorption. *ACS Appl. Mater. Interfaces* **2018**, *10*, 19615–19625. [CrossRef]
15. Xue, Y.; Cheng, W.T.; Cao, M.; Gao, J.Z.; Chen, J.Q.; Gui, Y.Y.; Zhu, W.M.; Ma, F.Q. Development of nitric acid-modified activated carbon electrode for removal of Co^{2+}/Mn^{2+}/Ni^{+} by electrosorption. *Environ. Sci. Pollut. Res.* **2022**, *29*, 77536–77552. [CrossRef]
16. Huynh, L.T.N.; Tran, T.N.; Ho, T.T.N.; Le, X.H.; Le, V.H.; Nguyen, T.H. Enhanced electrosorption of NaCl and nickel(II) in capacitive deionization by CO_2 activation coconut-shell activated carbon. *Carbon Lett.* **2022**, *32*, 1531–1540. [CrossRef]
17. Liu, C.; Wu, T.; Hsu, P.C.; Xie, J.; Zhao, J.; Liu, K.; Sun, J.; Xu, J.W.; Tang, J.; Ye, Z.W.; et al. Direct/alternating current electrochemical method for removing and recovering heavy metal from water using graphene oxide electrode. *ACS Nano* **2019**, *13*, 6431–6437. [CrossRef]
18. Wang, L.; Gao, Y.; Wang, X.M.; Chai, Y.J.; Sun, X.Q. Efficient and sustainable electro-sorption of rare earth by laser-induced graphene film. *Sep. Purif. Technol.* **2022**, *300*, 121853. [CrossRef]
19. Cao, R.Y.; Zhang, J.F.; Wang, D.; Sun, F.W.; Li, N.; Li, J.X. Electrodeposition cobalt sulfide nanosheet on laser-induced graphene as capacitive deionization electrodes for uranium adsorption. *Chem. Eng. J.* **2023**, *461*, 142080. [CrossRef]
20. Sakthivel, M.; Sivakumar, M.; Chen, S.M.; Hou, Y.S.; Veeramani, V.; Madhu, R.; Miyamoto, N. A facile synthesis of $Cd(OH)_2$-rGO nanocomposites for the practical electrochemical detection of acetaminophen. *Electroanalysis* **2017**, *29*, 280–286. [CrossRef]
21. Maile, N.C.; Shinde, S.K.; Koli, R.R.; Fulari, A.V.; Kim, D.Y.; Fulari, V.J. Effect of different electrolytes and deposition time on the supercapacitor properties of nanoflake-like $Co(OH)_2$ electrodes. *Ultrason. Sonochem.* **2019**, *51*, 49–57. [CrossRef] [PubMed]
22. Ren, J.J.; Zheng, L.C.; Su, Y.M.; Meng, P.P.; Zhou, Q.Y.; Zeng, H.; Zhang, T.; Yu, H.J. Competitive adsorption of Cd(II), Pb(II) and Cu(II) ions from acid mine drainage with zero-valent iron/phosphoric titanium dioxide: XPS qualitative analyses and DFT quantitative calculations. *Chem. Eng. J.* **2022**, *445*, 136778. [CrossRef]
23. Wang, R.F.; Deng, L.G.; Fan, X.J.; Li, K.; Lu, H.Q.; Li, W. Removal of heavy metal ion cobalt (II) from wastewater via adsorption method using microcrystalline cellulose–magnesium hydroxide. *Int. J. Biol. Macromol.* **2021**, *189*, 607–617. [CrossRef] [PubMed]
24. Rohit, R.C.; Jenifer, A.; Jagadale, A.D.; Kumbhar, V.S.; Lee, H.; Lee, K. Facile synthesis of Ce-doped α-cobalt hydroxide nanoflakes battery type electrode with an enhanced capacitive contribution for asymmetric supercapacitors. *J. Energy Storage* **2020**, *28*, 101227. [CrossRef]
25. Mark, C.; Biesinger, B.P.; Payne, A.P.; Grosvenor, L.W.M.; Lau, A.R.; Gerson, R.S.C. Resolving surface chemical states in XPS analysis of first row transition metals, oxides and hydroxides: Cr, Mn, Fe, Co and Ni. *Appl. Surf. Sci.* **2011**, *257*, 2717–2730.

26. Altamirano-Garcia, L.; Vazquez-Arenas, J.; Pritzker, M.; Luna-Sánchez, R.; Cabrera-Sierra, R. Effects of saccharin and anions (SO_4^{2-}, Cl^-) on the electrodeposition of Co-Ni alloys. *J. Solid State Electrochem.* **2015**, *19*, 423–433. [CrossRef]
27. Ma, H.L.; Wang, F.; Shen, M.H.; Tong, Y.; Wang, H.X.; Hu, H.L. Advances of $LiCoO_2$ in cathode of aqueous lithium-Ion batteries. *Small Methods* **2023**, 2300820. [CrossRef] [PubMed]

materials

MDPI

Article

Determination of Fluorine by Ion-Selective Electrode and High-Resolution Continuum Source Graphite Furnace Molecular Absorption Spectrometry with Respect to Animal Feed Safety

Zofia Kowalewska [1,*], Karolina Goluch [2], Waldemar Korol [2], Rafał Olchowski [3] and Ryszard Dobrowolski [4,*]

1 Faculty of Civil Engineering, Mechanics and Petrochemistry, Warsaw University of Technology, Łukasiewicza 17, 09-410 Plock, Poland
2 National Laboratory for Feedingstuffs, National Research Institute of Animal Production, Chmielna 2 Str., 20-079 Lublin, Poland; karolina.goluch@iz.edu.pl (K.G.); korol@clpp.lublin.pl (W.K.)
3 Department of Pharmacology, Toxicology and Environmental Protection, Faculty of Veterinary Medicine, University of Life Sciences, Akademicka Sq. 12, 20-950 Lublin, Poland; rafal.olchowski@up.lublin.pl
4 Department of Analytical Chemistry, Institute of Chemical Sciences, Faculty of Chemistry, Maria Curie-Sklodowska University, M. C. Sklodowska Sq. 3, 20-031 Lublin, Poland
* Correspondence: zofia.kowalewska@pw.edu.pl (Z.K.); rdobrow@poczta.umcs.lublin.pl (R.D.)

Abstract: Fluorine, depending on its concentration and chemical form, is essential or toxic to humans and animals. Therefore, it is crucial to be able to determine it reliably. In this study, fluorine was determined in animal feed after extraction with HCl (gastric juice simulation). The standard potentiometric method with a fluoride-selective electrode (ISE) and newly developed high-resolution continuum source graphite furnace molecular absorption spectrometry (HR-CS GFMAS) method was applied. Feed samples turned out to be a challenge for HR-CS GFMAS. Chemical interferences (formation of competing molecules, CaF, GaCl, and GaP, instead of the target GaF molecule) and spectral effects (including a phosphorous molecule spectrum and atomic lines) were identified. An additional difficulty was caused by reagent contamination with F and memory effects. Difficulties were eliminated/reduced. The quality of ISE analysis was multi-directionally verified (including comprehensive proficiency testing). A risk of inaccuracy at low F concentration, where the calibration relationship is nonlinear, was investigated. The results of both methods were consistent, which confirms the accuracy of the methods and informs that the extracted fluorine is in fluoride form. The results of extensive ISE tests conducted in Poland in 2021–2023 have shown that, in most cases, the fluoride content is significantly lower than the threshold values.

Keywords: fluorine determination; fluorides; ion selective electrode; ISE; potentiometry; HR-CS; molecular absorption spectrometry; graphite furnace; GaF; animal feed; calcium phosphate

1. Introduction

Fluorine is an element commonly found in the environment. In terms of occurrence, it is the 13th most abundant element in the earth's crust (0.059% by weight) [1]. Important minerals of fluorine are fluorapatite, fluorite, and cryolite [1–3].

Fluorine enters the secondary environment naturally through the weathering of rocks and minerals, volcanic gas emissions, forest fires, and, to a small extent, marine aerosols. Its anthropogenic sources include coal-fired energy, the ceramics industry, aluminum processing, and the production of phosphorus fertilizers. Agriculture also contributes to fluorine emissions from the use of phosphate fertilizers, sewage sludges, and pesticides [1–3].

In appropriate doses, fluorine is considered an element necessary for the proper development of animals and humans [1,4,5]. Deficiency of this element in the diet causes tooth decay, weakening of tooth enamel, and growth retardation, while its excess can lead to fluorosis, neurological deficits, worsening reproduction ability, as well as impaired thyroid and endocrine system functions [1,4,5]. Young animals are more susceptible to fluorosis because growing bones absorb fluoride more easily.

Animals absorb fluorine mainly as fluoride anions from feed and drinking water [1,4–6]. Typically, fluorine in tap water is of natural origin. The concentration of F is higher in groundwater, which is drawn from areas where rocks are rich in fluorine. Phosphate fertilizers may also contribute to excessive water pollution with F [2,3]. In other areas, tap water may be fluorinated to compensate for low fluorine levels. Fluorination is one of the most controversial topics in medicine, and some countries have decreased fluorine doses [1].

Plants can absorb F from the air through their leaves or from soil. The bioavailability of F in soil is usually low, but it depends on the plant species, the form of fluorine, and the soil type. Fluorine absorbed by plant roots is mostly retained there and does not reach the plant's above-ground parts [1]. A bigger problem in animal nutrition may be feed contaminated with soil with high fluorine content or animals grazing [1,4]. However, it is believed that fluorine from the atmosphere leaches into deeper soil layers, and therefore, surface soils have a lower F content than underground soils. The presence of native F in soils depends on its content in the parent rock. Agricultural soils can be high in F when phosphorus fertilizers are used [3]. The cause of excessive fluorine consumption by farm animals may also be excessive supplementation with phosphate rock, the mineral fluorapatite. For this reason, the supplement can be purified of F before application [1].

Fluorine in animals accumulates mainly in calcified tissues that humans do not normally consume. Reducing and controlling F in feed is primarily aimed at protecting animal health. Therefore, the European Union has established permissible levels of fluorine content in animal feed [7]. The limits depend on the feed material. The threshold value for most animal feed is 150 mg kg^{-1}. The threshold values for some materials, such as phosphate feed or krill, are much higher (2000 and 3000 mg kg^{-1}, respectively). In Europe, fluorine content limits in animal feed are generally not exceeded [4]. However, fluorine emissions into the biosphere are constantly increasing due to the widespread use of fluorine compounds [1,2,5]. Therefore, the content of this element in animal feed should be monitored.

Numerous methods have been applied for F determination, i.e., wet chemical (including titrimetric), electrochemical (including potentiometry and voltammetry), spectrophotometric, chromatographic (including ion chromatography, gas chromatography, and capillary electrophoresis) methods, as well as neutron activation analysis and total reflection X-ray fluorescence [8–13]. Due to the extreme properties of fluorine, the conventional analytical approach for element determination, the application of atomic spectrometry techniques, is impossible or very rare; instead, indirect methods are developed [11–13].

In the European Union, a standard method for determining fluorine in animal feedstuffs relies on potentiometric determination with a fluoride-selective electrode (ISE) following extraction using 1 M hydrochloric acid [14]. Thus, the determined form of fluorine is free fluoride anions. Various methods can be used to carry fluorides into solution (e.g., open ashing, alkali fusion, microwave-induced combustion, and pyrohydrolysis [8,9,15,16]). However, the simple usage of 1 M HCl is fully justified in the case of animal feed analysis as it simulates gastric juices [17].

A special electrode made of rare earth fluorides was used to sense fluoride activity in a solution for the first time in 1966 [18]. The method is widely applied, and it is still being investigated. For example, it was used for the analysis of various seeds [16], and it is under investigation as an official method for fluoride control in infant formula [19]. It was pointed out that the ISE electrode is highly selective to fluoride anion over other anions and works over the range of five orders of magnitude of fluoride anion concentration [18]. Unfortunately, the calibration curve is nonlinear for lower fluoride content, which worsens with long electrode's lifetimes [15]. The accuracy of the analysis can depend on the calibration procedure. The least accurate and most significantly underestimated results were obtained using a double standard addition method [20]. The ISE method needs to maintain pH in a narrow range, 5.2–5.5, to avoid OH^- interferences or transformation of fluoride anions into forms that are not detected by the method (HF or HF_2^-) [21,22]. Interferences can also come from cations (such as Fe^{3+} or Al^{3+}), which can react with fluoride anions and form complexes, depending on pH [22]. To overcome interference problems and maintain ionic strength, sodium acetate and sodium citrate are often added to the measured solution [15,17,22,23]. The effectiveness of various masking agents has been investigated lately, and the best results have been obtained for TISAB (total ionic strength adjustment buffer) [24]. Inconsistencies in the results of determining fluorides in tea infusions using ISE and ion chromatography methods were attributed to the difficulty of complete isolation of the fluoride ion signal in ion chromatography [22]. On the other hand, the good performance of citrates as the complexing agent was confirmed [22].

Currently, one of the more modern analytical techniques available is graphite furnace high-resolution continuum source molecular absorption spectrometry (HR-CS GFMAS). The technique was introduced in the first decade of the XXI century [8,25–27], and it has been widely developed in the second decade up to now [11,12]. The HR-CS GFMAS relies on generating a target molecule of an analyte in a heated graphite furnace. The appearing molecules absorb the radiation of a high-intensity xenon lamp at specific wavelengths, which is measured by a high-resolution system. To date, HR-CS GFMAS has not been used for the determination of fluorine in animal feed.

The goal of this work was to determine fluorine in animal feed using ISE and HR-CS GFMAS methods. In both cases, for the sample preparation step, the extraction with 1 M HCl, according to European Standard EN 16279:2012 [14], was selected. The same sample preparation procedure eliminates doubts connected with the efficiency of extraction.

2. Materials and Methods

2.1. Instrumentation and Its Operation

Fluoride measurement in the solutions by the ISE method was made with a combined Orion fluoride electrode (Thermo Fisher Scientific, Waltham, MA, USA) and a pH-meter CG 843 P (Schott, Hofheim, Germany). Each solution was mixed on a magnetic stirrer (Wigo, Pruszków, Poland).

A high-resolution continuum source absorption spectrometer contrAA 800 (Analytik Jena, Jena, Germany) was used to determine fluorine via molecular absorption. The radiation source was a high-intensity xenon lamp working in a hot spot mode and emitting continuum radiation. A double monochromator comprising a prism and an echelle grating enabled it to reach an excellent resolution of 1.4 pm per pixel. Of the 588 pixels of the CCD detector, 200 were available and covered the measured spectral range (211.1072–211.3874 nm). The range was centered at 211.248 nm, the center of the most sensitive rotational line of the target molecule, gallium monofluoride (GaF). The line belongs to the $0 \rightarrow 0$ vibrational transition within the $X^1\Sigma \rightarrow C^1\Pi$ electronic transition of this molecule. Seven pixels were selected for GaF molecule absorption measurements. An iterative mode, based on absorption minima (IBC mode), was used for background correction.

The gallium monofluoride was generated in a graphite furnace with an integrated PIN platform. Each new graphite furnace was initially covered with 230 µg of zirconium (three 25 µL and two 20 µL injections of 0.2% Zr solution, each followed by thermal pretreatment according to a routine heating cycle [28]).

Additionally, for the thermal stabilization of fluorine, a mixture of palladium and zirconium (5 µL of a solution containing 0.1% of Pd and 0.003% of Zr) was injected before the start of each graphite furnace heating cycle. This first injection also consisted of 6 µL of a 0.5% Ga solution. The heating program of graphite furnace is presented in Table 1. For thermal reduction of Pd, the temperature of 1100 °C was applied in step 5. The furnace was then cooled, and a second injection was executed involving 6 µL of 0.5% Ga solution, 5 µL of 0.4% sodium acetate (NaAc solution), and 6 µL of working solution. The role of NaAc was the neutralization of possibly formed volatile HF.

Table 1. Program of graphite furnace heating.

Step	Step Name	Temperature, °C	Heating Rate, °C s^{-1}	Hold Time, s	Argon Flow	Remarks
1	Drying	80	6	15	Max	Injection [1]
2	Drying	90	3	10	Max	
3	Drying	110	5	10	Max	
4	Pyrolysis	350	20	10	Max	
5	Pd reduction	1100	100	10	Max	
6	Drying	70	NP [3]	10	Max	
7	Drying	80	2	10	Max	Injection [2]
8	Drying	105	5	30	Max	
9	Drying	250	50	5	Max	
10	Pyrolysis	500	50	20	Max	
11	Gas adaptation	500	0	5	Stop	
12	Vaporization	1300	1500	5	Stop	
13	Cleaning	2500	500	9	Max	

[1] First injection took place before step "1" and was 5 µL of a solution containing 0.1% Pd and 0.003% Zr and 6 µL of a solution containing 0.5% of Ga (total volume 11 µL). [2] Second injection took place in step "7" and was 6 µL of a sample, 6 µL of a solution containing 0.5% of Ga, and 5 µL of a solution containing 0.4% of NaAc (total volume 17 µL). [3] No power.

2.2. Reagents, Standards, and Solutions

Sodium fluoride at a concentration of 1000 mg L^{-1} (Merck, Darmstadt, Germany) was used as a fluorine stock standard solution.

The following reagents were used in the ISE method and for sample preparation: hydrochloric acid 1 M was prepared by dilution of hydrochloric acid 35–38% (Avantor Performance Materials Poland S.A., Gliwice, Poland). The acetate buffer was prepared from sodium acetate trihydrate (Eurochem, Tarnów, Poland). Its pH was fixed by acetic acid 99.5–99.9% (Avantor Performance Materials Poland S.A., Gliwice, Poland) at 7.0. Citrate buffer was prepared from trisodium citrate dihydrate (Chempur, Piekary Śląskie, Poland) and perchloric acid 70% (Chempur, Piekary Śląskie, Poland). Distilled water was used to prepare all the solutions. In interference studies, 1000 mg L^{-1} standards of Fe, Al, and Ca (from Merck, Darmstadt, Germany) were used.

The following reagents were used in the HR-CS FMAS method: 99.999% gallium (III) nitrate (V) hydrate (from Sigma-Aldrich, Saint Louis, MO, USA), 10 g L^{-1} palladium modifier (from Merk, Germany), 2 g L^{-1} zirconium standard solution (the fix anal was diluted in 500 mL of water; from Sigma Aldrich Chemie GmbH, Steinheim, Germany), 5% ammonium dihydrogen phosphate (V) solution in water (a modifier for AAS form Teknolab, Drobak, Norway), 0.1% P standard as phosphoric (V) acid (from CPAchem, Bogomilowo, Bulgaria), 0.1% Ca solution in 2.5% nitric (V) acid (a Ca standard form GUM, Warsaw, Poland), 35–37% hydrochloric acid from Chempur, Piekary Śląskie, Poland) and sodium acetate (from VHR BDU Chemicals, Leuven, Belgium). Water, purified in a reverse osmosis process, was used in HR-CS GFMAS experiments.

The used reagents were at least of analytical grade.

2.3. Samples and Their Preparation

Fluorine was extracted from samples with 1 M hydrochloric acid according to the EN 16279 standard method [14]. In the ISE method, 0.5 g of sample was weighed into a plastic beaker. Moreover, 20 mL of 1 M HCl was added, and extraction was conducted for 20 min at room temperature (23 °C ± 1 °C). To ensure appropriate pH (5.5 ± 0.1) and ionic strength, 50 mL of acetate buffer and citrate buffer were added. The solutions were transferred to a 200 mL plastic flask and brought to volume with distilled water. The solutions were filtered before analysis. Most samples (exceptions could be made to calcium phosphate samples) were analyzed on the day of solution preparation.

The working calibration solutions were also prepared according to the EN 16279 standard method. The used calibration range covered 0.01–10 mg L^{-1} F content. All working solutions contained 1 M HCl, diluted 1:10 *v:v* (i.e., 0.1 M HCl), as well as acetate buffer and citrated buffer at the same concentration as samples. Measurements were performed from the lowest standard concentrations to the highest.

In the HR-CS GFMAS method, the initial dilution ratio was the same, 1:40 *m:v*, as in the ISE method (extraction from 1.25 g of a sample to 50 mL of the volume of solution). Due to organizational reasons, the HR-CS GFMAS analysis was performed at least a few days after the extraction.

Further pretreatment for HR-CS GFMAS depended on the solution's appearance and expected F content as follows:

- Solutions of calcium phosphate (V) were opalescent, but no sediment precipitation (even a few weeks after preparation) was observed; the solutions were diluted in the ratio 1:4000 or 1:20,000 *m:v*;
- Solutions of other samples were inhomogeneous; after intensive shaking, the precipitated sediment was separated by filtration; the filtrates were further diluted, and the dilution ratio depended on F content.

The dilution ratio of a single dilution step was not higher than 10. Thus, some dilutions comprise a few steps. For dilution and preparation of solutions for standard addition calibration, 1.5-volume Eppendorf vials were used.

In the standard addition method, the following two kinds of solutions were prepared:

- Type A: an extract from a given sample at a given dilution ratio;
- Type B: an extract from a given sample, like solution A, but with the addition of 0.1 mg L^{-1} of F.

In this study, especially, the following three samples were investigated:

- S1, calcium phosphate, containing a high concentration of F (approximately 1700 mg kg^{-1}),
- S2, a mixed feed for turkey, which represents samples with low content of F (approximately 12 mg kg^{-1} of F),
- S3, a spiked feed for chickens, representing samples with moderate F content (approximately 70 mg kg^{-1} of F).

3. Results and Discussion

3.1. Studies Using the Ion-Selective Electrode Method

Two calibration techniques were used for fluoride determination depending on the fluoride content of the sample, i.e., the interpolation technique when the concentration was within the nonlinear range of the curve and the standard addition technique when the concentration was within the linear range. An example of the calibration curve in the range of 0.01–10 mg L^{-1} is presented in Figure S1. Good linearity is achieved in the range 0.1–10 mg L^{-1} (the determination coefficient R^2 equals 1.0000), while at lower concentrations of fluorides, the calibration line curves (the determination coefficient R^2 for the 0.01–10 mg L^{-1} range equals 0.9925).

As recommended by the EN 16297 standard [14], the standard addition method involves adding 5 mL of a working standard to the sample solution. The concentration of the standard is adjusted to cause a potential drop of 5 to 40 mV. For example, in the case of a spiked feed containing approximately 70 mg kg^{-1} of fluorides in the initial sample, the added standard contained 3 mg L^{-1} of fluorides, and in the case of a phosphate sample containing approximately 1700 mg kg^{-1} of fluorides, the added standard contained 100 mg L^{-1} of fluorides.

In the standard addition method, the fluoride content was calculated using the following formula:

$$c_s = c_A \frac{0.1}{1.1 \times 10^{\frac{\Delta E}{S}} - 1}$$

c_s—fluoride concentration in the sample solution, mg L^{-1},
c_A—fluoride concentration of the added standard solution, mg L^{-1},
ΔE—potential difference (mV) after adding 5 mL of standard solution,
S—the slope of the calibration curve in the linear range [mV].

If the entire calibration range was utilized and the quadratic approximation of the calibration curve applied, the determination coefficient R^2 was equal to 0.9988.

The error of using linear calibration in the low concentration range was evaluated. For a sample containing 13.2 mg kg^{-1} of fluorides (approximately 0.03 mg L^{-1} in solution), the results of fluoride determination using linear fitting of the calibration curve would be 13.3 mg kg^{-1}. Thus, the difference in this case is slight (a relative difference of less than 1%). Unfortunately, at lower fluoride concentrations (e.g., 4 mg kg^{-1}, i.e., 0.01 mg L^{-1} in solution), using linear calibration would lead to significant result overestimation (i.e., the result of 5.2 mg kg^{-1}, i.e., 0.013 mg L^{-1} in solution and a relative difference of 30%).

The linear fitting of the calibration curve was applied for fluoride concentrations of 0.1 mg L^{-1} and more.

For the evaluation of the detection and determination limits, a sample with a low content of fluorides (a mixed feed for turkey) was analyzed seven times. At the average result of fluoride concentration equal to 4.81 mg kg^{-1}, the standard deviation was 0.10 mg kg^{-1}, and the relative standard deviation was 2.2%. The detection and determination limits, calculated as three and ten times the standard deviation, amounted to 0.3 and 1.0 mg kg^{-1}, respectively. The values are satisfactory from the point of view of the aim of this work.

The precision of the analysis was evaluated based on the differences between parallel results for samples with low or high fluoride content. The results of the analyses are presented in Tables S1 and S2, respectively. The fluoride content in the first group's samples was from 5.0 up to 6.1 mg kg^{-1}. The average result was 5.4 mg kg^{-1}, and the average difference was 0.04 mg kg^{-1} (relative absolute value of difference: 3.7%). The content of fluorides in the samples of the second group was from 1348 up to 4353 mg kg^{-1}. The average result was 2262 mg kg^{-1}, and the average difference was 28 mg kg^{-1} (average relative absolute value of differences: 2.9%). The precision is satisfactory.

For evaluating accuracy, a known amount of fluorides was added (as a solution) to a sample before extraction and recovery of analysis were determined. The recovery was defined as the ratio of the obtained result to the calculated (theoretical) result, expressed in percent. For calcium phosphate, the results of recovery evaluation were as follows: 109, 104, 106, 95, and 102% (103% on average). For mixed feed for turkey, the recovery was as follows: 81, 87, and 79% (82% on average).

Further recovery experiments were conducted but with the addition of potential interferents. Fluoride anions can form complex compounds with Fe^{3+} and Al^{3+} [17,22–24]. It is expected that the reactions do not occur in the presence of citrates. Therefore, the EN 16279 method [14] was designed to use the masking agent. A relatively large amount of potential interferents was added to the extraction solution to verify the lack of Fe and Al effects. The effect of additional calcium concentration was also studied, first of all to evaluate its influence on the extraction efficiency. The results of the investigation of the three samples mentioned in Section 2.3 are presented in Figure 1.

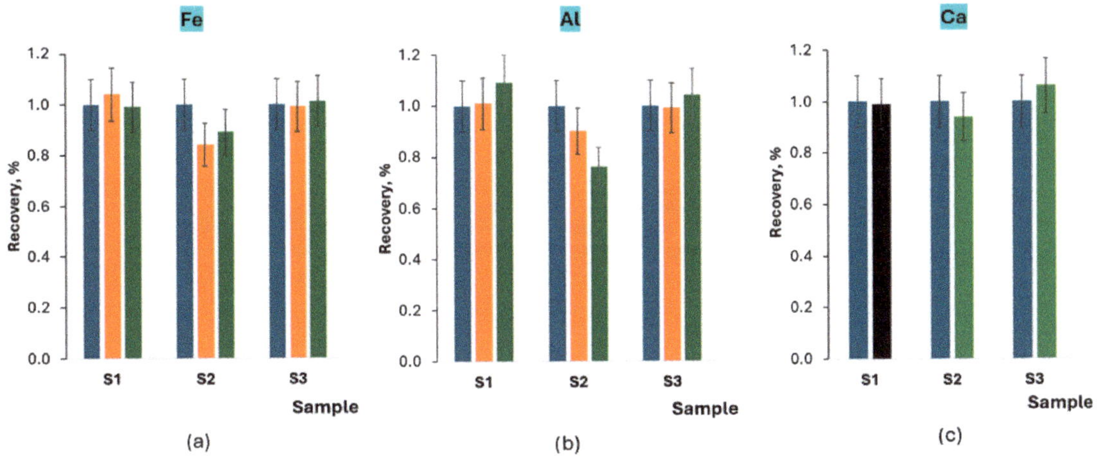

Figure 1. Influence of potential interferents on recovery of fluorine contained in samples S1, S2, and S3: Fe (**a**), Al (**b**) and Ca (**c**).

It can be concluded that Fe, Al, and Ca do not influence analysis results for samples S1 and S3, analyzed using the standard addition method and containing quite a high amount of fluorides. Some tendency to decrease results (up to 20%) can be observed in the case of Fe and Al added to sample S2. It can be caused by the consumption of some part of fluoride anions in complexation reactions or sample matrix influence, changing the slope of the calibration curve (which is not compensated with the standard addition method).

Proficiency testing results were used to evaluate data quality. In Table 2, the results of 13 proficiency testing programs devoted to the determination of fluorides in animal feed are presented. The investigations were carried out over the last 17 years. Various materials were investigated with various fluoride content (assigned values from 10 up to 1800 mg kg^{-1}). As the obtained z-score values show, the results of analyses of the National Laboratory for Feedingstuffs in Poland are satisfactory. The ISE method, as described in the current study, was applied.

It is essential to add that a relatively low number of laboratories participated in the proficiency testing (4–13), and almost all results were obtained using the same ISE method. Thus, the proficiency of laboratories was confirmed in the frame of the same analytical technique. Unfortunately, in the case of a systematic method error, it was not possible to reveal the error. Therefore, it can be concluded that accuracy was not entirely verified.

Table 2. Results of interlaboratory comparisons on fluoride determination in animal feed in 2006–2023.

PT Year	Type of Feed	Number of Laboratories/ Number of ISE Results	Assigned Value, mg kg^{-1}	Standard Deviation, mg kg^{-1}	NLF Result Obtained Using ISE Method, mg kg^{-1}	z-Score
2006 a	Feed phosphate	4/4	939	33.9	970	0.82
2006 a	Feed phosphate	4/4	1799	92.5	1956	0.96
2007 a	Feed phosphate	4/4	817	76.6	894	1.02
2011 a	Premixture for layer	4/4	16.1	2.83	18.4	0.82
2011 b	Compound feed for layers	10/9	9.88	5.40	6.18	−0.70
2013 b	Compound feed for chickens	11/10	21.3	4.13	20.7	−0.14
2014 b	Complementary feed for pigs	11/10	18.8	3.42	17.8	−0.30
2015 b	Compound feed for poultry	13/11	22.0	4.49	22.5	0.10
2016 b	Complementary feed for pigs	12/10	18.3	3.42	18.5	0.10
2017 b	Complementary feed for piglets	11/9	28.0	1.93	30.7	1.40

Table 2. *Cont.*

PT Year	Type of Feed	Number of Laboratories/ Number of ISE Results	Assigned Value, mg kg^{-1}	Standard Deviation, mg kg^{-1}	NLF Result Obtained Using ISE Method, mg kg^{-1}	z-Score
2020 b	Compound feed for broilers	8/6	15.0	2.61	17.0	0.76
2022 b	Compound feed for piglets	10/7	17.0	3.98	15.4	−0.40
2023 b	Fortified feed for poultry	8/6	75.8	14.2	69.6	−0.43

a—interlaboratory comparisons organized by the National Laboratory for Feedingstuffs (NLF, Lublin, Poland) as a part of the tasks of the reference laboratory; b—international proficiency test organized by the Österreichische Agentur für Gesundheit und Ernährungssicherheit International Analytical Group (AGES-IAG-Feedingstuffs, Vienna, Austria).

3.2. *Studies Using High-Resolution Continuum Source Graphite Furnace Molecular Absorption Spectrometry Method*

3.2.1. Selection of the Target Molecule

According to the literature [11,12,25–27,29–40], a few molecules, monofluorides of elements of IInd and XIIIth periodic table groups, were synthesized in the graphite furnace to determine fluorine via molecular absorption spectrometry. Some of such research was carried out in the past using classical AAS equipment of medium resolution [26,27,40], but real progress and wide application started with the introduction of high-resolution continuum source apparatus [11,12,25–27,29–39]. The device enables access to various wavelengths and efficient background correction, among others [25–27].

The HR-CS GFMAS enables the determination of the total amount of fluorine if only initial fluorine compounds are converted into the target monofluoride.

Using HR-CS GFMAS, gallium monofluoride [29–34] or calcium monofluoride [35–37] was mainly applied as a target molecule. The most crucial advantage of GaF is excellent sensitivity and low detection limits (characteristic mass of 7 pg and detection limit of 5 pg [30]); however, it is achieved using a complicated chemical modification. For the last reason, the application of another molecule, CaF, found popularity. Unfortunately, the sensitivity of such analysis was one to two orders of magnitude lower than in the case of GaF [35–37].

The conclusion on the relative sensitivity of F determination via absorption of GaF and CaF molecules in HR-CS GFMAS does not agree with the results of an alternative technique, high-resolution continuum source molecular spectrometry with flame as a chemical reactor. Careful adjustment of the measured spectral range and summing signals of many available pixels of the target molecule enabled obtaining similar (or even better) sensitivity and detectability for CaF [41] (and SrF [42]) as for GaF [13].

For the purpose of this work, it was decided to select GaF. It was expected that it would ensure achieving good sensitivity and detectability (as in the literature on HR-CS GFMAS). Therefore, it would allow sample dilution to decrease the content of the matrix in the graphite furnace. Another important reason is that animal feed could contain high and variable calcium concentrations that would result in variable sensitivity of F determination. On the other hand, animal feed is not expected to contain a significant amount of gallium. Additional difficulty could occur due to HCl application. It was found that a spectrum of CaCl, a molecule that can be formed if a measured solution contains HCl, overlaps the spectrum of the CaF molecule [37].

3.2.2. Selection of Spectral Conditions and Analytical Range

For the measurement of GaF absorption, the most sensitive line at 211.248 nm has been selected as in most of the previous papers [29–34]. This is the only single line of GaF molecule that appears in the measured spectral range. The line resembles common atomic single lines (used in AAS). As the line is relatively broad, seven pixels have been selected for F measurements. Iterative background correction designed for atomic absorption measurements has been used.

It is interesting that in the measured spectral range, a gallium atomic line is visible (211.197 nm, relative sensitivity 0.026). The appearance of this line indicates the presence of unbonded gallium in the graphite furnace, which is a good symptom (excess gallium is needed). The Ga 211.197 nm line and the 211.248 nm GaF line lie at a significant and "safe" distance from each other (do not overlap).

3.2.3. Pyrolysis/Vaporization Curves

The essential tool to recognize the behavior of an analyte in a graphite furnace is a series of measurements of an investigated solution with the stepwise increase of pyrolysis step temperature or measurement (i.e., vaporization) step temperature. In this way, respectively, pyrolysis or vaporization curves are obtained. Such experiments were performed for 0.1 mg L^{-1} standard solution and 0.1 mg L^{-1} standard solution with the addition of HCl, as well as solutions of samples S1–S3.

The results are presented in Figure 2. All pyrolysis curves were obtained using the 1300 °C temperature of the vaporization step (step number 12), and all vaporization curves were obtained using the 500 °C temperature of the pyrolysis step (step number 10).

Generally, the course of the curves for the five investigated solutions is quite similar. Significant fluorine losses certainly occur at pyrolysis temperatures above 700 °C. In the case of the standard solution of 0.1 mg L^{-1} and the highly diluted sample S1, the losses seem to occur even at lower temperatures. However, for these two solutions, the signals for pyrolysis temperatures of 300 °C and 500 °C are quite similar, taking into account approximately 10% relative difference of results of parallel measurements at given conditions.

Considering vaporization curves, the maximum signal is obtained for all investigated solutions using the 1200 °C temperature of step 12. The signal increase with the temperature from 1000 to 1200 °C is sharp, while the decrease of the signal at the temperature above 1200 °C is slower. It seems that the presence of HCl in the 0.1 mg L^{-1} standard solution or in the S2 and S3 sample matrix could slow down the signal decrease. Based on the pyrolysis–vaporization curves, the selected temperatures would be 500 °C for the pyrolysis step and 1300 °C for the vaporization step.

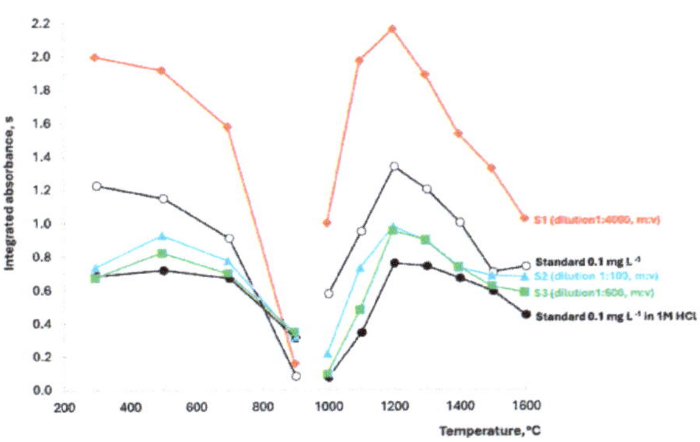

Figure 2. Pyrolysis–vaporization curves obtained for solutions of samples S1–S3 and 0.1 mg L^{-1} standard in a pure solution or as a solution containing 1 M of HCl. For better visibility, all absorbances for sample S1 were divided by two.

To better understand the processes in a graphite furnace, the signals measured using selected pyrolysis/vaporization temperatures are presented for 0.1 mg L^{-1} of F standard solution (Figure 3), S2 sample solution (Figure 4), S1 sample solution (Figure S2), S3 sample solution (Figure S3), and 0.1 mg L^{-1} of F standard solution in 1 M HCl (Figure S4). Each figure contains upper subfigures with time-resolved absorption (blue lines of GaF signal and red lines of BG signal) and lower subfigures with wavelength-resolved absorption in the measured spectral range. The lower situated, wavelength-resolved signals are presented at an enlarged scale to observe better baseline changes and the appearance of new phenomena (the result of this is cutting off the top of the high GaF peak). The scale of Y-axes in the upper subfigures is software-selected and varies, depending on the height of the signal. In contrast, the scale of Y-axes in lower figures was adjusted to always be the same for better comparison. The center of each wavelength-resolved absorbance picture is always the GaF molecular absorption line, while the Ga atomic absorption line has a maximum at pixel number 68 if it appears at all. The two absorption signals are exemplary and labeled in one of the subfigures of Figure 3.

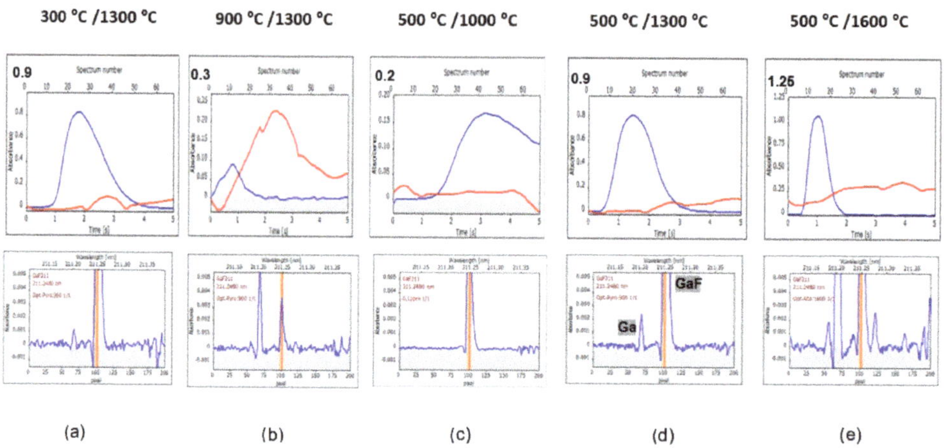

Figure 3. Signals of a 0.1 mg L^{-1} solution of fluorine as NaF, obtained using the following pyrolysis/vaporization temperatures: (**a**) 300 °C/1300 °C, (**b**) 900 °C/1300 °C, (**c**) 500 °C/1000 °C, (**d**) 500 °C/1300 °C, and (**e**) 500 °C/1600 °C. The main figure of each subfigure presents absorbance–time changes (the blue line is the analyte line, and the red line is the background line). The smaller figure of each subfigure denotes an enlarged wavelength-resolved absorption spectrum. The vertical orange line denotes the central pixel of the number 101.

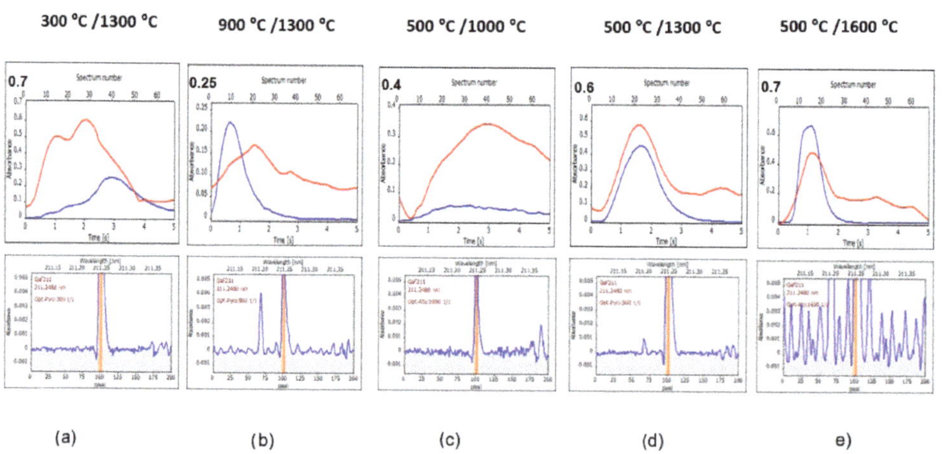

Figure 4. Signals of a solution of sample S2 (dilution 1:100 *m:v*, 1 M HCl), obtained using the following pyrolysis/vaporization temperatures: (**a**) 300 °C/1300 °C, (**b**) 900 °C/1300 °C, (**c**) 500 °C/1000 °C, (**d**) 500 °C/1300 °C, and (**e**) 500 °C/1600 °C. The main figure of each subfigure presents absorbance–time changes (the blue line is the analyte line, and the red line is the background line). The smaller figure of each subfigure denotes an enlarged wavelength-resolved absorption spectrum. The vertical orange line denotes the central pixel of the number 101.

Figure 3, Figure 4 and Figures S2–S4 show the following:

- Similarity of registered curves for 0.1 mg L^{-1} F standard and S1 highly diluted sample.
- Similarity of registered curves for 0.1 mg L^{-1} F with HCl and less diluted samples S2 and S3.
- One of the significant differences between these two groups is the higher background (BG) for the second group, especially distinctly seen for measurements using 300 °C/1300 °C and 500 °C/1000 °C.
- Another difference is a higher signal of Ga for the solutions of the first group (it is seen, for example, when Figure 2 and Figure S4 are compared, obtained for 0.1 mg L^{-1} of F and 0.1 mg L^{-1} of F with addition of 1 M HCl, respectively).
- For all the five investigated solutions, the same tendency of the GaF signal changes (appearance and shape) is observed with the changing temperatures; as is seen in the upper subfigures, with the increase of both pyrolysis and vaporization temperatures, the GaF signals move to the beginning of the measurement step and become narrow.
- Other phenomena appearing with the increase of both pyrolysis and vaporization temperatures are as follows: an increase of the Ga signal at 68 pixel and the appearance of a new structure background; the last phenomenon takes place especially using 1600 °C vaporization temperature; and the structure background observed using high 1600 °C vaporization temperature is a dense bush of lines in the case of samples S2 and S1 (Figure 4 and Figure S2, respectively).

The observed phenomena indicate interferences, which will be further considered in the following sections.

3.2.4. Influence of Hydrochloric Acid

Hydrochloric acid is indispensable in the applied extraction procedure [14]. It simulates the composition of the stomach environment and does not disturb potentiometric measurements. However, as was shown above, it is a cause of a background (time-resolved subfigures of Figure 3, Figure 4 and Figures S2–S4).

To confirm the HCl effect, a series of solutions with the same content of F and the increasing content of HCl was measured. As can be seen in Figure 5, the HCl effect can be connected with the appearance of an intensive background at the beginning of the vaporization curve (red curves at the time-resolved absorbance subfigures with maxima near the 11th spectrum, i.e., 0.7–0.8 s). The same BG is seen in green in the three-subfigures (wavelength- and time-resolved absorption) and the wavelength-resolved absorbance subfigures (the green signals measured only at the 11th spectrum overlap the blue signals, which is an average spectrum (averaging throughout the 5-s vaporization phase). A higher amount of HCl can change the shape of the GaF signal, which is shown in the time-resolved subfigure of Figure 4c. Also, a decrease of integrated absorbance takes place by approximately 10% and 30% for 0.5 M HCl and 1.5 MHCl, respectively, relative to the measurement without HCl.

The conclusions from the experiments are as follows:

- The HCl presence in the measured solution is a cause of the background appearing early in the vaporization step, which is separated in time from the GaF peak; at the highest concentration of HCl in a solution of a sample, 1 M, the spectral effect is some disturbance of baseline, which seems to be insignificant if Figure 3d and Figure S4d are compared (the selected pyrolysis/vaporization temperatures are equal to 500 °C/1300 °C).
- Due to the presence of HCl in the measured solution, a decrease of the Ga peak (comparison of Figure 3, Figures S4 and S5) and a decrease of the GaF signal occur. It can be explained as the formation of a concurrent compound, GaCl. The bonding energy of GaCl of 481 kJ mol^{-1} is relatively high, even though the bonding energy of GaF is 577 kJ mol^{-1} [43]. Therefore, less Ga is available for the formation of the target molecule, GaF, which would cause a decrease in the accuracy of the measurement and underestimation of F.

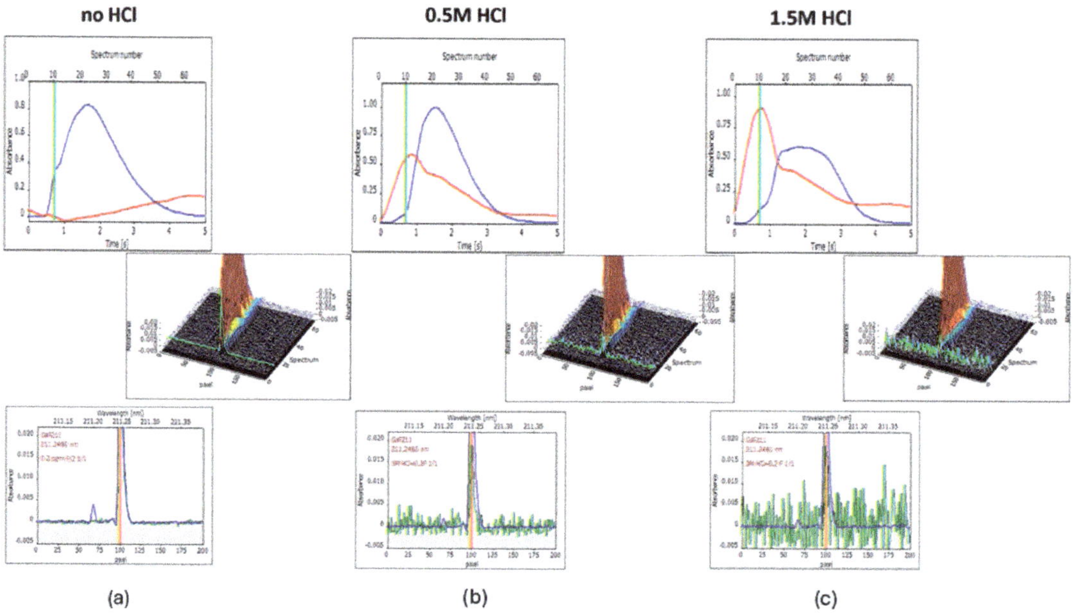

Figure 5. Influence of increased amount of HCl in solution on integrated absorbance and background: (**a**) no HCl, (**b**) 0.5 M HCl, and (**c**) 1.5 M HCl. Pyrolysis/vaporization temperatures: 500 °C/1300 °C. In middle parts of figures (**a**–**c**), 3-dimisional spectra are presented (wavelength- and time-resolved absorption), while the upper and lower subfigures contain 2-dimensional pictures, time-resolved absorption and wavelength-resolved absorption. The wavelength-resolved green signals were obtained at spectrum number 11 (time 0.8 s), which is marked in time-resolved spectra as a vertical light-green line.

3.2.5. Other Interferences

The spectra presented in Figure 3, Figure 4 and Figures S2–S4 contain additional signals, apart from the Ga atomic line at 68 pixel and the GaF line. The additional signals are significantly increased when the graphite furnace temperature is higher, i.e., in the options of pyrolysis/vaporization temperature 900 °C/1300 °C and, especially, 500 °C/1600 °C. Additional lines are always observed, even in the measurements of simple solutions of the F standard (Figure 3).

According to the software of the contrAA apparatus [28], at least 22 atomic lines and spectra of five molecules can be observed in the measured spectral range. The quoted atomic lines are of low sensitivity. These are mainly lines of elements requiring higher atomization temperatures (higher than 1600 °C). These elements are not expected to be present in a fluorine standard solution. No spectra of the six quoted molecules (CS, NO, OH, PO, and SiO) resemble the spectra obtained in this work.

In further experiments, model solutions containing 200 mg L^{-1} of Ca and 400 mg L^{-1} of P (approximately 1200 mg L^{-1} of $PO4^{3-}$ ion, coming from $NH_4H_2PO_4$) were investigated using 1300 °C or 1600 °C vaporization temperature (Figure 6).

The Ca effect at the 1300 °C vaporization temperature is negligible (the recovery relative to the result of measurement of a solution without Ca is equal to 102%), but at the 1600 °C, a distinct decrease of the GaF signal occurs (recovery of 79%). Introducing 200 mg L^{-1} of Ca to the measured solution does not lead to new spectral events. The Ca effect at 1600 °C vaporization temperature is probably of a chemical nature. It is the consequence of the concurrent reaction of F with Ca (the bonding energy of CaF of 527 kJ mol^{-1} is not much lower in comparison with the bonding energy of GaF [43]).

Probably, the effect does not occur at lower investigated vaporization temperatures because the temperatures are not high enough for the atomization of Ca.

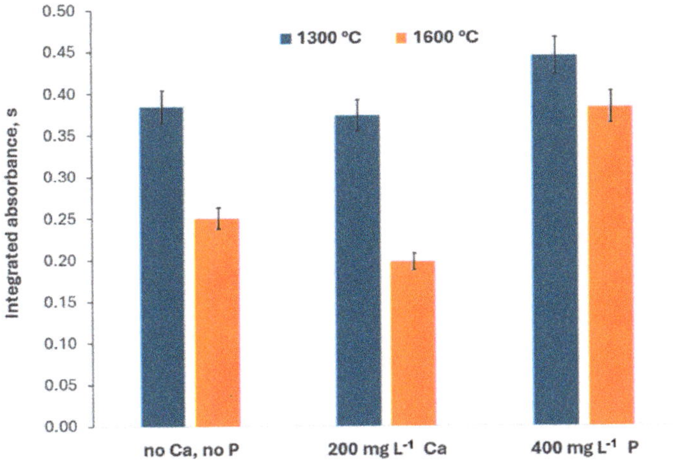

Figure 6. Signals of 0.1 mg L^{-1} of F, depending on the presence of 200 mg L^{-1} of Ca or 400 mg L^{-1} of P in the measured solution and the temperature of the measurement step.

As Figure 6 indicates, at the 1300 °C vaporization temperature, the presence of phosphate leads to a distinct increase in the measured GaF signal (recovery 122%). This effect is much higher using 1600 °C vaporization temperature (recovery 153%). It is interesting that in the spectra acquired in measurements of solutions containing $NH_4H_2PO_4$, an intensive structure background appears that resembles the spectra with a bush of lines in Figures 4 and S2, measured using 900 °C/1300 °C and 500 °C/1600 °C pyrolysis/vaporization temperatures. Figure S2 refers to a highly diluted (1:4000 *m:v*) solution of sample S1 (which is calcium phosphate used as animal feed), and Figure 4 refers to a low-diluted (1:100 *m:v*) solution of sample S2 (a feed for turkey). Various dilutions were applied due to the various fluorine contents. Nevertheless, the P content in the measured solutions was similar, approximately 50 mg L^{-1} in the S1 sample solution (calculation based on stoichiometry) and 70 mg L^{-1} in the S2 sample solution (calculations based on known P content in the original sample and the assumption of 100% extraction efficiency).

The next reagent was measured to verify the origin of the intensive "bush" background, a standard of P, H_3PO_4. Figure 7 shows spectra of H_3PO_4 (Figure 7a) and $Na_4H_2PO_4$ (Figure 7b–d,f). A reference spectrum from a software database attributed to the PO molecule [28] is presented in Figure 7c. The scale of Figure 7b is the same as in Figures 4 and S2. The spectrum of $Na_4H_2PO_4$ in Figure 7b is very intensive. Therefore, Figure 7e was prepared with the same spectrum but at a reduced scale.

The following can be seen in Figure 7:

- Spectra of H_3PO_4 and $NH_4H_2PO_4$ can be interpreted as the same "fingerprint spectrum"; the observed differences come from various relative intensities of the P molecule, the Ga atomic line (the 68th pixel), and the GaF molecule (the peak centered at the 101st pixel).
- The only spectrum measured at the higher vaporization temperature of 1600 °C (Figure 7f) shows (in comparison with Figure 7e measured at the 1300 °C vaporization temperature) much higher intensity of both the P-molecule spectrum and the Ga signal; on the other hand, the signal of GaF is less intensive (which corresponds well with Figure 2).
- The P-molecule spectrum directly overlaps the measured range of seven central pixels used for evaluation of the F signal.

- It is impossible to obtain a pure P-molecule spectrum because the graphite furnace is contaminated with fluorine; the contamination source is Ga salt, and the GaF signal for blank significantly decreases when the Ga solution is not injected (Figure 7d); some residual GaF signal still appears and makes it impossible to apply the LSBC (least squares background correction) method for elimination of interference by the P-molecule.
- The experiments illustrated by Figure 7d,e differ only in Ga solution application; the application of Ga in the experiment from Figure 7e causes an increase of GaF signal and a decrease of P-molecule signal; the Ga atoms could form the GaP molecule (bonding energy of 439 kJ mol^{-1} [43]).
- The spectra obtained for phosphates are similar to each other but differ from the reference spectrum attributed to the PO molecule [28]; there is literature data on the formation of the P_2 molecule in the graphite furnace, apart from the PO molecule [44,45]; it is not sure which molecule is formed in the current research; it is undoubtedly another molecule than the one used for software spectrum generation.

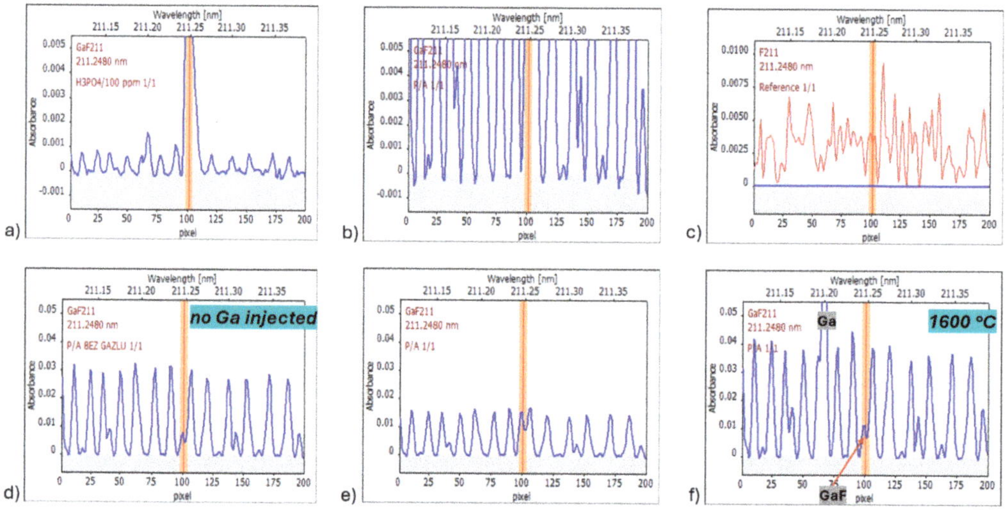

Figure 7. Wavelength-resolved absorbance in measurements of a solution of H_3PO_4 containing 32 mg L^{-1} of P (**a**) or a solution of $NH_4H_2PO_4$ containing 400 mg L^{-1} of P (**b**,**d**–**f**). Subfigures (**b**,**d**) differ only in the y-scale. Measurement temperature: 1300 °C (**a**,**b**,**d**,**e**) or 1600 °C (**f**). In case (**d**), the gallium solution was not injected. Subfigure (**c**) comes from the Aspect CS database [28], where it was attributed to the PO molecule.

The spectra obtained for samples S1 and S2 (Figure S2 and Figure 4, respectively) at vaporization temperature 1600 °C do not resemble exactly the P-molecule spectra from Figure 7. It seems that an additional effect occurs for the solutions investigated in Section 3.2.3 when a higher vaporization temperature of 1600 °C is applied. This is the atomization of an element/elements, resulting in their atomic lines' appearance. As the lines also appear for a simple standard solution of F (Figure 3e), the atomic lines cannot be attributed to the sample matrix. Probably, these are low-intensity lines of Ga, which is always injected into GF to form GaF molecules.

Thus, the wavelength-resolved spectra in Figure S2e and Figure 4e show the following:

- Overlapping signals of molecular absorption of GaF and a P-molecule.
- Signals of single atomic lines, including the Ga 211.2000 nm line (the 68th pixel).

However, the applied vaporization temperatures (even 1600 °C) are relatively low in terms of the temperature required for the atomization of most elements. It explains, for example, why the Fe signal is not seen (211.197 nm Fe line), although some investigated samples contain a lot of Fe. This Fe line was observed in flame experiments [13].

The registered atomic lines are distant from the wavelength range of absorbance by GaF and do not disturb the analyte signal directly. On the other hand, the overlap of the P-molecule spectrum on the GaF signal causes an overestimation of the results of F determination.

3.2.6. Strategy for F Measurements and Figures of Merit

Interferences due to hydrochloric acid, calcium, and phosphates were stated in previous sections. An additional issue is the need to keep a balance in the gallium application. The amount of applied gallium should be high enough to form the GaF molecule efficiently. Unfortunately, the available gallium salt (gallium (III) nitrate of 99.999% purity) is contaminated with fluorine. These problems were taken into account in the selection of measurement conditions.

The following was decided:

- To use the preliminary selected 1300 °C vaporization temperature; this temperature is high enough to form the GaF molecule efficiently (Figure 2), and some excess of Ga is visible as a 211.2000 nm atomic line at the 68th pixel in wavelength-resolved spectra of Figure 3d, Figure 4d and Figures S2d–S4d; simultaneously, the 1300 °C vaporization temperature is low enough to not cause atomization of calcium, which could bond fluorine and lead to underestimation of the F determination; also, at this temperature, the spectral effect of phosphates does not occur (lack of P-molecule spectra in Figure 4d and Figure S2d);
- To use standard addition calibration to overcome chemical interferences, the HCl effect and the phosphate effect, it is important to remember that animal feed can contain other possible interferents, which were not investigated here; standard addition calibration is a sure safeguard against such problems as well as a simple tool against differences in the physical properties of standards and sample solutions.
- To use a relatively high but accepted from the point of view of F level dilution ratio of initial solutions, it would diminish interferences; e.g., due to the high dilution of sample S1, being calcium phosphate, no P-molecule spectrum is visible in the wavelength-resolved spectrum in Figure S2d (the S1 sample is diluted 1:4000 *m:v* here); the high dilution of this kind of material containing a lot of F (above 1000 mg kg^{-1}) justifies resignation from standard addition calibration for this kind of samples.

Taking into account these decisions and the conditions specified in Section 2.1, a calibration curve was obtained for the range of concentrations of 0.05–0.4 mg L^{-1}. Satisfactory parameters of the curve are presented in Table 3. Excellent sensitivity was achieved, 2.1 pg (better than in published works [29–34]), which relates to concentration 0.0003 mg L^{-1} in the measured solution injected in the volume of 6 µL. However, due to contamination of Ga with F, the blank (0.030 mg L^{-1}) is a few orders of magnitude higher. The instrumental detection limit of 0.01 mg L^{-1} is satisfactory for the goal of this work.

Table 3. Figures of merit.

Parameter	Value of Parameter
Characteristic concentration, pg	2.1
Calibration curve parameters:	
– equation (range up to 0.4 mg L^{-1})	$y = 12.831x + 0.122$
– determination coefficient, R^2	0.9959
– correlation coefficient, R	0.9980

Table 3. *Cont.*

Parameter	Value of Parameter
Blank (H_2O):	
– average F concentration, mg L^{-1}	0.030
– standard deviation (s), mg L^{-1}	0.0032
– instrumental detection limit (3 s), mg L^{-1}	0.0096

The detection/determination parameters in the analysis of real samples are specified in Table 4. All samples were divided into two groups according to the expected F content and threshold F concentration.

Table 4. Dilution of samples and calibration range.

Analytical Parameter	Calcium Phosphate	Other Feed
Threshold F concentration, mg kg^{-1}	2000	150
Dilution from the threshold content of F in the sample to 0.1 mg L^{-1} of F in the measured solution	1:20,000, *m:v* (1.25 g/50 mL → 1:10 → 1:10 → 1:5)	1:1500, *m:v* (1.25 g/50 mL → 1:5 → 1:7.5)
Parameters recalculated for sample:		
– blank	600	45
– detection limit, mg kg^{-1}	200	15
– calibration range, mg kg^{-1}	600–8000	50–450
Dilution	-	1:200, *m:v* (1.25 g/50 mL → 1:5)
Parameters recalculated for sample:		
– blank		6
– detection limit, mg kg^{-1}		2
– calibration range, mg kg^{-1}		6–60
Dilution	-	1:50, *m:v* (1.25 g/50 mL → 4:5)
Parameters recalculated for sample:		
– blank		1.5
– detection limit, mg kg^{-1}		0.5
– calibration range, mg kg^{-1}		1.5–15

The first group is comprised of calcium phosphate, which can be highly diluted (1:20,000 *m:v*) and does not need standard addition calibration. The calibration range is from the determination limit of 600 mg L^{-1} in the initial sample (ten times the standard deviation multiplied by the dilution ratio) up to 0.4 mg L^{-1} in solution, which corresponds to 8000 mg L^{-1} of F in the initial sample. The concentration of 0.1 mg L^{-1} of F in solution relates to the threshold value of 2000 mg kg^{-1}.

The second group consists of other materials with a threshold F content of 150 mg L^{-1}. From the point of view of the decision on the fulfillment of the official requirements f, the sample dilution ratio would be 1500 *m:v*, which would result in 0.1 mg L^{-1} concentration in solution, related to the threshold value. The range of the method would be, from the determination limit, i.e., 50 mg kg^{-1} up to 450 mg kg^{-1}, which corresponds to 0.3 mg L^{-1} of F in solution (the upper concentration is less than 0.4 mg L^{-1} due to the use of a standard addition calibration, i.e., addition of 0.1 mg L^{-1} of F). However, the results of analyses of large numbers and variable types of animal feed of various origins, executed over a few years using the potentiometric method (over 80 samples—Sections 3.1 and 3.4), indicate low fluoride concentration, much lower than the threshold value. Therefore, also the HR-CS GFMAS method could be set to determine the extremely low content of fluorine. Example analysis ranges related to dilutions of 1:50 and 1:200 *m:v* are presented in Table 4. The detection and determination limits recalculated for the initial sample at the lowest recommended dilution (1:50) are 0.5 and 1.5 mg kg^{-1}, respectively.

The developed HR-CS GFMAS method was designed to determine fluorine in a wide range of concentrations in various materials using the same experimental conditions, volumes, and modifiers. The variable parameter is sample dilution.

The drawback of the method is the contamination of the graphite system with fluorine (coming from the available gallium salt). It is taken into account by subtracting the blank, but it narrows the calibration range of the method. The need for standard addition calibration limits the upper part of the calibration range.

Another drawback of the HR-CS GFMAS method are memory effects, which can be observed as increased GaF signals for pure deionized water measured after solutions with higher F content. The effect should be controlled by periodic measurements of blank.

The relative standard deviation calculated for parallel analyses is at the level of a few percent. For the determination of F in calcium phosphate, repeatability was evaluated. Nine pairs of results were obtained (Table S2; the results of each pair in repeatability conditions and particular pairs possibly in reproducibility conditions). The average result was equal to 1658 mg kg^{-1}, the average difference was 11 mg kg^{-1} and the average relative absolute value of differences was 2.5%.

The accuracy of the HR-CS GFMAS method was evaluated in the next section.

3.3. Comparison of ISE and HR-CS GFMAS Methods

The samples S1–S3 were analyzed using ISE and HR-CS GFMAS methods. The results of both methods shown in Table 5 are consistent with each other, which confirms the good accuracy of the methods. No underestimation of ISE results was observed, in contrast to the experiment with added Fe and Al, discussed in Section 3.1. Possibly, the original content of Fe and Al in the investigated samples is too low to cause a decrease in results.

The consistency of the results of the ISE method (analysis immediately after extraction) and the HR-CS GFMAS method (filtrate analysis from the extract at least a few days after extraction) indicates that there is no fluoride in the sediment emerging before HR-CS GFMAS measurement. The consistency of the results of both methods also informs that fluorine in extracts is in the fluoride form (in HR-CS GFMAS, the total amount of fluorine is determined, while in the ISE method only fluoride anions are detected).

As Table 5 shows, the precision of analysis of samples S1–S3 is comparable. The precision can also be compared, taking into account the results of calcium phosphate from Tables S2 and S3. The differences of two parallel results were evaluated. The average relative absolute value of differences for nine pairs of results obtained in reproducibility conditions was small (~3%) and similar for both methods.

Table 5. Results of fluorine determination, mg kg^{-1}. Analyses using both techniques were performed in different laboratories at an interval of at least a few days.

Sample	Proficiency Testing [1]	ISE		HR-CS GFMAS	
	Assigned Value ± Standard Deviation	Average Result ± Standard Deviation (Number of Parallel Analyses)	RSD, %	Average Result ± Standard Deviation (Number of Parallel Analyses)	RSD, %
S1 Calcium phosphate	-	1680 ± 43 (8)	2.6	1690 ± 50 (8)	2.9
S2 Mixed feed for turkey	11.2 ± 3.7	13.1 ± 1.3 (2)	9.9	12.5 ± 0.8 (3)	6.4
S3 Spiked feed for chicken	70 ± 26	64 ± 6 (2)	9.3	66 ± 7 (3)	9.8

[1] Proficiency testing organized by AGES-IAG-Feedingstuffs (Vienna, Austria).

A comparison of ISE and HR-CS GFMAS methods is further shown in Table 6 and Table S4.

The HR-CS GFMAS method has the potential to decrease the detection limit. Exemplarily, a higher volume of investigated solution can be injected, and further efforts to find less polluted gallium can be undertaken.

Undoubtedly, the ISE technique is faster and cheaper. On the other hand, in the HR-CS GFMAS, an autosampler is utilized.

Table 6. Comparison of features of ISE and HR-CS GFMAS methods (relative evaluation: "+" positive, "−"negative, "+/−" neutral).

Evaluated Parameter	ISE	HR-CS GFMAS
Sensitivity	+	+
Detectability	+/−	+
Precision	+	+
Total amount of F determination	−	+
Interferences: Ca/PO_4^{3-} effect	+	+/−
Interferences: Fe, Al effect	+/−	+
Interferences: HCl effect	+	−
Calibration range and parameters	+	+/−
Investment costs	+	−
Running costs	+	−
Speed of analysis	+	−
Automation *	−	+
Contamination risk	+	−

* Autosampler utilization.

3.4. Results of Fluorides Monitoring in Animal Feed Used in Poland in 2021–2023

The ISE method was widely applied for determining fluorides. Table 7 shows the results of fluoride testing in different types of feed samples collected in Poland in the years 2021–2023. Fluoride content in feed materials, with respect to mean values, varied from low content in plant and animal materials, respectively 4.5 mg kg^{-1} and 7.3 mg kg^{-1} to medium content in calcium carbonate (37.2 mg kg^{-1}) and high content in feed phosphates (2790 mg kg^{-1}). For feed materials of plant and animal origin, the fluoride levels found were 20–30 times lower than the maximum level of 150 mg kg^{-1}, without posing a risk to animals.

Table 7. Fluoride content in different types of animal feed samples collected in Poland in the years 2021–2023.

Kind of Animal Feed (Number of Samples of This Type)	Content of Fluorides, mg kg^{-1}	Mean Value, mg kg^{-1}	Median, mg kg^{-1}	Standard Deviation, mg kg^{-1}	Coefficient of Variation, %	Threshold Value, mg kg^{-1}
Feed materials of plant origin (34)	2.3–8.9	4.5	4.4	1.4	31	150
Feed materials of animal origin (18)	2.8–25.3	7.3	4.7	6.1	84	150
Calcium carbonate (13)	9.7–73.1	37.2	36.4	19.1	51	350
Calcium phosphate (5)	1620–4420	2790	2230	1250	45	2000
Compound feed for poultry (17)	5.2–20.7	12.0	11.5	4.7	39	150

4. Conclusions

It was determined that the calibration relationship in the ISE method is linear in the 0.1–10 mg L^{-1} fluoride concentration range. At the F concentration below 0.1 mg L^{-1} (0.1 mg L^{-1} in solution relates to 40 mg kg^{-1} in a sample), the calibration relationship is nonlinear, and interpolation, instead of the standard addition method, has to be applied for calibration. Model studies indicated the possibility of a certain underestimation of results of fluoride determination by the ISE method in samples with low fluoride content (recovery reduced to 90–80%). It may be due to a change in the slope of the calibration curve by the matrix components or the consumption of a certain amount of fluorides by cations (such as Fe^{3+} or Al^{3+}). These matrix effects will not be detected in proficiency testing, as almost all laboratories use the same ISE method. Therefore, it is essential to have an alternative method for comparison.

A new method for determining F in animal feed using a high-resolution continuum graphite furnace molecular absorption spectrometry (HR-CS GFMAS) was developed. The feed samples turned out to be a challenge for the HR-CSGFMAS method. The components of the measured solutions caused chemical interferences (formation of competing molecules of CaF, GaCl, and GaP) and spectral effects (a spectrum appearing in the measurement phase before the GaF signal, a spectrum of a P-containing molecule, and atomic lines, e.g., gallium lines). Other difficulties were the contamination of available gallium compounds with fluorine and memory effects.

The following strategy, undertaken in HR-CS GFMAS, turned out to be effective and enabled to eliminate/diminish interferences as follows:

- Application of a molecule providing the best sensitivity (GaF), which allows for higher dilution of samples.
- Relatively low measurement temperature, 1300 °C, which is sufficient to produce GaF molecules but too low to lead to the formation of interfering compounds.
- Using the standard addition method to eliminate chemical interferences (except for highly diluted samples, where the standard addition method is unnecessary).

Excellent sensitivity was achieved in the HR-CS GFMAS (characteristic mass of 2 pg), better than reported in the literature.

The quantification limits of ISE and HR-CS GFMAS methods are comparable and satisfactory (1.0 and 1.5 mg kg^{-1}, respectively), as well as precision (relative standard deviation at the level of a few percent). The consistency of the results obtained by the ISE and HR-CS GFMAS methods indicates both the absence of matrix effects and the presence of fluorine in extracts in fluoride anion form.

The ISE method has been positively verified in many proficiency testing programs (13 rounds for various materials containing 10–1800 mg kg^{-1} F).

Generally, it was confirmed that the ISE method is simple, fast, and cost-effective and is a good tool for routine control of fluoride content in the feed. The HR-CS GFMAS can be used alternatively. The application of both methods enables their cross-validation.

The results of extensive ISE tests conducted in Poland in 2021–2023 have shown that in the vast majority of cases, the fluoride content is more than an order of magnitude lower than the threshold values, which proves that animal feed does not pose a risk of excessive fluoride intake by animals or humans consuming this animal food.

Supplementary Materials: The following supporting information can be downloaded at: https://www.mdpi.com/article/10.3390/ma17122812/s1. Figure S1: Example calibration curve in ISE method The curve is entirely linear in the range 0.1 mg L^{-1}–10 mg L^{-1} (R^2= 1.0000). It starts to curve at the concentration of 0.1 mg L^{-1}. When the whole range of F concentration, 0.01–10 mg L^{-1}, R^2= 0.9925 for linear approximation and R^2= 0.9988 for quadratic approximation; Figure S2: Signals of a solution of sample S1 (dilution 1:600 *m:v*, 0.3 M HCl), obtained using the following pyrolysis/vaporization temperatures: (a) 300 °C/1300 °C, (b) 900 °C/1300 °C, (c) 500 °C/1000 °C, (d) 500 °C/1300 °C, (e) 500 °C/1600 °C. The main figure of each subfigure presents absorbance-time changes (the blue line is the analyte line, and the red line is the background line). The smaller figure of each subfigure denominates enlarged wavelength-resolved absorption spectrum. The vertical orange line denominates the central pixel of the number 101; Figure S3: Signals of a solution of sample S3 (dilution 1:600 *m:v*), obtained using the following pyrolysis/vaporization temperatures: (a) 300 °C/1300 °C, (b) 900 °C/1300 °C, (c) 500 °C/1000 °C, (d) 500 °C/1300 °C, (e) 500 °C/1600 °C. The main figure of each subfigure presents absorbance-time changes (the blue line is the analyte line, and the red line is the background line). The smaller figure of each subfigure denominates enlarged wavelength-resolved absorption spectrum. The vertical orange line denominates the central pixel of the number 101; Figure S4: Signals of a 0.1 mg L^{-1} solution of fluorine standard as NaF in 1 M HCl, obtained using the following pyrolysis/vaporization temperatures: (a) 300 °C/1300 °C, (b) 900 °C/1300 °C, (c) 500 °C/1000 °C, (d) 500 °C/1300 °C, (e) 500 °C/1600 °C. The main figure of each subfigure presents absorbance-time changes (the blue line is the analyte line, and the red line is the background line). The smaller figure of each subfigure denominates an enlarged wavelength-resolved absorption spectrum. The vertical orange line denominates the central pixel of the number 101; Table S1: Evaluation of intralaboratory-reproducibility of analysis of samples with low F content using ISE method; Table S2: Evaluation of intralaboratory-reproducibility of analysis of calcium phosphate using ISE method; Table S3: Evaluation of intralaboratory-reproducibility of analysis of calcium phosphate using HR-CS GFMAS method. Samples dilution 1:20,000, *m:v*; Table S4: Summary of HR-CS GFMAS and ISE methods.

Author Contributions: Conceptualization, Z.K., W.K. and R.D.; methodology, Z.K., W.K. and R.D.; software, Z.K., W.K. and K.G.; validation, Z.K., W.K., K.G. and R.D.; formal analysis, Z.K. and K.G.; investigation, Z.K. and K.G.; resources, Z.K. and K.G.; data curation, Z.K. and K.G.; writing—original draft preparation, Z.K.; writing—review and editing, Z.K. and K.G.; visualization, Z.K. and K.G.; supervision, Z.K., W.K., R.D. and R.O. All authors have read and agreed to the published version of the manuscript.

Funding: The Warsaw University of Technology partly financed the work of ZK in the frame of a grant from the Dean of the Faculty of Civil Engineering, Mechanics, and Petrochemistry (no. 504/04489/7192/44.000000). The research using the HR-CS GF MAS method was carried out on devices co-funded by the Warsaw University of Technology within the Excellence Initiative: Research University (IDUB) program.

Institutional Review Board Statement: Not applicable.

Informed Consent Statement: Not applicable.

Data Availability Statement: The raw data supporting the conclusions of this article will be made available by the authors on request.

Acknowledgments: The authors are grateful to Joseph H. Rule from Old Dominion University (Norfolk, VA, USA), and his wife Anna M. Rule for their help with the language editing during preparation of the manuscript.

Conflicts of Interest: The authors declare no conflicts of interest.

References

1. Han, J.; Kiss, L.; Mei, H.; Remete, A.M.; Ponikvar-Svet, M.; Sedgwik, D.M.; Roman, R.; Fustero, S.; Moriwaki, H.; Soloshonok, V.A. Chemical aspects of human and environmental overload with fluorine. *Chem. Rev.* **2021**, *121*, 4678–4742. [CrossRef] [PubMed]
2. Fuge, R. Fluorine in the environment, a review of its sources and geochemistry. *Appl. Geochem.* **2019**, *100*, 393–406. [CrossRef]

3. Ramteke, L.P.; Sahayam, A.C.; Ghosh, A.; Rambabu, U.; Reddy, M.R.P.P.; Popat, K.M.; Rebary, B.; Kubavat, D.; Marathe, K.V.; Ghosh, P.K. Study of fluoride content in some commercial phosphate fertilizers. *J. Fluor. Chem.* **2018**, *210*, 149–155. [CrossRef]
4. Alexander, J.; Autrup, H.; Bard, D.; Carere, A.; Costa, L.G.; Pierre, J. Opinion of the Scientific Panel on contaminants in the food chain on a request from the Commission related to fluorine as undesirable substance in animal feed. *EFSA J.* **2004**, *100*, 1–22.
5. Wu, S.; Wang, Y.; Iqbal, M.; Mehmood, K.; Li, Y.; Tang, Z.; Zhang, H. Challenges of fluoride pollution in environment: Mechanisms and pathological significance of toxicity—A review. *Environ. Pollut.* **2022**, *304*, 119241. [CrossRef] [PubMed]
6. Warren, J.J.; Levy, S.M. Current and future role of fluoride in nutrition. *Dent. Clin. N. Am.* **2003**, *47*, 225–243. [CrossRef] [PubMed]
7. Council of the European Union. *Directive 2002/32/EC of the European Parliament and of the Council of the 7th May 2002 on Undesirable Substances in Animal Feed;* Council of the European Union: Luxembourg, 2022; p. 10.
8. Mello, P.A.; Barin, J.S.; Duarte, F.A.; Bizzi, C.A.; Diehl, L.O.; Muller, E.I.; Flores, E.M.M. Analytical methods for the determination of halogens in bioanalytical sciences: A review. *Anal. Bioanal. Chem.* **2013**, *405*, 7615–7642. [CrossRef]
9. Tjabadi, E.; Mketo, N. Recent developments for spectrometric, chromatographic and electroanalytical determination of the sulfur and halogens in various matrices. *Trends Anal. Chem.* **2019**, *118*, 207–222. [CrossRef]
10. Yahyavi, H.; Kaykhaii, M.; Mirmoghaddam, M. Recent developments in methods of analysis for fluoride determination. *Crit. Rev. Anal. Chem.* **2016**, *46*, 106–121. [CrossRef]
11. Resano, M.; Garcia-Ruiz, E.; Aramendia, M.; Belarra, M.A. Quo vadis high-resolution continuum source atomic/molecular absorption spectrometry? *J. Anal. At. Spectrom.* **2019**, *34*, 59–80. [CrossRef]
12. Resano, M.; Aramendia, M.; Nakadi, F.V.; Garcia-Ruiz, E.; Alvarez-Llamas, C.; Bordel, N.; Pisoneiro, J.; Bolea-Fernandez, E.; Liu, T.; Vanhaecke, F. Breaking the boundaries in spectrometry. Molecular analysis with atomic spectrometric techniques. *Trend. Anal. Chem.* **2020**, *129*, 115955. [CrossRef]
13. Kowalewska, Z.; Brzezińska, K.; Zielinski, J.; Pilarczyk, J. Method development for determination of organic fluorine in gasoline and its components using high-resolution continuum source flame molecular absorption spectrometry with gallium fluoride as a target molecule. *Talanta* **2021**, *227*, 122205. [CrossRef] [PubMed]
14. *EN 16279:2012;* Animal Feeding Stuffs—Determination of Fluoride Content after Hydrochloric Acid Treatment by Ion-Selective Electrode Method (ISE). European Committee for Standardization: Brussels, Belgium, 2012.
15. Malde, K.; Bjorvatn, K.; Julsham, K. Determination of fluoride in food by the use of alkali fusion and fluoride ion-selective electrode. *Food Chem.* **2001**, *73*, 373–379. [CrossRef]
16. Crestani, P.C.; Pereira, T.C.; Larruscain, E.T.T.; Laureno, C.R.; Flores, E.M.M.; Duarte, F.A. A simple, low cost and fast sample preparation method for fluorine determination by ISE in leguminous seeds and oilseeds. *Food Chem.* **2023**, *429*, 136916. [CrossRef] [PubMed]
17. Torma, L. Collaborative study of ion selective electrode method for the direct determination of fluoride in feeds. *J. AOAC Int.* **1975**, *58*, 477–481. [CrossRef]
18. Fant, M.S.; Ross, J.W. Electrode for sensing fluoride ion activity in solution. *Science* **1966**, *154*, 1553–1554.
19. Erney, R.M.; Charles, B.K. Fluoride determination in milk, soy, and water-based products using ion-selective electrode and direct measurement technique: Single-laboratory validation, first action 2022.05. *J. AOAC Int.* **2024**, *107*, 103–111. [CrossRef] [PubMed]
20. Konieczka, P.; Zygmunt, B.; Namiesnik, J. Comparison of fluoride ion-selective electrode based potentiometric methods of fluorine determination in human urine. *Bull. Environ. Contam. Toxicol.* **2000**, *64*, 794–803. [CrossRef]
21. Spano, N.; Bortolu, S.; Addis, M.; Langasco, I.; Mara, A.; Pilo, M.I.; Sanna, G.; Urgeghe, P.P. An analytical protocol for the differentiation and the potentiometric determination of fluorine-containing fractions in bovine milk. *Molecules* **2023**, *28*, 1349. [CrossRef]
22. Balcerzak, M.; Janiszewska, J. Fluorides in tea products and analytical problems with their determination. *Crit. Rev Anal. Chem.* **2013**, *43*, 13–147. [CrossRef]
23. Janiszewska, J.; Balcerzak, M. Analytical problems with the evaluation of human exposure to fluorides from tea products. *Food Anal. Methods* **2013**, *6*, 1090–1098. [CrossRef]
24. De Silva, S.M.; Deraniyagala, S.; Walpita, J.K.; Jayaweera, I.; Diyabalanga, S.; Cooray, A.T. Masking ability of various metal complexing ligands at 1.0 mM concentrations on the potentiometric determination of fluoride in aqueous samples. *J. Anal. Methods Chem.* **2020**, *2020*, 6683309. [CrossRef] [PubMed]
25. Welz, B.; Becker-Ross, H.; Florek, S.; Heitmann, U. *High Resolution Continuum Source AAS;* Wiley-VCH: Weinheim, Germany, 2005.
26. Welz, B.; Lepri, F.G.; Araujo, R.G.O.; Ferreira, S.L.C.; Huang, M.D.; Okruss, M.; Becker-Ross, H. Determination of phosphorus, sulfur and the halogens using high-temperature molecular absorption spectrometry in flames and furnaces—A review. *Anal. Chim. Acta* **2009**, *647*, 137–148. [CrossRef] [PubMed]
27. Butcher, D.J. Molecular absorption spectrometry in flames and furnaces: A review. *Anal. Chim. Acta* **2013**, *804*, 1–15. [CrossRef] [PubMed]
28. *Aspect CS, 2.3.0.0; Software to contrAA 800;* Analytic Jena: Jena, Germany.
29. Heitmann, U.; Becker-Ross, H.; Florek, S.; Huang, M.D.; Okruss, M. Determination of non-metals via molecular absorption using high resolution continuum source absorption spectrometry and graphite furnace atomization. *J. Anal. At. Spectrom.* **2006**, *21*, 114–1320. [CrossRef]
30. Gleisner, H.; Welz, B.; Einax, J.W. Optimization of fluorine determination via the molecular absorption of gallium mono-fluoride in graphite furnace using high-resolution continuum source spectrometer. *Spectrochim. Acta Part B* **2010**, *65*, 864–869. [CrossRef]

31. Wurtenberger, I.; Gust, R. A highly sensitive method for invitro testing of fluorinated drug candidates using high-resolution continuum source molecular absorption spectrometry (HR-CS MAS). *Anal. Bioanal. Chem.* **2014**, *406*, 3431–3442. [CrossRef] [PubMed]
32. Ley, P.; Sturm, M.; Ternes, T.A.; Meermann, B. High-resolution continuum source graphite furnace molecular absorption spectrometry compared with ion chromatography for quantitative determination of dissolve fluoride in river water samples. *Anal. Bioanal. Chem.* **2017**, *409*, 6949–6958. [CrossRef] [PubMed]
33. Metzger, M.; Ley, P.; Sturm, M.; Meermann, B. Screening method for extractable organically bound fluorine (EOF) in river water samples by means of high-resolution continuum source graphite furnace molecular absorption spectrometry (HR-CS GF MAS). *Anal. Bioanal. Chem.* **2019**, *411*, 4647–4660. [CrossRef]
34. Gawor, A.; Tupys, A.; Ruszczyńska, A.; Bulska, E. An improved methodology for determination of fluorine in biological samples using high-resolution molecular absorption spectrometry via gallium fluorine formation in a graphite furnace. *Appl. Sci.* **2021**, *11*, 5493. [CrossRef]
35. Mores, S.; Monteiro, G.C.; da Silva Santos, F.; Carasek, E.; Welz, B. Determination of fluorine in tea using high-resolution molecular absorption spectrometry with electrothermal vaporization of the calcium monofluoride CaF. *Talanta* **2011**, *85*, 2681–2685. [CrossRef] [PubMed]
36. Borges, A.R.; Francois, L.L.; Welz, B.; Carasek, E.; Vale, M.G.R. Determination of fluorine in plant materials via calcium monofluoride using high-resolution graphite furnace molecular absorption spectrometry with direct solid sample introduction. *J. Anal. At. Spectrom.* **2014**, *29*, 1564–1569. [CrossRef]
37. Huang, M.D.; Becker-Ross, H.; Okruss, M.; Geisler, S.; Florek, S.; Richer, S.; Meckelburg, A. Direct determination of fluorine in niobium oxide using slurry sampling electrothermal high-resolution continuum source molecular absorption spectrometry. *Spectrochim. Acta Part B* **2014**, *94–95*, 34–38. [CrossRef]
38. Ozbek, N.; Akman, S. Method development for the determination of fluorine in water samples via the molecular absorption of strontium monofluoride formed in an electrothermal atomizer. *Spectrochim. Acta Part B* **2012**, *67*, 32–37. [CrossRef]
39. Ozbek, N.; Akman, S. Determination of fluorine in milk and water via molecular absorption of barium monofluoride by high-resolution continuum source atomic absorption spectrometer. *Microchem. J.* **2014**, *117*, 111–115. [CrossRef]
40. Flores, E.L.M.; Bouin, J.S.; Flores, E.M.M.; Dressler, V.L. A new approach for fluorine determination by solid sampling graphite furnace molecular absorption spectrometry. *Spectrochim. Acta Part B* **2007**, *62*, 918–923. [CrossRef]
41. Kowalewska, Z.; Brzezińska, K. Calcium monofluoride as a target molecule for determination of organic fluorine in gasoline and its components using high-resolution continuum source flame molecular absorption spectrometry. *Fuel* **2022**, *309*, 122197. [CrossRef]
42. Kowalewska, Z. Strontium monofluoride, a new molecule to measure fluorine using high-resolution continuum source flame molecular absorption spectrometry and its application for gasoline analysis. *Fuel* **2024**, *356*, 129501. [CrossRef]
43. Lide, R.D. (Ed.) *CRC Handbook of Chemistry and Physics*, 84th ed.; CRC Press: Boca Raton, FL, USA, 2004; pp. 9.53–9.57.
44. Resano, M.; Briceno, J.; Belarra, M.A. Direct determination of phosphorus in biological samples using a solid sampling-high resolution-continuum source electrothermal spectrometer: Comparison of atomic and molecular absorption spectrometry. *J. Anal. At. Spectrom.* **2009**, *24*, 1343–1354. [CrossRef]
45. Huang, M.D.; Becker-Ross, H.; Okruss, M.; Geisler, S.; Florek, S. Determination of phosphorous using high-resolution diphosphorus molecular absorption spectra produced in the graphite furnace. *Spectrochim. Acta Part B* **2016**, *115*, 23–30. [CrossRef]

materials

MDPI

Article

Improved Voltammetric Procedure for Chloride Determination in Highly Acidic Media

Rafał Maciąg [1,2,*], Wojciech Hyk [2], Tomasz Ratajczyk [2] and Mikołaj Donten [2]

1 PIT-RADWAR S.A., Poligonowa 30, 04-051 Warsaw, Poland
2 Faculty of Chemistry, University of Warsaw, Pasteura 1, 02-093 Warsaw, Poland; wojhyk@chem.uw.edu.pl (W.H.); tk.ratajczyk@uw.edu.pl (T.R.); donten@chem.uw.edu.pl (M.D.)
* Correspondence: rafal.maciag@pitradwar.com or rm.maciag@uw.edu.pl

Abstract: Cyclic voltammetry (CV) can be applied as a reliable method for the determination of chloride ions in a range from several to a couple hundred (about 200) ppm. Since the standard potential of chloride ion/gaseous chlorine is 1.36 V vs. normal hydrogen electrode (NHE), the efficient oxidation of Cl^- ion occurs at very positive electrode potentials, usually higher than +1.7 V vs. NHE. It is possible to observe this phenomenon only at noble-metal or inert carbon electrodes. Many other solutes, usually organic compounds, are often oxidized at this potential. This is the reason why the determination of Cl^- content directly from an increase in the oxidation current is not reliable and could lead to overestimated values. However, gaseous chlorine, generated at a positive potential dissolve in the analyzed solution, could be reduced in the reverse scan of a cyclic voltammetric curve. Optimization of the experimental procedure using statistical tools enables the development of an improved method for the direct quantification of chloride ions in acid copper electroplating baths. Under the proposed experimental conditions, the calibration curve (Cl_2 voltammetric reduction current vs. chloride ions concentration) is represented by the linear model for the concentration of chlorides ranging from 10 to 200 mg/dm^3. The developed method for analyzing chloride ions in an acid sulfate electroplating copper bath has many unique properties. It is fast; the time of a single analysis is less than 20 min. In automatic mode, it can be repeated up to 50 times a day. The method does not require processing of the sample of the analyzed bath before measurement. As a result, no additional chemical reagents are used, and the test sample can be returned to the plating bath.

Keywords: acid copper plating bath; chloride determination; voltammetry; electrodeposition; printed circuit boards

1. Introduction

Currently, acid copper electroplating is a key technology in the semiconductor industry and the production of printed circuit boards (PCBs) [1–3]. In both applications, the copper electroplating process is a commonly used method of producing interconnects. The acid copper electroplating baths contain copper sulfate, sulfuric acid, chloride ions, and organic additives such as accelerators (grain refiners), suppressors, and leveling agents [4,5]. The use of organic additives is essential because their absence results in a copper deposit with properties that are significantly degraded (e.g., coarse-grained structure and low ductility). Therefore, mechanical properties like ductility and tensile strength are insufficient for industrial applications [6]. The presence of organic additives like suppressors, accelerators, levelers, and chloride ions in the acid copper plating bath change the cathodic copper ion reduction. It creates a synergistic effect with all the organic components [7]. Suppressors and chloride ions are absorbed at the surface of the cathode, which lower the plating rate [8,9]. As a result, the copper coatings are fine-grained with equiaxial grain structure. That guarantees a high ductility and tensile strength, which are critical features for high quality and reliability of PCBs [10].

In the damascene process [11], the presence of chloride ions within the described concentration range is essential to produce defect-free copper deposits (free from voids, seams, dimples, bumps). When consumable (copper) anodes are used, chloride ions aid anode corrosion, setting up a uniform and adherent anode film. The facts mentioned earlier, regarding the influence of chloride ions on the cathodic reduction of Cu^{2+} ions, demonstrate that achieving consistent and high-quality copper deposition under industrial conditions requires strict control of the chloride ion concentration within a very narrow range. The method of chloride analysis for use in the electronic industry should be characterized by the shortest possible time to enable the highest possible frequency.

The methods of chloride ion determination in aqueous solutions are usually based on volumetric [12], potentiometric—Ion Selective Electrode (ISE) [12,13], liquid chromatography [12,14,15], and spectrophotometry techniques [12]. All of them give reliable results in media with low or moderate ion strength (diluted solutions of accompanying agents) and preferably neutral pH. These methods are effective in the determination of Cl^- ions in water from various sources, such as sea water, tap water, and selected waste water [12]. The application of methods listed above are difficult or even impossible for samples containing a high concentration of other ionic substances, especially acids or complexation agents. In this case, analyzed samples must be specifically adapted to technical requirements of applied techniques (titration, potentiometric, or liquid chromatography) [13,16]. These specific requirements usually eliminate the aforementioned methods as a direct analytical method for the determination of chloride ions in electroplating baths. It is especially difficult in the case of analyzing acidic copper plating baths. The commonly used Mohr titration procedure [17] requires a pretreatment of the analyzed solution and often gives results significantly different from the real value. These reasons induced our efforts to develop a simple direct method for the determination of Cl^- ions in the range of 10 to 100 ppm in highly acidic media containing high concentrations of copper sulfate. The natural simple solution of the problem is applying a voltammetric method. Chloride ion—gaseous chorine is a redox couple with a standard potential of 1.36 V [18]. This means the signal of oxidation in the Cl^- ion should be noticeable in an acidic aqueous solution. The main issue covered by this paper is the method of chloride determination adopting linear scan and cyclic voltammetry (CV) [19] as reliable analytical procedures efficient in the determination of chlorine in acidic media of high ionic strength.

2. Materials and Methods

2.1. Apparatus

Electrochemical measurements and electrodeposition were carried out with the electrochemical workstation OrigaFlex-OGF01A (OrigaLys ElectroChem SAS, Lyon, France, OrigaMaster 5, Version: 2.1.0.3). The temperature of the measured bath was kept at 25 ± 1.0 °C.

CV was chosen as an electrochemical method for determining the content of chloride ions in acid copper sulfate plating baths used for the production of Printed Circuit Boards (PCBs).

The chloride determination process was carried out in a specially designed voltammetric cell. The designing, manufacturing, and testing of a measuring cell was performed in-house. A prototype of the voltammetric cell is illustrated in Figure 1. The working volume of the measuring cell is approximately 70 $cm^3 \pm 10\%$. Commercially available completely inert plastic was selected as the building material for the production of a dedicated measuring cell. The reference electrode (RE) was prepared in this cell. In addition, this cell was also used to perform analytical measurements of the tested sample of the electroplating solution.

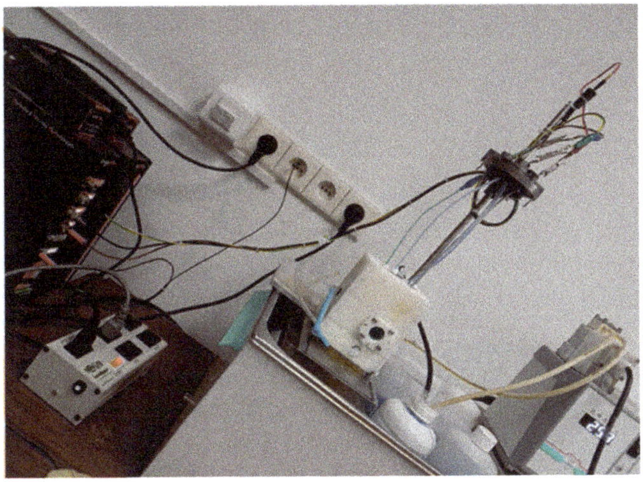

Figure 1. A prototype of the voltammetric cell.

Cylindrical polycrystalline platinum electrode with a surface about 0.38 cm^2 (dimensions—length 12 mm, diameter 1 mm) was employed as a working electrode (WE). The counter electrode (CE) was a polycrystalline platinum pipe-shaped foil, the surface of which was approximately 20 times larger than the surface of the working electrode. RE was a cylindrical polycrystalline platinum electrode with a surface about 0.38 cm^2 covered with a copper layer (copper RE (Cu^{2+}/Cu)). The copper film was produced by electrodeposition at a cathodic current density (Direct Current—DC) of 1.5 A/dm^2. During the electrodeposition of a copper layer on the surface of a polycrystalline platinum electrode, efficient mechanical agitation of the plating bath was applied.

Cathode current density was controlled by an OrigaFlex-OGF01A galvanostat/potentiostat. Deposition time was fixed at 5 min, which resulted in a 1.3 $\pm$ 5% µm thick layer. Two anodes were used in the system. This option was used to obtain a copper layer of the most uniform thickness for the entire surface of the platinum cylindrical cathode.

All the potentials recorded in this study were related to this RE (Cu^{2+}/Cu). For all CV measurements a three-electrode system was used. CV measurements of chloride content were performed between 0.6 V and 1.35 V at a scan rate of 0.1 V/s.

2.2. Analytical Measurement Procedure

A special analytical procedure has been developed for chlorides as a component of the acid copper sulfate plating baths. Quantitative measurements of chlorides were carried out using the CV technique.

The procedure for quantitative determination of chlorides consists of the following steps:

- Quasi-Reference Electrode (QRE) (Cu^{2+}/Cu) Preparing
- Preparation of the WE surface in order to obtain constant and repeatable measurement conditions (electrochemical treatment by application of a potential of +1.8 V vs. Cu^{2+}/Cu for 20 s)

- Quantitative measurement scan preparation (circulation period of the tested sample for 20 s, rest period of the tested sample for 20 s, application of 4 voltametric cycles in the range 0.6/1.35 V)
- Performing a quantitative measurement of chlorides (application of one voltametric cycle in the range 0.6/1.35 V vs. Cu^{2+}/Cu at 0.1 V/s).
- Ending of the measuring sequence (electrochemical treatment by application of a potential of +1.8 V vs. Cu^{2+}/Cu for 20 s)

2.3. Chemicals

The copper sulfate pentahydrate concentrate solution of semiconductor grade (Moses Lake Industries, Inc., Moses Lake, WA, USA) and H_2SO_4 (analytical reagent grade) were used for preparing the Virgin Makeup Solution (VMS), which is a plating solution without any added organics (i.e., zero accelerator, suppressor, leveler, organic and inorganic contamination). The additives used were ELECTROPOSIT™ M Additive (suppressor) and ELECTROPOSIT 1300 S from DuPont (Marlborough, MA, USA) formerly Dow (accelerator), and the chloride anions sourced from HCl (standard solution 1 mol/dm^3, Chempur, Piekary Śląskie, Poland).

The ELECTRODEPOSITION™ 1300 Acid Copper for PWB Metallization Applications plating process is designed for reliable through–hole plating of multilayer and double-sided PCBs. Deionized water (resistivity around 18 MΩ-cm, Millipore, MiliQ™, (Merck, Darmstadt, Germany)) was used in preparation of all solutions. The tested electrolyte composition was obtained from industry recommended values (*). The composition of the acid copper plating bath is presented in Table 1.

Table 1. Chemical composition of the tested copper plating bath.

Component *	Range *
Copper (II) (as $CuSO_4 \cdot 5H_2O$)	0.18–0.22 M (45.0–55.0 g/dm^3)
Sulfuric Acid	2.45–2.65 M (240.0–260.0 g/dm^3)
Chlorides	1.13–1.69 mM (40.0–60.0 mg/dm^3)
ELECTROPOSIT™ M Additive	40 cm^3/dm^3
ELECTROPOSIT 1300 S	1 cm^3/dm^3

(*) Technical data sheet, Electrodeposition ™ 1300 Acid Copper for PWB Metallization Applications.

3. Results

3.1. RE Selection (Prerparation) and Testing Its Potential Stability

A three-electrode system was used in the development of the voltammetric method of chloride determination. The system requires an RE. The use of commercially available REs (silver chloride or saturated calomel electrode) in chloride determination in acid sulfate baths, used in PCB production, was proven unsuitable. The commercial electrodes may contaminate the analyte with chloride ions; additionally, the electrode needs topping up periodically.

The need to maintain the concentrations of all working components of the acid copper plating bath in a very narrow working range, including the concentration of copper sulfate (within ±5% of the optimal value), makes it possible to use a system consisting of a cylindrical platinum electrode covered with a copper layer as the QRE [20].

The copper film was produced by electrodeposition at a cathodic current density of 1.5 A/dm^2. Two anodes were used in the system. This solution was used to obtain a copper layer of the most uniform thickness for the entire surface of the cylindrical cathode. The acid copper plating bath constituents for QRE and the plating parameters were given in Table 2.

Table 2. Chemical composition and plating parameters of the copper acid sulfate bath called ELECTRODEPOSITION™ 1300 Acid Copper for PWB Metallization Applications by DuPont, used for the production of electrical contacts in PCBs.

Component	Range
Copper (II) (as $CuSO_4 \cdot 5H_2O$)	0.18–0.22 M (45.0–55.0 g/dm^3)
Sulfuric Acid	2.45–2.65 M (240.0–260.0 g/dm^3)
Chlorides	1.13–1.69 mM (40.0–60.0 mg/dm^3)
ELECTROPOSIT™ M Additive	40 cm^3/dm^3
ELECTROPOSIT 1300 S	1 cm^3/dm^3
Temperature	298 K (25 °C)
Cathodic Current Density (DC)	1.7 A/dm^2

Three deposition times were selected: 2 min, 5 min, and 15 min. The average thickness increased in proportion to the electrolysis time. For 2 min, the thickness was approximately 0.5 µm. However, for 5 min it was about 1.3 µm and for 15 min it was about 4 µm. Quasi-reference electrodes generated for 5 min were selected for Open Circuit Potential (OCP) tests.

The QRE (Cu^{2+}/Cu) potential stability was quantified using open circuit cell potential [21] measured against silver chloride electrode (with saturated KCl solution) at 298 K (25 °C).

Three types of copper plating solutions were tested. The first one was a VMS, that was an acid copper plating bath containing optimal concentrations of inorganic components and free from any organic or inorganic contamination. The second solution type was a newly prepared acid copper plating bath containing optimal concentrations of organic and inorganic components and free from any organic or non-organic contamination. For the third type of solution, a sample was taken from the production plating bath after aging increased by 150 Ah/L. Chemical composition of the sample of the acid copper bath used for the OCP stability measurements are presented in Table 3.

This selection of the acid copper plating bath sample used in industrial conditions resulted from the fact that with the process of electroplating, the accumulation of a series of by-products of plating solution additives will directly affect the hole filling ability and deep plating ability of the plating solution and the uniformity of copper coating thickness. Improper growth control process will directly lead to loose, rough, porous, or broken copper coating surface crystals, resulting in excessive hole filling depression value, insufficient deep plating, and decreased uniformity of copper coating thickness [4].

To conduct studies on the stability of the QRE potential using the OCP method, the following procedure was proposed. In the first stage of the study, a QRE was made for each bath sample (described in Table 3), and then its stability was tested in a freshly prepared plating bath (describe in Table 3 as the newly prepared acid copper plating bath). In the second stage, the QRE was re-manufactured for each tested sample of the plating bath (except for the VMS sample). This time, the OCP method measurement was conducted in the same solution from which the QRE was made.

Table 3. Chemical composition of the samples of the copper acid sulfate bath called ELECTRODEPOSITION™ 1300 Acid Copper (For PWB Metallization Applications) by DuPont, used for the OCP stability measurements.

Components of the Acid Copper Plating Bath	Concentration of Components of the Acid Copper Plating Baths Used for OCP Measurements		
	The Virgin Makeup Solution	The Newly Prepared Acid Copper Plating Bath	The Production Acid Copper Plating Bath
$CuSO_4{\cdot}5H_2O$	0.20 M (51.0 g/dm^3)	0.20 M (51.0 g/dm^3)	0.21 M (52.4 g/dm^3)
H_2SO_4	2.60 M (255.0 g/dm^3)	2.55 M (255.0 g/dm^3)	2.55 M (250.0 g/dm^3)
Cl^-	1.41 mM (50 mg/dm^3)	1.41 mM (50 mg/dm^3)	1.30 mM (46 mg/dm^3)
ELECTROPOSIT™ M Additive	0 cm^3/dm^3	Unknown **	Unknown **
ELECTROPOSIT™ 1300 S	0 cm^3/dm^3	Unknown **	Unknown **

(**)—the content of organic additives was monitored by Hull Cell analysis (plating time: 10 min, DC amperage: 1.5, temperature: 298 K (25 °C), mixing bath: air agitation, polished brass test plate).

Since the obtained data are very similar to each other, regardless of the stage of life of the electroplating bath, for a better presentation of the obtained data, it was decided to present the results obtained for the bath at two diametrically different stages of its life in the form of a graph. Figures 2 and 3 present examples of test results for the stability of the QRE (Cu^{2+}/Cu) potential generated from the acidic copper plating bath at various stages of the bath's life. The rest of the obtained data showing the stability of the QRE potential was presented in numerical form in Table 4.

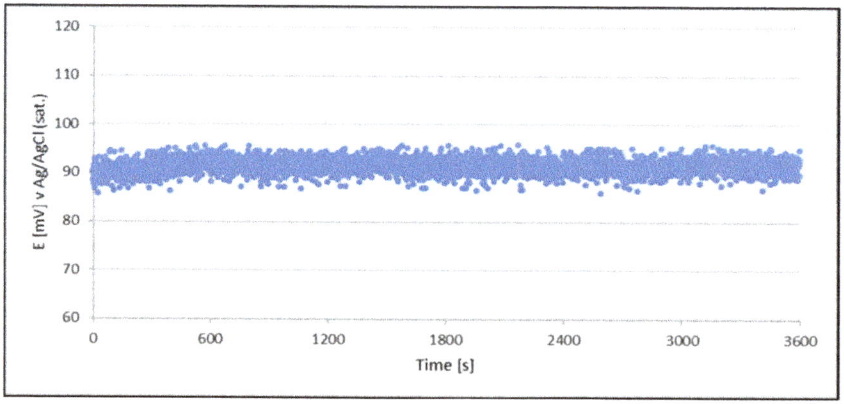

Figure 2. OCP variation with time for a QRE (Cu^{2+}/Cu) made from a sample of a newly prepared acid copper plating bath The QRE was tested in a newly prepared acid copper plating bath.

Data obtained from OCP measurements indicate that the condition of acid copper plating solution contaminated due to aging of bath did not affect the stability of the potential quasi-reference electrodes.

The results clearly indicate the potential raise over initial 80–100 s, which then remains stable until the end of the measurement period (1 h). The initial change in potential happens due to adsorption–desorption processes occurring on the surface until equilibrium was reached. During the process of cathodic reduction of copper (II) ions, their concentration (oxidized form) directly at the cathode surface is lower than in the bulk. After interrupting the electrolysis process, the system needs time to re-establish equilibrium, which, according to the Nerst equation, must lead to an increase in the recorded potential.

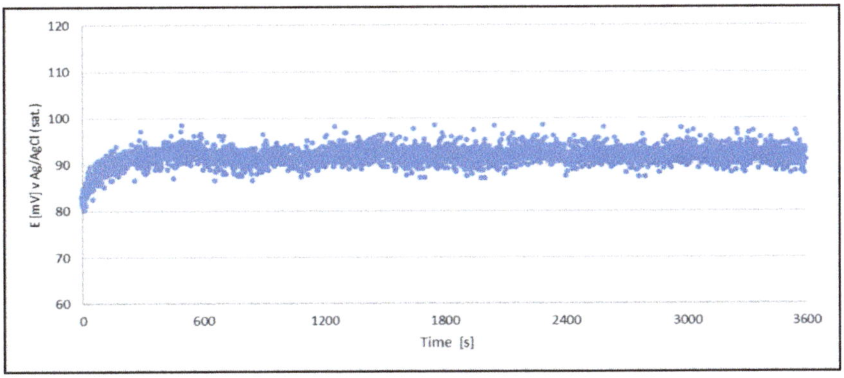

Figure 3. OCP variation with time for a QRE (Cu^{2+}/Cu) made from a sample of a production acid copper plating bath. The QRE was tested in the production acid copper plating bath too.

Table 4. Recorded range of potential changes during OCP measurements for the produced quasi reference electrodes.

Type of Acid Copper Bath Used for Preparing Quasi-Reference Electrode	OCP Measurements for a Newly Prepared Acid Copper Plating Bath Average Value [mV]	OCP Measurements for a Production Acid Copper Plating Bath Average Value [mV]
The Virgin Makeup Solution	92.0	N/A [a]
The newly prepared acid copper plating bath	91.6	N/A [a]
The production acid copper plating bath	92.6	91.8

a—N/A it means that there is no data.

For the OCP measurements performed, the average recorded potential was 92 mV. These data are collected in Table 4.

The QRE (Cu^{2+}/Cu) stability studies have demonstrated its suitability. The electrode delivers a stable potential over a period required for chloride determination (approximately 5 min). Additionally, each measurement could be performed using a freshly prepared quasi-reference electrode.

Figure 4 is showing the relation between electrical potential, various concentrations of the copper in the ELECTROPOSIT™1300 Acid Copper bath and temperature of the solution.

The working range for copper concentration in the ELECTROPOSIT™1300 Acid Copper bath by DuPont was from 0.18 to 0.22 mol/dm^3. During the experiments, the temperature was maintained between 296 and 300 K. Assuming an average temperature of 297 K, the potential could be calculated using the Nernst equation (for a concentration of 0.2 mol/dm^3 of Cu $(II)_{aq}$, after conversion to activities). The calculated potential was 95 mV.

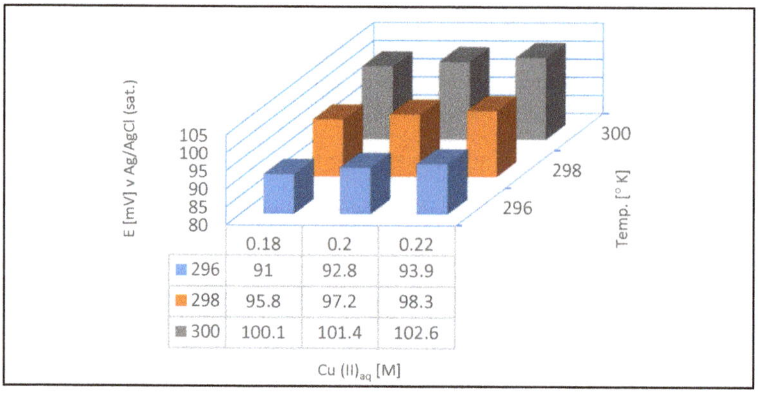

Figure 4. Potential changes depend on variations in copper concentration and temperature, which were permissible within the operating range for DuPont's ELECTROPOSIT™1300 Acid Copper technology.

The small difference between the theoretical value and the actual formal potential determined from OCP measurements may result from the assumptions made when calculating activity coefficients. This approach was necessary because the ionic strength of the examined galvanic bath was approximately 3.

3.2. Characterization of the Method

The method presented in the article bases on oxidation and reduction signal of Cl_2/Cl—redox couple. Others reported in the literature procedures were indirect ones and based on interactions of chloride ions with other systems such as the oxidation and reduction of silver [22], ferrocenemethanol [23], or Co(II)-based anion receptors [24]. None of the methods can be used in highly acidic media such as an acidic copper plating bath.

The technique proposed in the paper is based on the direct oxidation signal of the Cl^- ion and it was designed for an acidic copper sulfate bath, but it can be utilized in other media, especially highly acidic.

The proposed method of determination of Cl^- concentration was based on the electrochemical behavior of redox couple chlorine/chloride ion. The standard potential of this system in aqueous media was determined as 1.36 V vs. NHE and did not depend on the pH of the solution. The electrochemical window of the Pt electrode covers the potential in acidic solution [25,26]. Thus, the voltammetric method could be applied at low pH of the analyte. Also, the method was not affected by high concentration of many ions which could not be oxidized or reduced at similar to Cl_2/Cl^- redox couple potentials. This was a substantial advantage of the procedure because other methods, such as titration, ion chromatography, ion selective potentiometry, and even coulometry are strongly affected by pH and ion strength of the analyte. So, the listed methods could not be directly applied, for example, in analysis of acid copper electroplating baths. In the developed procedure, chloride ion concentration could be determined directly from oxidation current of Cl^-, observed at very positive potentials, or reduction current of gaseous chlorine, generated during oxidation of Cl^- ions. Both signals were potentially useful for analysis. At the potential range from 0.4 V to oxygen evolution, no other significant current signals could be observed.

Figure 5 displays cyclic voltammograms for chloride oxidation/chlorine reduction. These cyclic voltammograms were recorded for a freshly prepared (unused) acid copper plating electrolyte (containing organic additives at optimal concentration levels). During the initial forward scan (from 0.6 to 1.35 V), an increasingly oxidation potential was applied.

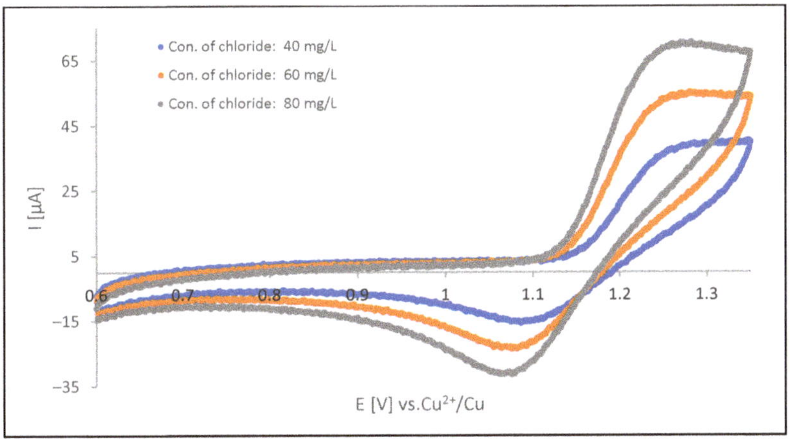

Figure 5. Cyclic voltammograms recorded at acid copper plating bath. Signals at about 1.28 V correspond to Cl^- oxidation signals at about 1.08 V corresponds to Cl_2 reduction. All three CV curves were registered at sweep rates of 0.1 V/s for each direction of potential sweep. These voltammograms were recorded at a polycrystalline Pt electrode (temperature 298 K).

Intensity (current) of both of the signals depends on concentration of chloride ions present in the analyzed solution. The linearity of the oxidation current is compared to the concentration dependence apparent; however, the signal could be disturbed by the oxidation current of organic additives and contaminants of the plating solution. Thus, the oxidation signal could be used only in uncontaminated solutions and was not very reliable in production baths. In such a situation, the analysis could be performed by measuring the reduction current of gaseous chlorine in the reverse half-cycle. Stabilization of this signal can be performed by repeating the oxidation reduction cycle up to obtaining steady-state conditions. In our experimental case, cycling was performed in a potential range of 0.60 to 1.35 V, with scan rate of 0.1 V/s. The 5th cycle was taken for measurement.

Analyses of chloride ion/chlorine redox couple voltammogram indicate that the system is reversible in meaning the existence of electroactive reduced and oxidized forms in the solutions. However, separation of oxidation and reduction peaks suggests that not transport of the electroactive species, but the kinetics of the electrode reactions is what limits the redox process. Moreover, the kinetic limit of the current does not interfere with the linearity of the correlation plot current vs. chloride ion concertation.

3.3. Calibration of the Method

The calibration plot was carried out for solutions identical with acid copper plating solution used for PCB fabrication (composition reported in Table 1). Figure 6 displays calibration plot of peak current vs. Cl^- concentration.

In the proposed method of Cl^- ion determination, the top limit of its suitability was determined by solubility of gaseous Cl_2 in the analyzed solution. According to the literature data, the limit of chlorine concentration in the analyzed bath was circa 48 mmol/dm^3 [27,28]. Above this limit, the calibration plot should reach a plateau. At the reported experimental conditions, the calibration plot was linear up to concentration of Cl^- of 200 mg/dm^3, which is compatible with the operational range of the chloride ion concentration in an acid copper electroplating bath. Modification of the experimental condition, particularly the scan rate, should shift the range of linearity of calibration plot to higher or lower values.

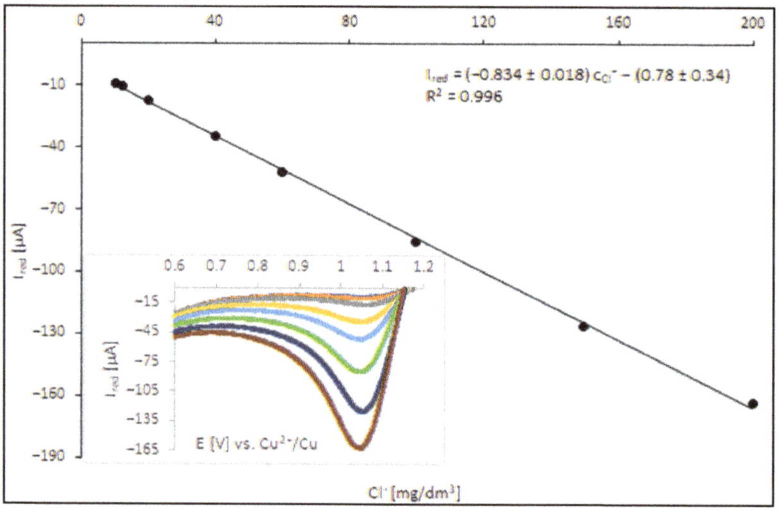

Figure 6. The calibration curve for chloride concentration ranging from 10 to 200 mg/dm^3. The colors of the curves correspond to the following chloride ion concentrations: dark blue—10 mg/dm^3; orange—12 mg/dm^3; gray—20 mg/dm^3; yellow—40 mg/dm^3; blue—60 mg/dm^3; green—100 mg/dm^3; navy blue—150 mg/dm^3, brown—200 mg/dm^3.

Calibration was a crucial stage in developing an analytical method for the quantification of chloride ions in acidic media. It results in the construction of a linear relationship between directly measured voltammetric signals—the limiting currents of the Cl_2 reduction process (I_{red})—and the corresponding chloride concentrations (c_{Cl}) in the prepared matrix samples spiked with a known amount of chloride ions standard. Since both correlated variables were affected by various measurement errors, their experimental values were associated with standard uncertainties. In the case of the dependent variable (voltammetric current), the standard uncertainty ($u(I_{red})$) was estimated using the A method [29], i.e., based on the magnitude of the random dispersion (repeatability) of the measurement results (four replicated independent voltammetric measurements). In the case of an independent variable (chloride concentration), the estimated standard uncertainties ($u(c_{Cl})$) were combinations of several independent sources of uncertainty resulting from the adopted sample preparation scheme, i.e., dilution of the certified reference material (CRM—hydrochloric acid). At least three sources of uncertainty could be distinguished: concentration of the CRM analyte, volume of the CRM sample (pipette limiting error), and the total volume of the system (volumetric flask limiting error). The combination of these uncertainty sources yields the combined relative uncertainty in the chloride concentration at the level of 1%.

Table 5 presents a complete list of the numeric data employed for the construction of the calibration curve. In the calculation process a linear relationship $I_{red} = a \cdot c_{Cl}^{-} + b$, where a and b were the slope and the intercept of the straight line was assumed. To obtain the most representative estimates of the values of coefficients a and b and their standard uncertainties (u(a) and u(b)), it was required to employ a general weighted linear regression scheme for calculations, taking the uncertainties of both scales into account and the significant heterogeneity of uncertainties in the measured values of the dependent variable. In the weighted regression calculation scheme the extent to which a given experimental point determines the course of the calibration line was quantified by its weight coefficient. Uncertainties of both correlated variables contribute to the weight coefficient in the following form (Equation (1)):

$$W_i = [u^2(I_{red,i}) + a^2u^2\,(c_{Cl^-,i})]^{-1} \tag{1}$$

where i numbers indicate the experimental points employed for the regression calculation.

Table 5. Experimental results for the construction of the calibration curve.

No.	c_{Cl^-} [mg/dm^3]	$u(c_{Cl^-})$ [mg/dm^3]	Ired [μA]	u(Ired) [μA]	Weight, W
1	10	0.10	−9.10	0.058	97.3
2	12	0.12	−10.733	0.088	56.2
3	20	0.20	−17.50	0.32	7.63
4	40	0.40	−34.66	0.47	3.04
5	60	0.60	−51.85	0.55	1.82
6	100	1.0	−85.13	0.93	0.644
7	150	1.5	−126.9	1.1	0.355
8	200	2.0	−163.53	0.19	0.355

The calculated values of weights are listed in Table 5.

Under the proposed experimental conditions (described in the previous section), the obtained regression model for the concentration dependence of the voltammetric current can be represented by the following equation (Equation (2)):

$$I_{red} = (-0.834 \pm 0.018)\ c_{Cl^-} - (0.78 \pm 0.34) \tag{2}$$

with the correlation coefficient $r = -0.9998$.

It was found that the linear dependence between the Cl_2 reduction current and the chloride ions concentration was valid for the concentration of chlorides ranging from 10 to 200 mg/dm^{-3}. The upper level of the proposed concentration range was considerably above the typical Cl^- concentration found in electroplating baths. It was also important to note that any change in the voltammetric conditions (e.g., scan rate, the type of RE etc.) requires recalibration of the developed analytical method.

A calibration curve is also a convenient tool for the estimation of the method quantification limit (LOQ) according to the following formula (Equation (3)):

$$LOQ = p \times u(b)/a \tag{3}$$

where u(b) is the standard uncertainty of the intercept and p represents a multiplication constant (usually set to 10).

The LOQ parameter for the proposed method was found to be 2 ppm.

The proposed method was used for the determination of chloride concentration in copper plating bath used for fabrication of PCBs. In this case, there is no simple direct method for the determination of Cl^- level in the bath solution. Our effort was turned to determine this component in a range usually expected in the copper plating bath solution, which is 10 to 100 ppm (mg/dm^3) [30,31]. Commonly applied argentometric titration is a time-consuming process and requires modification of the sample [13]; thus, it cannot be considered a procedure for continuous monitoring of the chloride ions in the copper plating bath. Additionally, such a procedure generates a significant amount of waste which has to be treated. Our voltammetric method requires collection only of a small portion of the bath solution, performing the analytical procedure, and returning the analyzed portion back into the monitored plating bath. Such proceeding could be applied because the sample of the solution used for analysis was not modified during the determination process.

The newly developed electrochemical method of chloride ion determination (based on cyclic voltammetry) in acid copper plating bath was compared to a titration method, commonly used for analysis of copper plating baths.

The determination of Cl^- ion concentration was performed for two solutions. The first one was a newly prepared bath sample of recommended composition shown in Table 6. The bath was prepared using 0.1 M hydrochloric acid standard solution as a chloride source, which was transferred using automatic micropipette into a 1 dm^3 A-class volumetric flask. The second sample was collected from the plating bath which was working for a significantly long time. The bath was operating properly but was not strictly controlled for the concentration of the constituents, especially organic additives.

Table 6. Composition of copper plating solutions utilized for verifying the new method of chloride determination.

	Concentration of Components of the Acid Copper Plating Baths	
Components of the Acid Copper Plating Bath	**A Newly Prepared Acid Copper Plating Bath**	**A Production Acid Copper Plating Bath**
$CuSO_4{\cdot}5H_2O$	0.21 M (52.3 g/dm^3)	0.22 M (54.9 g/dm^3) ***
H_2SO_4	2.60 M (255.0 g/dm^3)	2.55 M (250.0 g/dm^3) ***
Cl^-	1.41 mM (50 mg/dm^3)	1.11 mM (39 mg/dm^3) ***
ELECTROPOSIT™ M Additive	40 cm^3/dm^3	unknown
ELECTROPOSIT™ 1300 S	1 cm^3/dm^3	unknown

(***) values determined in commonly used analytical methods (titrations).

The comparison results of analyses performed by voltammetric and titration techniques are shown in Table 7. The voltammetrically determined Cl^- ion contents indicate significant differences compared to the modified argentometric titration results. However, the reliability of the data were confirmed by additional analyses performed for intentionally spiked bath solutions. For both of them, the concentration of chlorides was increased two times, by two consecutive spikes, each of 25 mg/dm^3 (ppm). The proportional response observed for the implemented method proves its reliability and shows, beyond doubt, that it can be applied not only in the form of comparison result with calibration plot but also using standard addition method (single or multiple). It means that the developed chloride ion determination method may be directly used for working plating line (continuous monitoring) and for electrolyte samples (analyses performed in offline mode), as well.

Table 7. Chloride content for two different acid copper plating baths obtained by two different analytical methods.

C	Chloride Concentrations in the Tested Acid Copper Plating Baths						
	Parameters			Results			
	P1 [mg/dm^3]	P2 [mg/dm^3]	P3 [mg/dm^3]	R1 [mg/dm^3]	R2 [mg/dm^3]	R3 [mg/dm^3]	R4 [mg/dm^3]
A	50	+25	+25	55	77	105	32
B	unknown	+25	+25	24	47	76	39

Where C (a type of the tested acid copper plating baths), A (a newly prepared acid copper bath), B (a production acid copper plating bath), P1 (assumed initial conc.), P2 (first spike), P3 (second spike), R1 (initial conc. determined by newly developed analytical voltammetric method), R2 (results of the first spike), R3 (results of the second spike), R4 (initial concentration determined by standard method (modified argentometric titration)).

4. Discussion

The described voltammetric method of determination of chloride ion content in aqueous solutions may be an attractive alternative for commonly used methods such as application of ISEs and argentometric titration. The advantage of voltammetric determination of concentration of Cl^- ion is its good performance in solutions of high ionic strength and high concentration of acid. In such experimental conditions, usage of ISE is impossible and titration is not reliable due to problems with detection of the titration end point. Thus, the presented method could be applied in determination of concentration of chloride ion in acidic copper plating baths.

The analytical procedure was fit to conditions and the range of concentration of analyte in commercially used electrolytes, i.e., 10–200 ppm, however the range of determined concentration may be changed by modification of experimental parameters, e.g., scan rate. The performed investigation indicated that the method gives reliable results in newly prepared and heavily used solutions. An additional advantage of the method is possibility of its automation integration with industrial copper plating process. It was found that the method was also stable for off-line analyses based on standard addition (simple or multiple). In summary, the proposed analytical method is comparable to other methods commonly used in analysis of chloride ion and additionally gives good results in aggressive media such as acid copper plating baths.

5. Conclusions

The difficulty of accurate determination of chloride concentration in highly acidic media was successfully solved applying a voltammetry technique. The described method is simple, fast, and reliable which was proved applying it for Cl^- analysis in an acid sulfate copper plating bath. A complete analytical system containing stable and robust RE-based on Cu (II)/Cu redox couple and inert Pt WE was proposed and proven for this analytical technique. The constructed system does not require any maintenance even for months of operation. Since the sample solution is not modified during analytical procedures, it can be returned to plating bath. Thus, the method does not generate chemical waste as titration (up to now the only one efficient way to determine Cl^- level in highly acidic media). Additionally, the described method is easy to integrate with other electrochemical procedures used for plating bath monitoring and is fast (less than 20 min for whole cycle). In automatic mode it could be repeated up to 50 times a day. The method is extremely stabile and in appropriate cleaning and conditioning Pt working electrode, e.g., by oxidation during vigorous oxygen evolution, repeatable results can be obtained through couple of months. The appropriate tests support the statement.

Author Contributions: Conceptualization, review, and supervision, M.D.; experimental work and writing, R.M.; proofreading, methodology, and MS editing, T.R.; metrology and statistic, W.H. All authors have read and agreed to the published version of the manuscript.

Funding: This research was funded by Polish Ministry of Science and Higher Education, grant number DWD/3/4/2019.

Institutional Review Board Statement: Not applicable.

Informed Consent Statement: Not applicable.

Data Availability Statement: The original contributions presented in the study are included in the article; further inquiries can be directed to the corresponding author.

Acknowledgments: The authors would like to thank Kazimierz Wikiel (Technic Inc.) for their valuable substantive comments.

Conflicts of Interest: Author Rafał Maciąg was employed by the PIT-RADWAR S.A. The remaining authors declare that the research was conducted in the absence of any commercial or financial relationships that could be construed as a potential conflict of interest.

Abbreviations

VMS	Virgin Makeup Solution
OCP	Open Circuit Potential
PCB	Printed Circuit Board
QRE	Quasi-Reference Electrode

References

1. Wang, C.; Peng, C.; Xiang, J.; Chen, Y.-M.; He, W.; Su, X.-H.; Luo, Y.-Y. Research and Application of Copper Electroplating in Interconnection of Printed Circuit Board. *J. Electrochem.* **2021**, *27*, 257–268. [CrossRef]
2. Zou, H.-B.; Tan, C.-L.; Xiong, W.; Xi, D.-L.; Liu, B.-Y. Introduction of Development and Application Technology of Organic Additives for Acid Copper Electroplating. *J. Electrochem.* **2022**, *28*, 2104531. [CrossRef]
3. Guo, L.; Li, S.; He, Z.; Fu, Y.; Qiu, F.; Liu, R.; Yang, G. Electroplated Copper Additives for Advanced Packaging: A Review. *ACS Omega* **2024**, *9*, 20637–20647. [CrossRef]
4. Sun, Z.; He, G.; Zhao, Z.; He, X.; Zhao, H. Research Progress of Electroplated Micropore for Interconnection Technology. *J. Phys. Conf. Ser.* **2023**, *2430*, 012011. [CrossRef]
5. Wong, K.Y.; Liew, P.J.; Lau, K.T. Copper Filling of Printed Circuit Broad (PCB) Industry: A Review. *Sci. Int.* **2022**, *34*, 131–137.
6. Konstantin, G.; Kück, H.; Münch, R. A novel manufacturing technology for tensile test specimens for the characterization of copper in plated through holes (PTH). In Proceedings of the 2013 IEEE 15th Electronics Packaging Technology Conference (EPTC 2013), Singapore, 11–13 December 2013. [CrossRef]
7. Zheng, L.; He, W.; Zhu, K.; Wang, C.; Wang, S.; Hong, Y.; Chen, Y.; Zhou, G.; Miao, H.; Zhou, J. Investigation of poly (1-vinyl imidazole co 1, 4-butanediol diglycidyl ether) as a leveler for copper electroplating of through-hole. *Electrochim. Acta* **2018**, *283*, 560–567. [CrossRef]
8. Ren, P.; An, M.; Yang, P.; Zhang, J. Unveiling the synergistic inhibition of Cl− copper plating: Pivotal roles of adsorption and desorption. *J. Electroanal. Chem.* **2021**, *898*, 115624. [CrossRef]
9. Zhang, Y.; An, M.; Yang, P.; Zhang, J. Recent Advances in Electroplating of Through-Hole Copper Interconnection. *Electrocatalysis* **2021**, *12*, 619–627. [CrossRef]
10. *IPC-6012B*; Qualification and Performance Specification for Rigid Printed Boards August 2004, Section 3.2.6.8. Available online: https://www.innovative-circuits.com/docs/IPC-6012B.pdf (accessed on 20 September 2024).
11. Carpio, R.; Jaworski, A. Review—Management of Copper Damascene Plating. *J. Electrochem. Soc.* **2019**, *166*, D3072–D3096. [CrossRef]
12. Wu, D.; Hu, Y.; Liu, Y.; Zhang, R. Review of Chloride Ion Detection Technology in Water. *Appl. Sci.* **2021**, *11*, 11137. [CrossRef]
13. Mustafa, J.; Abdullah, M.M.; Husain, S.; Muhaisen, H.M.H.; Umar, K.; Luqman, M. Development of a Highly Selective Chloride Ion Potentiometric Sensor Utilizing a Novel N, N-Ethylene-bis-(Salicylideneaminato) Zinc (II) Molecule for Clean Environment. *Sci. Adv. Mater.* **2024**, *16*, 450–458. [CrossRef]
14. Pillion, D.J.; Meezan, E. Liquid-chromatographic determination of chloride in sweat from cystic fibrosis patients and normal persons. *Clin. Chem.* **1985**, *31*, 1155–1157. [CrossRef]

15. Newton, B. Analysis of copper plating baths. In Proceedings of the Twenty Third IEEE/CPMT International Electronics Manufacturing Technology Symposium (Cat. No.98CH36205), Austin, TX, USA, 21 October 1998; pp. 358–361. [CrossRef]
16. Kaiser, E.; Rohrer, J. Determination of Chloride in Acid Copper Plating Bath, The Application Notebook—February 2002. Available online: https://cdn.sanity.io/files/0vv8moc6/chroma/d625fd1620ad91a70511337796727e89487afd08.pdf/article-11767.pdf (accessed on 18 September 2024).
17. Braun, T.; Bujdosó, E.; Schubert, A. Literature of Analytical Chemistry: A Scientometric Evaluation. CRC Press: Boca Raton, FL, USA, 2019.
18. Kaczur, J.J. Oxidation Chemistry of Chloric Acid in NOx/SOx and Air Toxic Metal Removal from Gas Streams. *Environ. Prog.* **1996**, *15*, 245–254. [CrossRef]
19. Linear Sweep and Cyclic Voltametry: The Principles. Available online: https://www.ceb.cam.ac.uk/research/groups/rg-eme/Edu/linear-sweep-and-cyclic-voltametry-the-principles (accessed on 18 September 2024).
20. Kahlert, H. *Handbook of Reference Electrodes*; Springer: Berlin/Heidelberg, Germany, 2010; pp. 291–308. [CrossRef]
21. Overview of Reference Electrodes and Alternative Reference Electrodes Brief Discussion about Standard and Pseudo Reference Electrodes, Document #: DRK10053 (REV001 I APR 2016). Available online: https://www.google.com/url?sa=t&source=web&rct=j&opi=89978449&url=https://www.pineresearch.com/shop/wp-content/uploads/sites/2/2016/10/DRK10053-Overview-of-Reference-Electrode-Operation-and-Alternative-Reference-Electrodes-REV001.pdf&ved=2ahUKEwiQ3NCFyIuIAxXsgP0HHVNAJRsQFnoECBkQAQ&usg=AOvVaw0frOpN-TZ4CSpDWMwm-BLB (accessed on 18 September 2024).
22. Arai, K.; Kusu, F.; Noguchi, N.; Takamura, K.; Osawa, H. Selective Determination of Chloride and Bromide Ions in Serum by Cyclic Voltammetry. *Anal. Biochem.* **1996**, *240*, 109–113. [CrossRef]
23. Bujes-Garrido, J.; Arcos-Martínez, M.J. Development of a wearable electrochemical sensor for voltammetric determination of chloride ions. *Sens. Actuators B Chem.* **2017**, *240*, 224–228. [CrossRef]
24. Malecka-Baturo, K.; Daniels, M.; Dehaen, W.; Radecka, H.; Radecki, J.; Grabowska, I. Voltammetric Sensing of Chloride Based on a Redox-Active Complex: A Terpyridine-Co(II)-Dipyrromethene Functionalized Anion Receptor Deposited on a Gold Electrode. *Molecules* **2024**, *29*, 2102. [CrossRef]
25. Yokoyama, Y.; Fukutsuka, T.; Miyazaki, K.; Abe, T. Origin of the Electrochemical Stability of Aqueous Concentrated Electrolyte Solutions. *J. Electrochem. Soc.* **2018**, *165*, A3299–A3303. [CrossRef]
26. Barón-Jaimez, J.; Joya, M.R.; Barba-Ortega, J. Bismuth electrodes, an alternative in stripping voltammetry. *J. Phys. Conf. Ser.* **2013**, *466*, 012025. [CrossRef]
27. Young, C.L. Sulfur Dioxide, Chlorine, Fluorine and Chlorine Oxides. In *Solubility Data Series*, 1st ed.; Pergamon: Oxford, UK, 1983; Volume 12, ISBN 9781483156675.
28. Density of Sulfuric Acid and Sulfur Trioxide. Available online: https://wissen.science-and-fun.de/chemistry/chemistry/density-tables/density-of-sulfuric-acid-and-sulfur-trioxide/ (accessed on 18 September 2024).
29. *JCGM 100:2008*; Evaluation of Measurement Data—Guide to the Expression of Uncertainty in Measurement. First edition 2008 Corrected version 2010, the Joint Committee for Guides in Metrology. Available online: https://www.bipm.org/documents/20126/2071204/JCGM_100_2008_E.pdf/cb0ef43f-baa5-11cf-3f85-4dcd86f77bd6 (accessed on 18 September 2024).
30. Barthelmes, J. Acid Copper Plating with Insoluble Anodes—A Novel Technology in PCB Manufacturing. *Trans. IMF* **2000**, *78*, 135. [CrossRef]
31. Dow, W.-P.; Yen, M.-Y.; Liu, C.-W.; Huang, C.-C. Enhancement of filling performance of a copper plating formula at low chloride concentration. *Electrochim. Acta* **2008**, *53*, 3610–3619. [CrossRef]

materials

MDPI

Article

Comparative Analysis of Metal Electrodeposition Rates towards Formation of High-Entropy WFeCoNiCu Alloy

Tomasz Ratajczyk * and Mikołaj Donten

Faculty of Chemistry, University of Warsaw, 02-093 Warsaw, Poland
* Correspondence: tk.ratajczyk@uw.edu.pl

Abstract: This study presents a calculation and comparison of Fe, Co, Ni and Cu deposition rates in the tungsten codeposition process based on the electrodeposition of numerous tungsten alloys. Eight different tungsten alloys containing from two to five metals were electrodeposited in constant conditions in order to compare the exact reduction rates. The calculated rates enabled control of the alloy composition precise enough to obtain a high-entropy WFeCoNiCu alloy with a well-balanced composition. The introduction of copper to form the quinternary alloy was found to catalyze the whole process, increasing the deposition rates of all the components of the high-entropy alloy.

Keywords: tungsten; iron; cobalt; nickel; copper; alloys; high-entropy alloy; electrodeposition; induced codeposition; deposition rates

1. Introduction

The electrodeposition of tungsten alloys is an important research niche, yielding various applicative materials due to their corrosion resistance or catalytic properties. Dating from 1931 [1], research on tungsten electrodeposition has led to the development of several kinds of tungsten alloys with metals such as Fe, Co, Ni [2] and Cu [3], which act as *inducing metals* for the reduction in tungstates [4]. Besides tungsten, such alloys must contain one or more inducing metals. There has been no single original work recently that has compared the electrodeposition efficiencies of the most basic binary alloys of tungsten, let alone of materials consisting of three or more components. Most original papers cover only selected binary or ternary alloys, focusing mainly on their applicational properties [5–10]. The range of possible applications of tungsten alloys obtained via induced codeposition is broad indeed. The most common applications of electrodeposited tungsten alloys include protective and decorative layers, especially as anti-corrosive protective coatings for steel surface enrichment, a substitute for hard chrome coatings [5–7]. Another important application for electrodeposited tungsten alloys is obtaining catalytic layers, mainly for hydrogen evolution [8–10]. All the alloys studied in the mentioned works consist of tungsten and one or two other inducing metals. A trend to approach the electrodeposition of materials containing more metallic components is visible in the most recent publications [11–13].

An attempt at obtaining an alloy of higher complexity in terms of qualitative composition and controlling it quantitatively should rely on a comparison of deposition rates for the individual metals in the system, mostly the inducing metals. However, the mentioned background makes the comparison difficult and uncertain. The minimum variables necessary to calculate the rates are current density j, current (faradaic) efficiency CE and exact alloy composition, or at least the atomic content of tungsten $at\%$ W for binary alloys. In such cases, the deposition rate v of a certain element X on the normalized area of the cathode, in the z-electronic process, can be expressed as $v = j \cdot CE \cdot \mathrm{F}^{-1} \cdot z^{-1} \cdot at\%\ X$ [$\mathrm{mol} \cdot \mathrm{s}^{-1} \cdot \mathrm{m}^{-2}$]. Although CE is a quantity that is crucial to describe the deposition process, numerous authors skip it in their publications. Furthermore, most works that contain the needed information present it not as exact numbers, but rather as graphs, which focus on the influence of a particular parameter on the outcome. Many of these parameters, which affect the deposition rates, make it difficult to compare the results of various researchers, especially since each of them conducts the experiments under different conditions, i.e., bath composition (both qualitative and quantitative) and temperature. Thus, despite almost a century of research on tungsten codeposition, the numerical results cannot be easily unified.

In Table 1, several sets of data have been collected concerning the deposition rates v of Fe, Co and Ni in binary alloys with W. The numbers were calculated from j, *CE* and *at*% W contained in tables or gathered from plots (in the latter case, the source is mentioned with an asterisk). In the absence of more general criteria of choice, the highest achieved rate of Fe/Co/Ni deposition was chosen for each source. All the v values [$\mathrm{mol \cdot s^{-1} \cdot m^{-2}}$] were multiplied by 10^5 for legibility.

Table 1. Rates of Fe, Co and Ni codeposition with tungsten, according to cited papers. Numbers from sources marked with "*" have been retrieved from graphs.

Alloy	WFe			WCo				WNi	
Source	[11] *	[12]	[13]	[14]	[15] *	[16] *	[17]	[18]	[19] *
j [mA·cm^{-2}]	100	100	100	35	36	50	15	100	15
CE [%]	27.8	15	39.7	15.3	74	53	8	3.75	11.1
at% W	6.3	29.4	38.6	38.2	22.3	30.5	31	34.7	46.4
at% Fe/Co/Ni	93.7	70.6	61.4	61.8	77.7	69.5	69	65.3	53.6
v Fe/Co/Ni	135	36.6	101	17.1	107	95.4	4.29	12.7	4.63

Although the occurrence of major discrepancies between the rates achieved by separate researchers cannot be denied, the results do overlap partly. Most notably, the v of Ni is usually more than an order of magnitude lower compared to Fe and Co. The difference in j applied by various authors is the most obvious cause of the inconsistency in the results, but even for the same j, the numbers vary due to the divergency of the other plating conditions, such as bath composition and temperature. Nevertheless, the papers cover some important cases of the influence of those parameters on the v. Plots of *at*% W and *CE* against j are quite common [14–16,18–20], even though some authors stay at a constant value of j, described as optimal [17,21]. Plots of induced metal deposition rates against j tend to increase monotonically due to the apparent increase in current passing through the circuit. The increase is concave down though, mostly because of the *CE* decrease and simultaneous increase in the *at*% W in the alloys. Moreover, applying a pulse current was found to possibly increase the *at*% W and *CE*, with a higher v of cobalt at higher frequencies of the pulses [17]. Other variables found in the prior papers include plating bath temperature [22] or tungstate concentration [23], which showed no clear correlations with the v of nickel.

Introducing ions of an additional inducing metal to the bath, which leads to the deposition of ternary or multinary alloys, has a certain influence on deposition rates for particular metals. The presence of Cu(II) ions in the solution [24,25] inevitably lowers the *at*% of other metals, but increases the *CE* of the whole deposition process, especially in the case of WNiCu alloys. The interpretation of the data contained in previous works is even more difficult when it comes to mixing more than one iron-group metal into the alloy, as most papers on the electrodeposition of ternary tungsten alloys do not cover any comparison to binary alloys. Possible combinations of tungsten electrodepositions along with the inducing metals may eventually lead to the formation of multinary alloys, consisting of four (e.g., WFeCoNi) or five (e.g., WFeCoNiCu) substantial components. Alloys containing at least five metals above 5 at% can be classified as *high-entropy alloys*, according to the original definition coined by Yeh [26]. Materials similar to the quintenary WFeCoNiCu alloy described hereby were published in recent papers [11,12], but they differ substantially in the qualitative composition [11] or alloy morphology and structure [12], a result of utilizing different plating baths.

As mentioned before, more precise control of the deposition rates of particular metals is an essential requirement for obtaining W alloys with more than one inducing metal. To obtain an alloy consisting of three or more metals, the deposition rates for particular metals have to be calculated for the same system at constant plating parameters. The main aim of the following study is to organize the current knowledge on these deposition efficiencies, which may facilitate more precise tuning of the applicational properties of electrodeposited tungsten alloys, and enable further investigation on deposition of multinary alloys of tungsten. Thus, the following work is intended to quantify the rates of Fe, Co, Ni and Cu codeposition with W and to evaluate the ratios between these rates, as a prelude to research on the electrodeposition of multi-component alloys of tungsten, based on the electrodeposition of nanocrystalline WFeCoNiCu high-entropy alloy.

2. Materials and Methods

As this study concerns the comparison of deposition rates of the inducing metals in alloys with tungsten, several different materials were obtained at varying concentrations of inducing metal ions, while keeping all the essential conditions constant. All the samples were deposited from a tungstate–citrate aqueous plating bath. All the reagents were analytical grade. Before each experiment, 400 mL portions of the plating baths were freshly prepared by mixing concentrated solutions of the inducing metal salts into a new portion of the tungstate–citrate bath. The solutions were not stirred nor moved in any manner during the electroplating. The composition of the tungstate–citrate bath utilized in the study is enumerated in Table 2. A fresh portion of the plating bath prepared in such a manner has a pH of 8.0–8.2 and was not observed to change substantially during the experiments.

Table 2. Composition of the plating bath.

Reagent	Concentration
Sodium citrate	270 mM
Sodium tungstate	260 mM
Boric acid	170 mM
Phosphoric acid	90 mM
Nonoxynol-10 (surfactant additive)	70 ppm
Butynediol (brightening additive)	50 ppm
Distilled water (Millipore Milli-Q)	—
Nickel sulphate Cobalt sulphate Ferric ammonium citrate Copper sulphate	various concentrations, totaling 27 mM, explained further in detail

The usage of ferric ammonium citrate deserves some individual attention. For the electrodeposition of iron–tungsten alloys, baths containing iron(II) salts are commonly used, such as ferrous sulphate. The ease of oxidation of iron(II) to iron(III) makes such baths unstable, so they would have to be either freshly prepared for each experiment or constantly monitored. Replacing Fe(II) with Fe(III) solves this problem, so the Fe(III) ion concentration is known more precisely. The choice of this particular ferric compound was justified by its high solubility and by Fe(III) ion already being in an appropriate citrate complex.

As noted before, the concentrations of particular inducing metal salts varied throughout the experiment. The utilized baths can be divided into four series, A, B, C and D, respecting their composition. The number of repetitions is mentioned in the brackets next to the series names. In series A (12 samples, 4 samples of each alloy) were baths containing one inducing metal salt, Fe(III), Co(II) or Ni(II), in a concentration of 27 mM, which produced the binary alloys WFe, WCo or WNi, respectively. Series B (12 samples, 4 samples of each alloy) consisted of baths with two (13.5 mM) or three (9 mM) inducing metal ions at equal concentrations totaling 27 mM altogether, leading to the deposition of the ternary alloys WFeCo, WfeNi and WcoNi, and the quaternary alloy WfeCoNi. Whereas the results of the deposition from bath series A and B allow a comparison of the deposition rates of particular iron-group metals in the same conditions, the remaining two series were set up to examine the influence of Cu on deposition rates in a high-entropy alloy. Series C (WfeCoNi, 11 samples) and D (WfeCoNiCu, 9 samples) consisted of baths designed to generate, respectively, quaternary and quinternary alloys possibly close to equal *at*% of the metals, i.e., equal deposition rates. The exact ion concentrations used in the C and D bath series are shown in Table 3. The experiments concerning the quaternary and the quinternary alloys were repeated a relatively large number of times, taking the system complexity into account.

Table 3. Concentrations of metal ions in C and D bath series.

Bath Series	C Fe(III)	C Co(II)	C Ni(II)	C Cu(II)
C	1.55	1.70	23.75	—
D	1.55	1.70	23.75	1.55

During the electrodeposition, the plating bath was separated from 0.25 M sodium sulphate solution as an anode electrolyte. Two inert Ti/RuO_x anodes were placed symmetrically around the cathode. All the samples were deposited with a constant current at a current density $j = 50\ mA\cdot cm^{-2}$. A potentiostat/galvanostat, EG&G PAR 173A, was utilized as the source of the current. All the experiments were conducted at a constant temperature of 65 °C. The coatings were plated on 0.999 Ag foil, cut into rectangles so that the area of the substrate was $A = 2.4\ cm^2$. The duration of the deposition was $t = 1800$ s. Before deposition, all the substrates were cleaned by annealing in a CH_4–O_2 flame, then mechanically scrubbed with a detergent and finally washed with ethanol. Each time, the clean Ag substrate was weighed on an analytical balance with a precision ±0.1 mg. After the deposition, each sample was weighed again in order to calculate the mass of the deposit Δm.

The composition of the deposited alloys was measured using an FE SEM Zeiss Merlin (Jena, Germany) with a Bruker Quantax 400 EDS analyzer (Billerica, MA, USA) with a 10 mm^2 detector. The EDS spectra were acquired at 15 kV beam energy and 400 pA beam current as an average from a (0.3 × 0.4) mm site in the very center of the sample. An example of the EDS spectrum for a WFeCoNiCu alloy sample is depicted in Figure 1a, with the characteristic lines labeled with element symbols. The numerical results were calculated from the acquired spectra using the original Bruker software (Quantax Esprit v.1.9.4) and normalized to 100% after omitting oxygen, the presence of which indicates surface oxidation of the alloy, and finally carbon, which might contaminate the surface before measurement. The oxygen content did not exceed a few percentage points in any sample. The metal content was quantified from the spectra using the K-series lines of Fe, Co, Ni and Cu, and the L-series lines of W (circled in Figure 1a). The mass content $wt\%X$ and atomic content $at\%\ X$ of the metals in the samples can be easily converted into each other, utilizing the molar masses M of all the metals present in the alloy. The morphology of the alloy surface was imaged with a secondary electron detector with the same beam parameters as the EDS measurements.

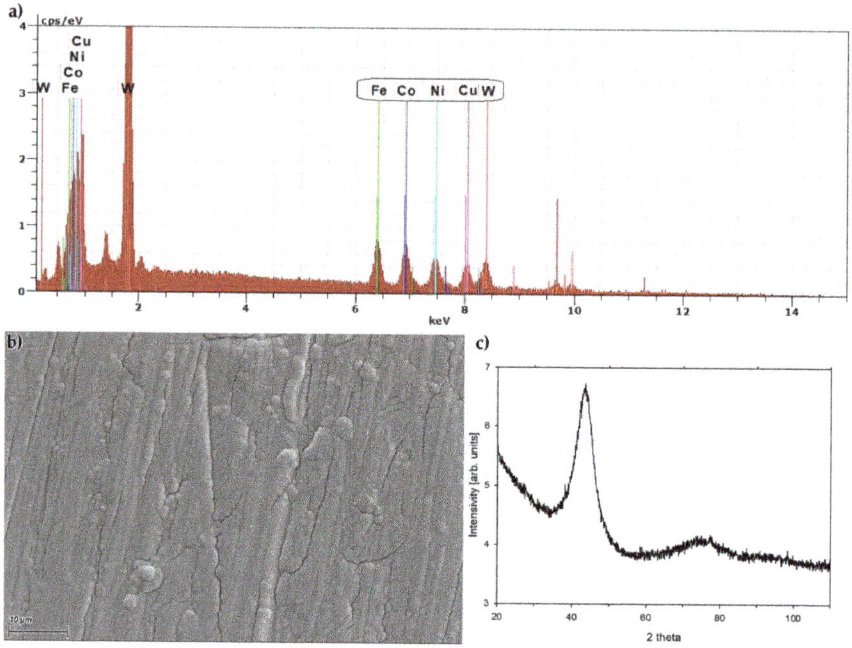

Figure 1. (**a**) EDS spectrum acquired for a WFeCoNiCu alloy sample. (**b**) SEM image of WFeCoNiCu coating surface. Scalebar in the bottom left corner. (**c**) XRD pattern for WFeCoNiCu alloy sample.

The XRD pattern was acquired with an X-ray diffractometer, Bruker D8 Discover, equipped with a collimated Cu Kα (1.54 Å) radiation source. The data were collected in the 20–110° 2θ range, with a 0.01° step-size, in a locked–coupled mode using a Vantec 1D-linear detector. The alloy sample prepared for the XRD analysis was deposited on a 15 × 15 mm piece of Ag foil as a 8 µm thick coating. The crystallite size was calculated using the Scherrer equation.

For each metal X (Fe, Co, Ni or Cu), the deposition rate [mol $\cdot$ cm^{-2} $\cdot$ s^{-1}] calculated for a particular sample v is $X = \Delta m \cdot wt\%X \cdot A^{-1} \cdot M^{-1} \cdot t^{-1}$, assuming that the electrodeposition process is not affected by changing A or t. The final results, i.e., the values of the v of the inducing metals, *at*% of tungsten and other metals, and *CE* of the whole process, were calculated as an average within the four mentioned bath series. All the following values of v have been multiplied by 10^5 for legibility of the data.

3. Results

All the obtained alloys formed shiny gray metallic coatings that adhered well to the substrates. Only the WFe coating was observed to turn brownish after a longer exposition to air, which indicates that this alloy is more prone than the others to corrosive oxidation. The samples of the quinternary WFeCoNiCu alloy appeared to be slightly brighter in comparison to the others, which may be caused by the well-levelled surface morphology of the coating.

Figure 1b depicts an image of the WFeCoNiCu alloy surface, acquired with SEM. The surface consists of compact metallic structures, arranged in lines, along with some spherical granules grown on the surface. The surface is slightly cracked in quite a regular manner. The morphology of the WFeCoNiCu sample is typical of electrodeposited tungsten alloys and resembles other alloys such as WNiCu [25]. The presence of the cracks on the alloy surface is most probably caused by high tensile stress, as well as by abundant hydrogen evolution on the cathode surface during the electrodeposition, which is also the cause for the appearance of the linear grooves, parallel to the vertical direction during the electrolysis.

A diffractogram acquired for the WFeCoNiCu alloy sample is shown in Figure 1c. It consists mainly of an intense, broad peak at 2θ = 43.3°, and an additional reflection, weaker and even broader, at 2θ ca. 76°. The first one is typically the only peak observed for the electrodeposited alloys of tungsten with iron-group metals containing more than 18 at% W [2,15,19,27]. The full width at half maximum of this peak is 5.5°, which translates to a crystallite size of 1.62 nm, which means that the alloy is nanocrystalline, close to amorphous. A second, weaker reflection was also observed for ternary electrodeposited tungsten alloys containing copper [25]; thus it can be stated that the novel WFeCoNiCu alloy has an analogous structure to the previously researched tungsten–iron-group-metal–copper alloy coatings.

In Table 4, the numerical results for the binary alloys are shown, i.e., experiment series A. The presented data include the deposition rates of the inducing metals, atomic content of tungsten and faradaic efficiency of the deposition processes. All the rates here and subsequently are expressed in SI units [mol $\cdot$ m^{-2} $\cdot$ s^{-1}] and multiplied by 10^5 for legibility.

Table 4. Comparison of deposition rates for binary alloys (series A).

Alloy	WFe	WCo	WNi
Me	Fe	Co	Ni
v Me	24.8 ± 3.0	9.6 ± 2.5	2.60 ± 0.69
v W	7.20 ± 0.98	3.1 ± 1.1	0.87 ± 0.21
at% W	22.5 ± 1.3	24.0 ± 2.0	25.3 ± 2.0
CE [%]	24.9 ± 3.9	7.3 ± 2.2	2.01 ± 0.50

The efficiency of the deposition depends strongly on which metal was codeposited with the tungsten, and so do the deposition rates for particular metals. More precisely, the deposition rate of cobalt for the WCo alloy is about 4 times higher than that of nickel in the WNi alloy, and the deposition rate of iron in the WFe is about 10 times higher than that of nickel. In general, the tungsten content in all of these binary alloys is ca. 24%, but it tends to be slightly higher when the deposition is slower. Thus, out of these three alloys, the WNi has the highest W content and the WFe has the lowest in the present study.

In experiment series B, ternary and quaternary alloys were deposited, utilizing baths containing equal concentrations of the inducing metal salts. Therefore, compared to series A, the concentration of a particular metal was two times lower for a ternary alloy and three times lower for the quaternary WFeCoNi alloy. Table 5 contains the relevant data for the alloys deposited in series B.

Table 5. Comparison of deposition rates for ternary and quaternary alloys (series B).

Alloy	v Fe	v Co	v Ni	v W	*at%* W	CE [%]
WFeCo	12.93 ± 0.37	7.47 ± 0.89	—	5.73 ± 0.28	21.96 ± 0.83	17.01 ± 0.72
WFeNi	12.0 ± 3.3	—	0.86 ± 0.11	4.44 ± 0.69	26.0 ± 2.0	12.4 ± 2.7
WCoNi	—	7.3 ± 1.0	0.60 ± 0.11	2.75 ± 0.50	25.6 ± 1.1	6.2 ± 1.0
WFeCoNi	6.8 ± 1.9	5.4 ± 1.0	0.447 ± 0.057	4.07 ± 0.76	24.43 ± 0.88	10.9 ± 2.3

The results from Tables 4 and 5 are presented below as bar charts in Figures 2 and 3. Figure 2 is a straight representation of the values of v for the Fe, Co and Ni deposition in particular binary, ternary and quaternary alloys. The values of the deposition rates, as given in Figure 2, are easily comparable visually within the categories of binary, ternary or quaternary alloys. However, as already mentioned, the particular inducing metal ion concentrations for the ternary metal deposition were divided by half, so the constant 27 mM concentration is shared by the two inducing metal ions in the bath, and so on with the bath for the quaternary alloy and three inducing metal ions. Thus, Figure 3 is presented as well, where the rates have been normalized to 1 mM metal ion concentrations, that is, divided by 27 for the binary alloys, by 13.5 for the ternary alloys and by 9 for the quaternary WFeCoNi alloy deposition. Presented in such a way, the differences between the values of the deposition rates can be compared for the particular inducing metals, especially in terms of divergencies from the linearity of v as a function of the metal ion concentration.

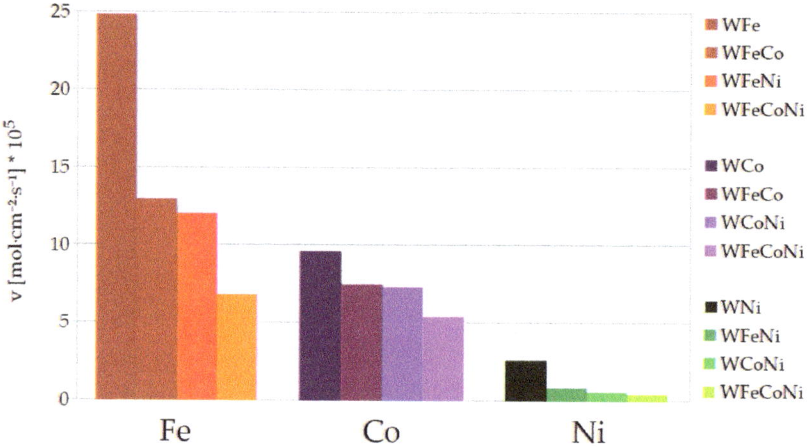

Figure 2. Comparison of Fe, Co and Ni deposition rates in binary, ternary and quaternary alloys.

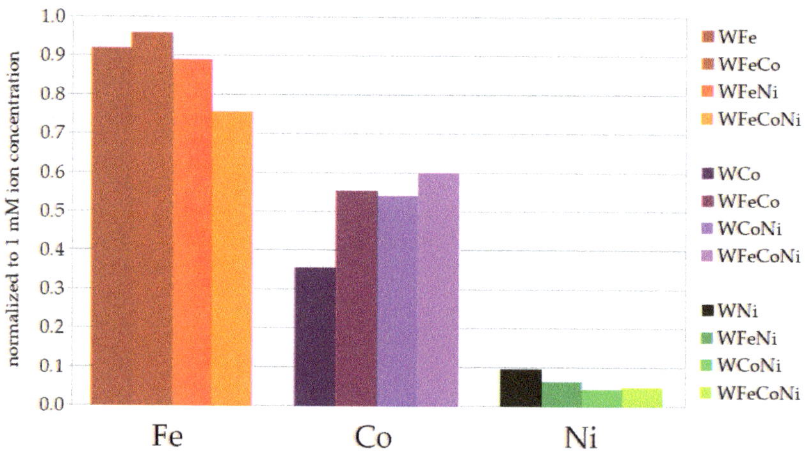

Figure 3. Metal deposition rates normalized to 1 mM metal ion concentration.

In general, during the deposition of the ternary alloys, iron still deposits at the highest rate, whilst nickel deposits the most slowly. In comparison to the WFe, the deposition rate of Fe is about 2 times lower for the WFeCo and WFeNi alloys, and approximately 3 times lower for the WFeCoNi, so it seems to be close to proportional to the iron concentration in the plating bath. However, the deposition rates for the remaining metals change in a different manner. As for Co, the deposition rates in the ternary alloys are noticeably higher than half the rate for the WCo; also, the deposition rate in the WFeCoNi is higher than a third of that for the binary alloy. The opposite can be observed for Ni: the rates of Ni deposition in the ternary and quaternary alloys are much lower than would be expected if they were proportional to the nickel concentration in the baths. For that reason, in the quaternary WFeCoNi alloy, the deposition rates of Fe and Co approach each other, so they are both more than 10 times higher than the rate of Ni deposition. The tungsten content in the ternary alloys containing nickel is close to that in the binary WNi, and for the WFeCo, it is very close to tungsten content in the WFe. In the WFeCoNi alloy, the tungsten content is somewhat in the middle, closest to *at*% W in the WCo alloy. In most cases, the faradaic efficiencies for the ternary and quaternary alloys are close to the mean values of the efficiencies for the binary alloys with corresponding metals.

Table 6 contains analogous data for experiment series C and D, i.e., the quaternary WFeCoNi and quinternary WFeCoNiCu alloys. As mentioned earlier, these baths contained adjusted concentrations of the metal ions, so to obtain alloys consisting of possibly equal *at*% of the inducing metals, also included in Table 6. Due to the v of nickel being more than 10 times lower than that of the other metals, the Ni(II) concentration had to be an order of magnitude higher than that of the other metals (see Table 3). Also, the proper concentration of Cu(II) was determined in a sequence of subsidiary tests between series C and D.

Table 6. Comparison of deposition rates for quaternary and quinternary alloys (series C and D).

Alloy	v Fe	v Co	v Ni	v Cu	v W	CE [%]
WFeCoNi	1.71 ± 0.22	1.94 ± 0.17	1.59 ± 0.23	—	2.25 ± 0.15	4.95 ± 0.33
WFeCoNiCu	2.30 ± 0.42	2.22 ± 0.37	1.99 ± 0.37	2.16 ± 0.34	2.74 ± 0.38	7.7 ± 2.8
Alloy	***at*% Fe**	***at*% Co**	***at*% Ni**	***at*% Cu**	***at*% W**	
WFeCoNi	22.8 ± 2.0	25.9 ± 1.9	21.3 ± 3.3	—	30.05 ± 0.42	
WFeCoNiCu	20.1 ± 1.9	19.4 ± 1.4	17.3 ± 1.7	19.3 ± 1.8	23.96 ± 0.76	

The values of the deposition rates contained in Table 6 are also depicted in Figure 4 in a visually comprehensive manner.

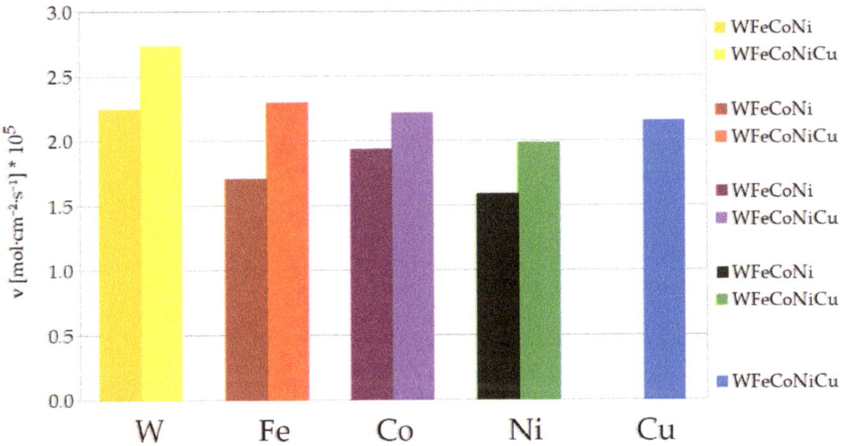

Figure 4. Influence of Cu on deposition rates in multinary alloys.

Apparently, the deposition rates of each inducing metal in both alloys are indeed close to equal, within standard deviation intervals. When comparing the WFeCoNi series B and C, it should be noticed that the tungsten content in the latter exceeds 30%, which is much higher than in any previous alloy in series A or B. On the other hand, the faradaic efficiency of the deposition process reaches only 5%, which is lower than for any other alloy except the binary WNi.

For the quinternary WFeCoNiCu alloy (series D), the tungsten content goes back to 24%, but all the rates of metal deposition are about 1.25 times higher than those without copper, including the v of tungsten. The deposition rate of Fe seems to be the most affected, and Co the least, although it is still clearly higher. Obviously, the copper itself deposits simultaneously with the other metals. Altogether, the introduction of copper makes the faradaic efficiency 1.5 times higher. Taking the standard deviation ranges into account, the obtained high-entropy WFeCoNiCu alloys have well-balanced atomic composition with 24% tungsten and ca. 19% of each inducing metal.

4. Discussion

A quinternary WFeCoNiCu alloy has been successfully synthesized utilizing the electrodeposition process. The morphology and crystalline structure of the alloy were shown to be comparable with previously researched binary and ternary tungsten alloys, especially the ternary alloys of tungsten, an iron-group metal and copper [25]. The novel quinternary alloy appears to be a homogeneous nanocrystalline material, close to amorphous. The XRD pattern (Figure 1c) proves that the WFeCoNiCu alloy contains no inclusions of other systems, such as metal oxides or solid solution phases of non-tungsten alloys. The issue of homogeneity of the electrodeposited tungsten alloys was already covered more exhaustively in our previous studies [27,28].

Out of the three most researched binary alloys of tungsten, i.e., with Fe, Co or Ni, the WFe is deposited with the highest faradaic efficiency and the WNi with the lowest, which was also observed in the cited papers. Overall, the achieved deposition rates are somewhat lower than the rates found in the literature (see Table 1), which may be caused by various factors such as discrepancies in the current densities, bath composition and temperature. However, for precise control of the composition of a quinternary alloy, knowledge of the ratios between the rates of electrodeposition in the same conditions is way more important than maximizing the rates themselves.

The current efficiencies are low, especially for nickel-containing alloys, compared to the much higher efficiencies of alloys such as WFe or WFeCo, which are an order of magnitude higher. Nevertheless, the low efficiency of the WNi alloy deposition does not stop it from being widely researched, due to its practical applications [6,10]. The particularly low metal deposition leads to abundant hydrogen evolution, which could cause issues with bath alcalization. For the current study, this issue was solved by utilizing fresh batches of the plating bath, but in other settings, the bath pH can be easily monitored and lowered in the case of the emergence of such a need.

In general, higher deposition rates of the inducing metals coincide with higher deposition rates of tungsten and a higher overall efficiency of the deposition process. According to the most recent mechanism proposal for tungsten codeposition [29], the $[(Me)(WO_4)(H)(Cit)]^{n-}$ type complex is a direct precursor of the electrodeposited tungsten alloy for *Me* being Ni or, most probably, other iron-group metals, namely, Co and Fe. Thus, the efficiency of the deposition process should be proportional to the concentration of the particular complex in the plating bath solution. Taking this into account indicates that the stability of these complexes present in the tungstate–citrate baths is increasing in the sequence *Me*: Ni < Co < Fe.

The copresence of two or three inducing metal ions in one plating bath at equal concentrations causes disproportionate changes in the deposition rates of the inducing metals. Most notably, the Ni deposition rate goes down in the presence of other inducing metals, which also could be explained by competition between the metal ions to form the adequate precursor complex.

Tuning the bath composition in a such manner that the metal deposition rates are equal inevitably requires that the metal ions in the solution be mostly nickel (about 88% of inducing metal ions in the bath). Hence, the faradaic efficiency is relatively low. However, even though the metal ions are mostly nickel, the efficiency of the deposition from such a bath is 2.5 times higher compared to the deposition of the binary WNi alloy. Also, in the quaternary alloy deposited from such a bath, the *at%* W is undoubtedly highest, exceeding 30%, which shows that this balanced setting is optimal in terms of the tungsten content. The addition of copper ions into the plating bath not only introduces copper into the alloy, but also catalyzes the deposition of every other metal, which leads to a remarkable boost in current efficiency. Questions about the exact role of copper in the mechanism of induced codeposition are yet to be answered in future studies.

Taking all the above into consideration should enable relatively precise control of Fe, Co, Ni and Cu content in the electrodeposited tungsten alloys of higher complexity, especially in the high-entropy quinternary WFeCoNiCu alloy. The deposition rates of the studied metals depend on their ion concentrations in a manner linear enough to modify the metal content predictably. Further iterations of adjusting the inducing metal ion concentrations may easily allow tuning of the composition of the high-entropy alloy, either for obtaining an excess of a chosen metal or to have the material composition well balanced. Due to the completely different behavior of tungsten species within the supposed codeposition mechanism, controlling the W content in the alloy would be a completely different task; thus was not included in the present study.

Author Contributions: Conceptualization, review and supervision, M.D.; experimental work and writing, T.R. All authors have read and agreed to the published version of the manuscript.

Funding: This research received no external funding.

Institutional Review Board Statement: Not applicable.

Informed Consent Statement: Not applicable.

Data Availability Statement: The original contributions presented in the study are included in the article; further inquiries can be directed to the corresponding author.

Acknowledgments: The authors would like to thank Pawel W. Majewski (University of Warsaw, Faculty of Chemistry) for acquisition of the XRD pattern.

Conflicts of Interest: The authors declare no conflicts of interest.

References

1. Fink, C.G.; Jones, F.L. The Electrodeposition of Tungsten from Aqueous Solutions. *Trans. Electrochem. Soc.* **1931**, *59*, 461. [CrossRef]
2. Tsyntsaru, N.; Cesiulis, H.; Donten, M.; Sort, J.; Pellicer, E.; Podlaha-Murphy, E.J. Modern Trends in Tungsten Alloys Electrodeposition with Iron Group Metals. *Surf. Eng. Appl. Electrochem.* **2012**, *48*, 491–520. [CrossRef]
3. Bacal, P.; Indyka, P.; Stojek, Z.; Donten, M. Unusual example of induced codeposition of tungsten. Galvanic formation of Cu-W alloy. *Electrochem. Commun.* **2015**, *54*, 28–31. [CrossRef]
4. Brenner, A. Electrodeposition of alloys. In *Principles and Practice*; Academic Press: New York, NY, USA; London, UK, 1963.
5. Brenner, A.; Burkhead, P.; Seegmiller, E. Electrodeposition of Tungsten Alloys Containing Iron, Nickel, and Cobalt. *J. Res. Natl. Bur. Stand.* **1947**, *39*, 351. [CrossRef]
6. Krasikov, A.V.; Merkulova, M.V.; Markov, M.A.; Bykova, A.D. Tungsten-rich Ni-W coatings, electrodeposited fromfrom concentrated electrolyte for complex geometry parts protection. *J. Phys. Conf. Ser.* **2021**, *1758*, 012019. [CrossRef]
7. Park, J.-H.; Hagio, T.; Kamimoto, Y.; Ichino, R. Electrodeposition of a Novel Ternary Fe-W-Zn Alloy: Tuning Corrosion Properties of Fe-W Based Alloys by Zn Addition. *J. Electrochem. Soc.* **2020**, *167*, 132508. [CrossRef]
8. Fan, C.; Piron, D.L.; Sleb, A.; Paradis, P. Study of Electrodeposited Nickel-Molybdenum, Nickel-Tungsten, Cobalt-Molybdenum, and Cobalt-Tungsten as Hydrogen Electrodes in Alkaline Water Electrolysis. *J. Electrochem. Soc.* **1994**, *141*, 382–387. [CrossRef]
9. Laszczyńska, A. Electrodeposited alloy electrodes in the Co-W-Mo system as highly efficient catalysts for hydrogen production from alkaline water electrolysis. *Mater. Chem. Phys.* **2024**, *314*, 128876. [CrossRef]
10. Vernickaite, E.; Tsyntsaru, N.; Sobczak, K.; Cesiulis, H. Electrodeposited tungsten-rich Ni-W, Co-W and Fe-W cathodes for efficient hydrogen evolution in alkaline medium. *Electrochim. Acta* **2019**, *318*, 597–606. [CrossRef]
11. Dehestani, M.; Sharafi, S. Gholam Reza Khayati. The effect of pulse current density on the microstructure, magnetic, mechanical, and corrosion properties of high-entropy alloy coating Fe–Co–Ni–Mo–W, achieved through electro co-deposition. *Intermetallics* **2022**, *147*, 107610. [CrossRef]
12. Bian, H.; Wang, R.; Zhang, K.; Zheng, H.; Wen, M.; Li, Z.; Li, Z.; Wang, G.; Xie, G.; Liu, X.; et al. Facile electrodeposition synthesis and super performance of nano-porous Ni-Fe-Cu-Co-W high entropy alloy electrocatalyst. *Surf. Coat. Technol.* **2023**, *459*, 129407. [CrossRef]

13. Shojaei, Z.; Khayati, G.R.; Darezereshki, E. Review of electrodeposition methods for the preparation of high-entropy alloys. *Int. J. Miner. Metall. Mater.* **2022**, *29*, 1683. [CrossRef]
14. Bobanova, Z.I.; Grabco, D.Z.; Danitse, Z.; Mirgorodskaya, Y.; Dikusar, A.I. Electrodeposition and Properties of an Iron–Tungsten Alloy. *Surf. Eng. Appl. Electrochem.* **2007**, *43*, 239–247. [CrossRef]
15. Tsyntsaru, N.; Bobanova, J.; Ye, X.; Cesiulis, H.; Dikusar, A.; Prosycevas, I.; Celis, J.-P. Iron–tungsten alloys electrodeposited under direct current from citrate–ammonia plating baths. *Surf. Coat. Technol.* **2009**, *203*, 3136–3141. [CrossRef]
16. Lietzke, M.H.; Holt, M.L. Codeposition of Iron and Tungsten from an Aqueous Ammoniacal Citrate Bath. *J. Electrochem. Soc.* **1948**, *94*, 252–261. [CrossRef]
17. Donten, M.; Stojek, Z. Pulse electroplating of rich-in-tungsten thin layers of amorphous Co-W alloys. *J. Appl. Electrochem.* **1996**, *26*, 665–672. [CrossRef]
18. Ibrahim, M.A.M.; El Rehim, S.S.A.; Moussa, S.O. Electrodeposition of noncrystalline cobalt–tungsten alloys from citrate electrolytes. *J. Appl. Electrochem.* **2003**, *33*, 627–633. [CrossRef]
19. Tsyntsaru, N.; Cesiulis, H.; Budreika, A.; Ye, X.; Juskenas, R.; Celis, J.-P. The effect of electrodeposition conditions and post-annealing on nanostructure of Co–W coatings. *Surf. Coat. Technol.* **2012**, *206*, 4262–4269. [CrossRef]
20. Sun, S.; Bairachna, T.; Podlaha, E.J. Induced Codeposition Behavior of Electrodeposited NiMoW Alloys. *J. Electrochem. Soc.* **2013**, *160*, D434–D440. [CrossRef]
21. Cesiulis, H.; Baltutiene, A.; Donten, M.; Stojek, Z. Increase in rate of electrodeposition and in Ni(II) concentration in the bath as a way to control grain of amorphous/nanocrystalline Ni-W alloys. *J. Solid State Electrochem.* **2002**, *6*, 237–244. [CrossRef]
22. Sridhar, T.M.; Eliaz, N.; Gileadi, E. Electroplating of Ni_4W. *Electrochem. Solid-State Lett.* **2005**, *8*, C58–C61. [CrossRef]
23. Younes, O.; Gileadi, E. Electroplating of High Tungsten Content Ni/W Alloys. *Electrochem. Solid-State Lett.* **2000**, *3*, 543–545. [CrossRef]
24. Vernickaite, E.; Tsyntsaru, N.; Cesiulis, H. Electrochemical co-deposition of tungsten with cobalt and copper: Peculiarities of binary and ternary alloys coatings formation. *Surf. Coat. Technol.* **2016**, *307*, 1341–1349. [CrossRef]
25. Bacal, P.; Donten, M.; Stojek, Z. Electrodeposition of high-tungsten W-Ni-Cu alloys. Impact of copper on deposition process and coating structure. *Electrochim. Acta* **2017**, *241*, 449–458. [CrossRef]
26. Yeh, J.W.; Chen, S.-K.; Lin, S.-J.; Gan, J.-Y.; Chin, T.-S.; Shun, T.-T.; Tsau, C.-H.; Chang, S.-Y. Nanostructured highentropy alloys with multiple principal elements: Novel alloy design concepts and outcomes. *Adv. Eng. Mater.* **2004**, *6*, 299–303. [CrossRef]
27. Donten, M. Bulk and surface composition, amorphous structure, and thermocrystallization of electrodeposited alloys of tungsten with iron, nickel, and cobalt. *J. Solid. State Electrochem.* **1999**, *3*, 87–96. [CrossRef]
28. Pisarek, M.; Janik-Czachor, M.; Donten, M. Local characterization of electrodeposited Ni–W amorphous alloys by Auger microanalysis. *Surf. Coat. Technol.* **2008**, *202*, 1980–1984. [CrossRef]
29. Eliaz, N.; Gileadi, E. The Mechanism of Induced Codeposition of Ni-W Alloys. *ECS Trans.* **2007**, *2*, 337–349. [CrossRef]

materials

MDPI

Article

Electrochemical Formation and Characterization of Functional Ag-Re Coatings

Oksana Bersirova [1,2,*], Valeriy Kublanovsky [2], Svetlana Kochetova [2] and Olena Bondar [2]

[1] Faculty of Chemistry and Geosciences, Vilnius University, 03225 Vilnius, Lithuania
[2] V.I. Vernadsky Institute of General and Inorganic Chemistry NAS of Ukraine, 03142 Kyiv, Ukraine
* Correspondence: bersibol@gmail.com

Abstract: Silver-white, matte, smooth, and durable deposits of silver-rhenium, with thicknesses ranging from 2.0 to 13.7 μm and containing 0.15 to 13.5 wt.% Re, were obtained with a current efficiency of 66–98% from a developed dicyanoargentate–perrhenate bath based on a borate–phosphate–carbonate silver-plating electrolyte. This study was focused on the influence of bath composition, the $[Ag(I)]:[ReO_4^-]$ ratio, surfactant additives, applied current density, temperature, and stirring, on the alloys' composition, structure, morphology, microhardness, adhesion, and porosity. A voltammetric analysis was conducted, considering the influence of ethanolamines on electrode processes. In baths with triethanolamine (TEA), coatings similar to a silver matrix with rhenium doped in mass fractions are likely achievable. Monoethanolamine (MEA) is recommended due to its process-activating properties. All coatings were nanocrystalline (τ = 28.5–35 nm). For deposits containing less than 10 wt.% Re, characteristic silver XRD peaks were observed, while for other deposits, additional peaks attributed probably to Re(VII) and Re(VI) oxides. A linear relationship $H_v - \tau^{-1/2}$, typical for Hall–Petch plots, was obtained, confirming that grain boundaries play a crucial role in mechanical properties of coatings. The conditions for stable electrochemical synthesis of promising functional Ag-Re coatings of predetermined composition (0.7–1.5 wt.% Re) were proposed for practical use in power electronics and energy sectors for manufacturing electrical contacts operating across a wide temperature range. This was realized by deposition from an Ag-rich bath in the area of mixed electrochemical kinetics, at potential values corresponding to the region of half the limiting current: j = 2.5–6 mA cm^{-2}, t = 19–33 °C.

Keywords: silver; rhenium; electrodeposition; monoethanolamine; triethanolamine; functional coating; contact material

1. Introduction

Silver-based composites and electrolytic alloys are prominent among thin-layer materials [1–3]. Electrolytic alloys of silver with rhenium [4,5], tungsten [6,7], and molybdenum are of interest both practically (e.g., as high-powered electrical contacts) and scientifically (because co-deposition of these metals allows the incorporation of refractory metals into alloys, which are difficult or impossible to obtain in pure form from aqueous solutions).

The interest in electrolytic alloys of silver with rhenium arises because of the combination of the unique electrical and catalytic properties of silver with the mechanical properties of rhenium [8,9]. The high melting point and exceptional resistance to recrystallization of rhenium have prompted its use in high-temperature environments like gas turbine superalloys, which represent its largest market [10,11]. Depositing Re alloys with metals like Ni,

 https://doi.org/10.3390/ma18091893

Co, Fe, In, Cr, Pd, Rh, Ag, or Au results in a significantly increased cathode efficiency (up to 80% or more) and a higher discharge potential for H^+ ions [12–14].

According to [9], Kemp first proposed to use rhenium for electrical contacts due to its wear resistance in slip rings and switching contacts on motor commutators. Rhenium offers an alternative to traditional hard contact metals like Pt and Rh. Despite its industrial potential, rhenium has not gained widespread popularity in connector engineering. Its limited application in electric contacts is because of two factors: the high cost of rhenium and its tendency to form Re_2O_7 oxide, with a melting point of 569 K and a boiling point of 635 K [15]. This can lead to mass loss in contact materials due to erosion from the electric arc, though it also imparts high resistance to sticking [16]. However, it is indicated that is has sporadic but predominantly successful uses in specific applications, such as external engines of marine vessels, due to its resistance to seawater [15,17]. Recent research has focused on rhenium incorporating into contact materials to develop new generations of nanocrystalline materials like Ag-Re using high-energy grinding and careful plastic consolidation [16,18–20]. Findings related to the manufacturing technology and initial research on the physical, mechanical, and electrical properties of such materials was presented in [15].

For applications in power electronics and energy, specifically for manufacturing electrical contacts functioning across a broad temperature range (from −196 °C to +250 °C), functional silver-based alloys doped with rhenium, chromium, zirconium, and rare earth elements (cerium, lanthanum, yttrium) for applying wear-resistant and corrosion-resistant coatings using cold gas-dynamic spraying were developed [21]. Key to enhancing manufacturability and achieving the desired performance of the nanostructured silver-based alloy are, firstly, a judicious selection of the components of alloy, and, secondly, the creation of a nanocrystalline coating structure with superior corrosion and thermal properties. Moreover, the alloy must be lightly alloyed; to maintain the low electrical resistance of silver (the lowest among all metals 0.304×10^{-8} Ohm·m), alloying elements should not exceed 12 wt.%. The authors determined that cold resistance at −196 °C is achieved by adding 0.7 to 1.5 wt.% rhenium to the alloy. Increasing rhenium content above 1.5 wt.% reduced temperature stability (from +250 to +160 °C), while decreasing it below 0.7 wt.% diminished cold resistance at lower temperatures (down to −196 °C), leading to loss of mechanical strength in the coating (delamination, cracking). The electrochemical deposition method, with its ability to control factors like solution composition, current density, hydrogen ion concentration, and temperature, offers flexible and efficient control over the deposition process. However, despite rhenium-containing electrolytic alloys being highly promising, only a few studies on the electrodeposition of silver-rhenium coatings were carried out [22–25]. It was shown that multicomponent electrolytic coatings possess improved solderability, strength, and electrocatalytic activity, often displaying synergistic effects due to the properties of the constituent components. The developed coatings can be used in the electrical industry to produce contact groups for various purposes, including solid-lubricating wear-resistant coatings for friction pairs in vacuum, extreme temperature, and pressure conditions, and coatings with enhanced surface conductivity and reduced contact resistance.

Silver coatings, known for their low cost, high electrical conductivity, and corrosion resistance, are applied to both low-voltage and high-voltage contacts. High-voltage contact coatings must withstand electrical erosion and the elevated temperatures generated during contact opening and closing. Applying a finishing layer of silver-rhenium electrolytic coating [26] can improve the precision, reliability, wear, and corrosion resistance of contacting surfaces.

In the context of wear-resistant coatings for friction pairs, the electrochemical method for applying metal solid film lubricants is preferable due to several advantages over traditional processes using metal matrices. These include improved accessibility in physical areas, better control ensuring homogeneity and uniformity of coating, and enhanced adhesion reliability under various mechanical factors. Turns et al. [27] investigated a Ag-Re alloy plating process from a cyanide bath and electrodeposited Ag-3% Re alloy. The obtained 97% Ag-3% Re coatings may serve as solid lubricants withstanding high temperatures and loads. In addition, the highly porous plating serves dual functions: holding a high amount of MoS_2 as grease and, once compacted under load, continuing to provide effective lubrication. Practical applications include aircraft wing hinge pins and the elimination of fretting corrosion and long-life contacts for rotary switches. Other baths for antifriction coatings consist of $K_4(Fe(CN)_6)$, $NaReO_4$, K_2CO_3, and monoethanolamine; a dissipative power of 53–57%, CE of 43–57%, and a rhenium content of 2–3% were determined [25].

To increase the rhenium content in the coatings [24], a self-regulating electrolyte was proposed having the following composition (in g L^{-1}): $KAgCN_2$—20–30; $NaReO_4$—1–3; NH_4NO_3—15–20; and triethanolamine—2–3, pH 9.5–10.0, at room temperature. The cathode current density was set at 0.5–10 mA cm^{-2} without stirring and 10–40 mA cm^{-2} with stirring. The dissipative power of the electrolyte was 60–63%, and the current efficiency (CE) was 65–97%. A maximum of 20% rhenium in the coating was achieved from the stirred solution and at a high cathode current density of 40 mA cm^{-2}. Under these conditions, the CE reached 85%, enabling the growth thickness of a coating up to 55 μm. So, developing new bath compositions for electrochemical co-deposition of silver-rhenium based on complex dicyanoargentate solutions is considered as most preferential.

As indicated in [6,7], the electrodeposition of thin-layer composite silver coatings via co-depositing of oxides of refractory metals is highly promising. The electrochemical reduction of ReO_4^- was studied at pH ranging from 0 to 7 [28]. The reduction of ReO_4^- to metallic Re involves multiple stages of sequential reduction of the intermediate products. It was found that low pH values and perrhenate concentrations favor the formation of the intermediate Re^{3+}, which aids in forming metallic Re. Thus, the pathway of reduction is as follows: $ReO_4^- \rightarrow ReO_3 \rightarrow Re^{3+} \rightarrow Re$.

Either at higher pH or ReO_4^- concentration, it tends to form on the cathode the intermediate ReO_2, which promotes the hydrogen evolution reaction and significantly impedes the reduction of low-valent intermediates. The corresponding reduction process is $ReO_4^- \rightarrow ReO_3 \rightarrow ReO_2$, Re_2O_3. X-ray studies have indicated that NH_4ReO_4 plays a crucial role in the reduction process, but high concentration of NH_4ReO_4 is detrimental to obtain Re in metallic form because rhenium oxides ReO_3 and ReO_2 are formed [28].

It was determined [26] that co-deposited rhenium with silver in the film exits in form of ReO_x. Deposition was conducted from a dicyanoargentate solution with the composition, g L^{-1}: $KAgCN_2$—20; NH_4ReO_4—1–5; NH_4NO_3—15; triethanolamine—5. The authors [26] achieved a maximum rhenium content in deposits of 0.7 wt.% using non-stationary reverse current, stirring, and ultrasound. These coatings exhibit high microhardness, wear, and corrosion resistance.

Controlled electrochemical formation of functional coatings from metals and alloys, including silver and its alloys with rhenium, having specific structures, can be achieved based on the understanding of the relationship between the electrochemical kinetics of electrode processes and electrodeposition [29–32]. Recently, from these perspectives, both the process of electrochemical deposition of silver [33–37] and the process of co-deposition of rhenium in alloys from complex baths [38–40] have been studied. The main principles of forming coatings of binary and ternary rhenium-containing alloys for various functional purposes have been established. However, there are practically no studies in the literature dedicated to the process of electrochemical synthesis of functional Ag-Re coatings.

A non-toxic dicyanoargentate–borate–phosphate–carbonate (BPC) silver plating electrolyte has been developed [29]. BPC electrolyte does not contain "free" cyanide-ions, and its stability is ensured by a borate–phosphate–carbonate buffer. The mechanism of silver recovery from BPC electrolytes has been studied [29]. The electrolyte allows for high-quality coating production with strong adherence to the substrate even after prolonged bath operation. It has been effectively incorporated into the technology for functional silver plating of micro- and nanoelectronics components. Therefore, for the electrochemical deposition of silver coatings with rhenium, we consider the previously developed complex buffer BPC electrolyte as the most promising for developing co-deposition bath compositions. The key ingredients of the new bath for Ag-Re alloy electrodeposition could be the functional silver plating BPC electrolyte and ammonium perrhenate as the rhenium source because it is commonly used for electrodeposition of rhenium and its alloy [41,42].

This study presents a novel and original approach to the formation of silver-rhenium coatings, based on the determination of a stable electrodeposition process range within the mixed electrochemical kinetics regime. This approach contributes to the advancement of the field of electrolytic deposition of metals and alloys with a controlled composition.

This work aimed to investigate the process of electrodeposition of silver-rhenium coatings from a dicyanoargentate–perrhenate bath, developed based on a borate–phosphate–carbonate buffer electrolyte for functional silver plating, and to characterize the structure and properties of obtained coatings.

2. Experimental Section

2.1. Bath Composition and Procedure

The study of electrodeposition of Ag-Re coatings was conducted using dicyanoagentate-perrhenate plating solutions. These solutions were prepared by adding ammonium perrhenate to borate–phosphate–carbonate (BPC) silver plating electrolyte [29], which contains $KAg(CN)_2$, and buffers such as KH_2PO_4, K_2HPO_4, H_3BO_3, K_2CO_3, and KOH (pH 7.0 ± 0.1). The choice of ammonium perrhenate as the rhenium ion source is due to its higher solubility in the operational temperature range compared to $KReO_4$ (3.25×10^{-2} vs. 5.08×10^{-3} g mol/100 g H_2O at room temperature) [43]. The specific compositions of these electrolytes are detailed in Table 1.

Electrodeposition occurred from baths with a constant silver concentration ($KAg(CN)_2$— 20 g L^{-1}) but varying concentrations of perrhenate ions (bath #1, 2, 5, 6, 7). The $[Ag^+]:[ReO_4^-]$ ratio was 10:1 (bath #2 and 4) and 2:1 (bath #5, 6, 7). Additionally, experimental studies were conducted in solutions where the silver concentration was reduced and $[Ag^+]:[ReO_4^-]$ ratio was 1:1.5 (bath #8); 1:5 (bath #3) and 1:10 (bath #9 and 10); as shown in Table 1.

The electrolysis conditions were also diversified, including a range of current densities (3, 5, 7.5, 10, 15 mA cm^{-2}) and temperatures (18, 20, 30, 40, 50, 60 °C), both without and with stirring using a magnetic stirrer.

All electrolytes were prepared using double-distilled water and analytical grade reagents. The process began with dissolving a sample of potassium dicyanoargentate ($KAg(CN)_2$) in double-distilled water. Then, a borate–phosphate–carbonate (BPC) buffer was added to this solution, enhancing the ionic strength and stabilizing ion states in both the electrolyte volume and cathode layers during electrolysis. The pH of the BPC electrolyte was maintained at 7.0 ± 0.1 and adjusted using boric acid. The required amount of ammonium perrhenate was added to the dicyanoargentate solution containing the BPC buffer, with careful pH monitoring. In experiments studying the effect of surfactants, specific concentrations (0.11 M $KAg(CN)_2$ and 0.055 M $KReO_4$) were used, with incremental additions of monoethanolamine (MEA) and triethanolamine (TEA) ranging from 0.1 to 0.5 mL (Table 1). For baths with ethanolamine additives, ammonium perrhenate was replaced with potassium perrhenate. $KReO_4$ was chosen to avoid additional cations and to use silver and rhenium salts with the same cation. The literature suggests that ammonium rhenium salt can slightly increase deposit porosity compared to potassium or sodium salts [44].

Table 1. The composition of dicyanoargentate–perrhenate baths used for the deposition of Ag-Re coatings.

Bath, (mol L^{-1})	**#1**	**#2**	**#3**	**#4**	**#5**	**#6**	**#7**	**#8**	**#9**	**#10**
NH_4ReO_4	-	0.011	0.022	0.0022	-	-	-	-	-	-
$KReO_4$	-	-	-	-	0.055	0.055	0.055	0.065	0.09	0.11
$KAg(CN)_2$	0.11	0.11	0.0044	0.022	0.11	0.11	0.11	0.044	0.009	0.011
KH_2PO_4	0.05	0.05	0.002	0.01	0.05	0.05	0.05	0.025	0.014	0.005
K_2HPO_4	0.3	0.3	0.01	0.06	0.3	0.3	0.3	0.15	0.085	0.03
K_2CO_3	0.025	0.025	0.001	0.005	0.025	0.025	0.025	0.001	0.0055	0.0025
H_3BO_3	0.2	0.2	0.007	0.04	0.2	0.2	0.2	0.1	0.05	0.02
C_2H_7NO (MEA)	-	-	-	-	-	0.033; 0.066; 0.099; 0.132; 0.160	-	-	-	-
$C_6H_{15}NO_3$ (TEA)	-	-	-	-	-	-	0.015; 0.030; 0.045; 0.060	-	-	-
$[Ag^+]:[ReO_4^-]$ ratio	-	10:1	1:5	10:1	2:1	2:1	2:1	1:1.5	1:10	1:10
pH (at 20 °C)	7.00	7.65	7.80	7.60	7.21	8.78 (0.16 M MEA)	8.15 (0.06 M TEA)	8.18	8.42	8.49

2.2. Electrochemical Measurements

Experiments were conducted in a 50 mL standard electrochemical glass cell within a temperature range of 20 to 60 °C, using a UTU-4 thermostat (ZEAMiL HORIZONT, Cracow, Poland) with an accuracy of ±0.1 °C. An electric magnetic stirrer at 120 rpm was used for mixing the electrolyte.

For voltammetric studies, a platinum wire, 1 mm in diameter and sealed in glass, served as the working electrode. Both platinum and electrochemical silver were used as working electrode materials. To prepare the silver electrode, the platinum wire's end was coated with a 1.5-micron thick electrolytic silver layer at a polarization of 0.150 V. All potential measurements were relative to a Ag/AgCl reference electrode in a saturated potassium chloride solution, with all potentials reported relative to this silver chloride reference electrode.

Voltammograms were recorded using a PC Controlled Potentiostat, type EF453 "Electroflex" (ELECTROFLEX, Budapest, Hungary), at temperatures of 20, 25, 30, 35, and 40 °C. Potentiodynamic polarization dependencies were captured over a broad range of scan rates from 1 to 100 mV s^{-1}. Each curve was recorded multiple times, ensuring good reproducibility of the results. When comparing curves from parallel experiments conducted under identical conditions, a complete alignment was observed. Prior to experimentation, the electrode was immersed in the solution for 30 min to establish an equilibrium potential. The initial sections of the polarization curves, obtained at a scan rate of 1 mV s^{-1}, were converted into semilogarithmic coordinates for determining the kinetic parameters of the electrode process. The activation energy was calculated from the slope of the inverse temperature dependence of the exchange currents.

The electrodeposition was performed in galvanostatic mode; the electrolysis times were set to either 30 or 60 min. A copper plate with a 2 cm^2 area served as the working electrode for coating electrodeposition. Silver wire or glassy carbon were used as anodes.

Current efficiency was calculated based on the mass of the obtained deposit, the passed charge, and the chemical composition of the deposit determined by the EDX method. The calculation utilized the following equation:

$$CE = \frac{\Delta m}{It} \Sigma \frac{wt\%_i \cdot n_i \cdot F}{M_i} \cdot 100\%, \tag{1}$$

where Δm is the measured weight of the deposit (g), t is the deposition time (s), I is the total current passed (A), $wt\%_i$ is the weight fraction of the element (Ag or Re) in the deposit, n_i is the number of electrons transferred per atom each metal (n_i = 1 and 7 for Ag and Re, respectively), M_i is the atomic mass of that element (M_i = 107.6 and 186.2 g mol^{-1} for Ag and Re, respectively), and F is Faraday's constant (96,485 C mol^{-1}).

The coating thickness δ was calculated based on the weight gain of the sample after application, the chemical composition of the deposit (wt.%), and the density of each metal according to the formula:

$$\delta = \frac{\Delta m}{\rho * S} = \frac{\Delta m}{\Sigma(\rho_i * wt\%_i) * S}, \tag{2}$$

where ρ_i is the density of the element (Ag and Re) (kg dm^{-2}); $wt\%_i$—mass fraction of silver and rhenium, respectively; Δm—is the measured weight of the alloy mass (g); S—cathode area (cm^2).

2.3. Surface Morphology, Chemical Composition and Crystallographic Structure Characterization

The morphology and chemical composition of the Ag-Re deposits were analyzed using SEM and EDX methods. This involved a scanning electron microscope (JSM 6700F, JEOL Ltd., Tokyo, Japan) equipped with an energy dispersive spectrometer (JED-2300), operating at an accelerating voltage of 20 kV, a beam current of 0.75 nA, and a beam size of 1 μm. For EDX analyses, the calculation time was set to 1 min.

The structure of the electrodeposited silver and Ag-Re alloys was investigated by X-ray diffraction (XRD) methods (Rigaku MiniFlex II, Tokyo, Japan). XRD patterns were produced with Cu Kα radiation (1.5406 Å) in 2θ scanning mode from 20 to 90° with a step of 0.01°. For most precise θ values for peaks and line broadening determination, the routines of the peak fitting using Gaussian- or Lorenzian-type functions were performed.

The crystallite sizes were calculated using the Scherrer formula:

$$\tau = K\lambda / \beta\tau \cos\theta \tag{3}$$

where $\beta\tau$ is the peak broadening (in rad); K is the Scherrer factor—the empirical constant 0.9; τ is the crystalline size (in Å).

2.4. Microhardness, Adhesion and Porosity Measurements

The Vickers microhardness of the samples was measured with a microhardness tester PMT-3 (Vostok-7, Moscow, Russia) under a load of 10 g and a load exposure of 10 s. Arithmetic mean values of H_V show the trend between the samples.

The adhesion of the coatings was tested using the "scotch tape test" according to ISO 2409 standard [45] methodology.

The porosity of the coatings was assessed by applying filter paper moistened with a solution containing (g L^{-1}): $K_3[Fe(CN)_6]$—3, NaCl—10.

Each test was carried out multiple times, ensuring good reproducibility of the results.

3. Results and Discussion

3.1. Voltammetric Studies

The study of deposition kinetics makes it possible to establish the patterns of chemical and electrochemical reactions depending on time, current density, concentration of additives, and other factors affecting the rate of both the overall process and intermediate stages [6,46]. Such studies are necessary to consider the patterns of changes in the rate of electrode processes, which are influenced by the passage of electric current through the "electrode-ionic system" boundary, charging of the electrical double layer, and the distribution of ions in the volume of the electrolyte.

3.1.1. Results of Studies of Stationary Potential

The measured stationary potentials for silver electrode in different baths for the deposition of Ag-Re coatings are shown in Table 2.

Table 2. The stationary potential of a silver electrode in different compositions of electrolytes for the deposition of Ag-Re coatings (temperature 20 °C).

Bath	**#1**	**#2**	**#3**	**#4**	**#5**	**#6**	**#7**
E_{Ag}, V (Ag/AgCl)	+0.200	−0.022	−0.122	+0.024	+0.007	+0.170	+0.110
E_{Pt}, V (Ag/AgCl)	−0.0040	−0.0145	−0.030	+0.004	+0.004	+0.150	+0.0038

The introduction of ammonium perrhenate into the BPC silvering electrolyte based on potassium dicyanoargentate affects the stationary potential of the silver electrode, changing it to the cathodic side by 222 mV to −0.022V (vs. Ag/AgCl) in the case of adding 0.011 M NH_4ReO_4, (bath #2) and by 193 mV to +0.007 (vs. Ag/AgCl) when adding five times more perrhenate (bath #5). The influence of the concentration of dicyanoargentate ions can also be traced. Thus, decreasing the concentration from 0.11 M (bath #2) to 0.022 M $KAg(CN)_2$ (bath #4) shifts the potential to the cathodic side by 176 mV, and to 0.004 M $KAg(CN)_2$—by 322 mV to −0.122 V (vs. Ag/AgCl) (bath #3).

The addition of 0.06 M triethanolamine (bath #7) has little effect on the stationary electrode potential, which remains the same as in a solution of the same concentration of silver and rhenium ions, but without surfactant additives (bath #5). Unlike triethanolamine, monoethanolamine depolarizes the electrode from +0.007 V (in bath #5) to +0.170 V, although the concentrations of both silver and rhenium in solution remain unchanged. In the dicyanoargentate–perrhenate solution, the stationary potential of the silver electrode is the same with the stationary potential in the BPC silver plating electrolyte of the same concentration of silver ions.

3.1.2. Results of Studies of Potentiodynamic Polarization Curves

Linear sweep voltammetry studies were carried out in the developed dicyanoargentate–perrhenate plating solutions showed in Table 1. Figure 1 presents polarization curves for pure silver (bath #1) and those obtained from baths with different $[Ag^+]:[ReO_4^-]$ component ratios: 10:1 (bath #2, #4); 2:1 (bath #5); and 1:5 (bath #3).

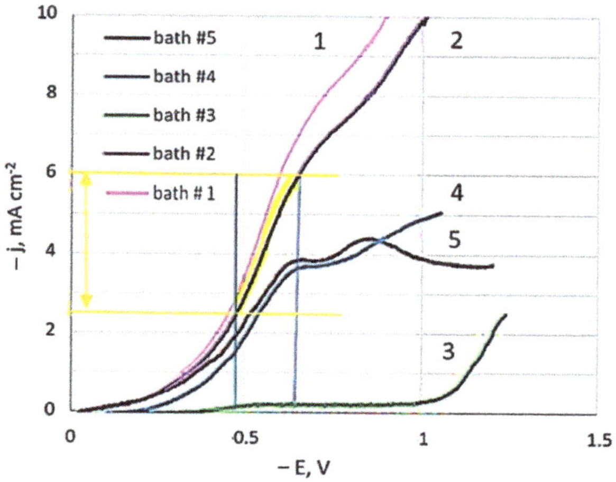

Figure 1. Cathode polarization curves depending on the concentration of baths components. The temperature is 20 °C. Potential sweep rate of 1 mV s^{-1}.

In the study of polarization curves at a potential sweep rate of 1 mV s^{-1} in a dicyanoargentate–perrhenate electrolyte, a limiting current plateau of 6.8 mA cm^{-2} (bath #2) was observed. The polarization curve in the Ag-rich bath, with rhenium concentration ten times less than silver (bath #2), closely resembled that of the BPC silver bath without perrhenate (bath #1), albeit slightly lower. Limiting current potentials were consistent across all curves. However, adding rhenium, even at a lower concentration compared to silver, slightly reduced the limiting current values. Comparing baths with equal silver concentration but five times higher rhenium concentration (curve 5 vs. curve 2) revealed even lower limiting current values. With the same silver to rhenium ratio (10:1) but diluted fivefold (curve 4 vs. curve 2), a significant decrease in limiting current values was noted. Bath #3, having 25 times less silver than the initial bath #2, showed the minimal limiting currents among all studied solutions. Curves 4 and 5 distinctly displayed two sections of limiting currents, while the second limiting current was not as pronounced in the other curves.

Voltammograms in semilogarithmic coordinates were used to calculate kinetic characteristics (apparent transfer coefficients, exchange current densities) at different temperatures. Adding NH_4ReO_4 salt to the BPC electrolyte reduced the exchange current value, leading to lower effective current density. The apparent transfer coefficient in bath #2 was 0.170 ± 0.005, indicating minimal influence of the electrode electric field (double layer energy) on activation energy [47]. Formal exchange currents increased nearly tenfold with temperature rising from 20 to 40 °C. The activation energy, calculated from the exchange currents against reverse temperature, was 65 kJ mol^{-1} in bath #2, closely matching the value of 65.6 kJ mol^{-1} determined for electroreduction of silver from a BPC electrolyte without ammonium perrhenate (bath #1).

Typical potentiodynamic voltammograms obtained at different potential sweep rates is depicted in Figure 2. A current peak is present on the non-stationary curves, and its value increases with the sweep speed. The potential of this current peak remains essentially constant at −0.65 V, relative to the Ag/AgCl reference electrode.

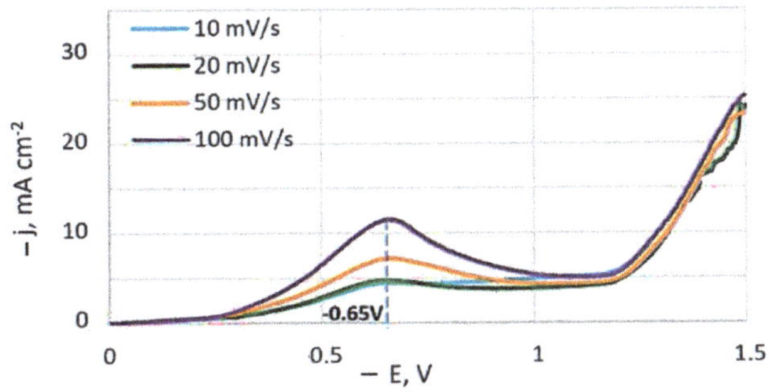

Figure 2. Potentiodynamic cathodic curves obtained at a temperature of 20 °C from bath #4. Potentials are given in relation to the Ag/AgCl reference electrode.

At slower charge transfer stages, the concentration of depolarizer on the electrode surface is governed by the rate constant. When a potential sweep is applied, the sweep rate also influences this concentration. Figure 3a shows that the relationship between the height of the peak cathode current and the square root of the potential sweep rate (at 20 °C) is generally linear. However, deviations from linearity at low sweep speeds are observed, which can be attributed to the effects of natural convection.

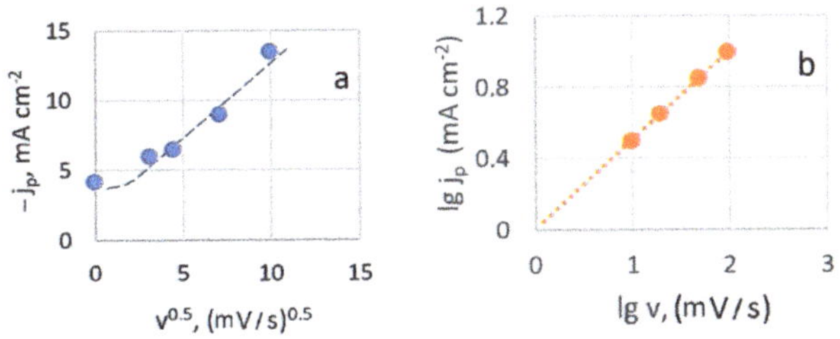

Figure 3. Dependence of the peak current on the scan speed in criterion coordinates: peak current is the square root of the scan speed (**a**) and logarithmic dependence (**b**).

This suggests that the diffusion process is irreversible and indicates that the electroactive complex is likely restored from an adsorbed state under these conditions. However, when the current peak dependence on sweep rate is plotted in logarithmic coordinates (Figure 3b), the calculated slope value $(\lg j_p/\lg v)_{t,C}$ is 0.5. This rate dependence typically indicates that processes are limited by charge transfer rate, diffusion, and kinetics of a chemical reaction (mixed kinetics). It excludes catalytic processes and those involving adsorption complexities.

3.1.3. The Effect of Surfactants on the Electrode Process

The impact of organic nitrate surfactant additives, specifically monoethanolamine (MEA) and triethanolamine (TEA), was examined. Although the use of these additives in baths for forming functional Ag-W (or tungsten oxides) and Ag-Re (or rhenium oxides) coatings are referenced [6,7,26], their influence on the electrode process has not been detailed.

In developing the optimal composition of BPC silver electrolytes, various additives like ethylenediamine, glycine, methylamine, ethylamine, dimethylamine, dicyclohexylamine, EDTA, KNO_3, NH_4NO_3, $K_4P_2O_7$ among others, were studied. It was found that ethylenediamine and nitrates are more promising for facilitating anodic processes [29]. However, the effects of MEA and TEA had not been explored, prompting this investigation.

A noticeable difference in the influence of these surfactant additives on the electrode process was observed. Figure 4a,b displays the polarization curves for silver-rhenium co-deposition in solutions with surfactant additives. The numerals adjacent to the curves indicate increasing amounts of added surfactants. It was observed that an increase in MEA concentration led to higher maximum discharge currents of electroactive ions (Figure 4b), while TEA addition resulted in decreased currents (Figure 4a).

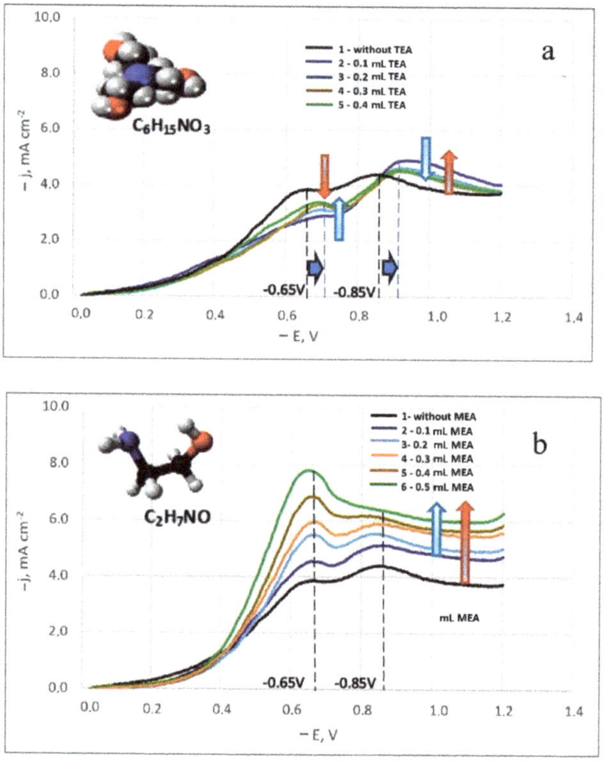

Figure 4. Potentiodynamic cathodic curves in dicyanoargentate–perrhenate silver electrolytes with TEA (**a**) and MEA (**b**) additives. The temperature is 50 °C. Potential sweep rate is 100 mV/s.

Although the equilibrium potential in these electrolytes remained unchanged, a shift in the current peak potentials was noted in the solutions containing TEA addition. For both the first and second peaks, the potential shifted by 50 mV towards more negative potentials (Figure 4a). Furthermore, the first limiting current values for all studied TEA concentrations were lower than those in a similar composition solution without additives (bath #5). With increasing TEA concentration, the first limiting currents increased slightly. For the second peak current, the values were higher than in bath #5, but they decreased slightly with increasing TEA concentration. This inhibition of the electrode process by TEA can be attributed to its potential adsorption on the surface of the coated substrate, obstructing the surface and hindering the reduction of electroactive dicyanoargentate ions due to possible competitive adsorption.

With the addition of MEA, there was a significant increase in the first peak currents, proportional to the MEA concentration. Initially, the second peak currents also increased slightly before the second limiting current peak leveled off (Figure 4b). For all concentrations studied, the curves were higher than the polarization curve obtained in bath #5. In an electrolyte solution, MEA can function as an additional ligand. Since the primary ligand, the $[CN]^-$ group, is bound, MEA^+ ligands help prevent the passivation of silver anodes by forming an intermediate silver(I) complex with monoethanolamine ($[Ag(MEA)]^+$ $pK_1 = 3.1$, $[Ag(MEA)_2]^+$ $pK_2 = 3.5$ [48]), followed by the formation of a strong complex $[Ag(CN)_2]^-$ $pK = 21.1$ [29,35] in the solution. The electrode process activation by MEA is likely due to its complexing abilities with silver, potentially leading to the formation of a complex polyligand silver compound.

This behavior of surfactant additives in the developed dicyanoargentate–perrhenate electrolyte can be leveraged to create various types of coatings. It is recommended to use MEA as a surfactant in the dicyanoargentate–perrhenate electrolyte due to its process-activating properties. In baths with TEA, coatings akin to a silver matrix with rhenium doped in fractions of a mass percent are likely achievable. For further studies on silver-rhenium electrodeposition, the maximum concentrations of MEA and TEA were chosen. The following sections present results regarding the deposition composition, structure and morphology, and some properties.

3.2. Electrodeposition of Ag-Re Coatings

3.2.1. Dependency of Deposit Composition and Morphology on Current Density and Hydrodynamic Electrolysis Condition

EDX analysis of samples obtained from Ag-rich bath #2 ($[Ag^+]$:$[ReO_4^-]$ ratio of 10:1) indicated that the resulting deposits contain up to several mass percent of rhenium. An increase in deposition current density from 3 to 15 mA cm^{-2} led to a decrease in rhenium content in the deposit (Figure 5a). This trend was consistent across samples deposited at different temperatures (20 to 60 °C), regardless of stirring (Figure 5b). The highest rhenium content in the deposit, 1.35 wt.%, was achieved with electrodeposition at a current density of 3 mA cm^{-2}, at room temperature, without stirring.

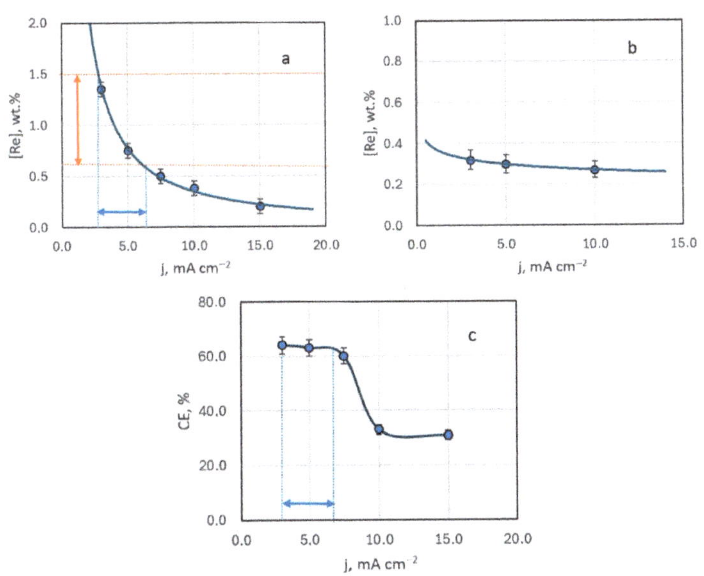

Figure 5. The mass fraction of rhenium in the deposit (**a**,**b**) and current efficiency (c) depending on the deposition current density. Electrolysis temperature, °C: 20—(**a**,**c**); 60—(**b**).

It was observed that the oxygen content in most samples was approximately three times higher than the rhenium content, suggesting that rhenium in the coatings exists predominantly in the form of ReO_x. This raises the possibility of forming composite "silver–rhenium oxides" coatings, although further research is required to confirm this.

Current efficiency also decreased with increasing deposition current density. Up to 10 mA cm^{-2}, the efficiency remained relatively constant, but at higher currents, it dropped by almost half and then stabilized (Figure 5c).

According to our previous studies [29], the range for stable production of functional coatings with a specified composition should correspond to the conditions of mixed electrochemical kinetics and is achieved at potentials corresponding to half the limiting current area.

Analysis of the voltammetric polarization curve (bath #2) allowed us to identify the corresponding range of operating current densities for the studied system (Figure 1). At 20 °C, this range of deposition current densities was 2.5–6 mA cm^{-2}. We highlight it in Figure 5a. It has been experimentally established that the deposits synthesized under these conditions contain strictly 0.7–1.5 wt.% Re. Such coatings are very interesting from a practical point of view. Since the electrodeposited silver alloys are lightly alloyed (up to 12 wt.% Re), the low resistance of the silver base is maintained. They can be promising for replacing electrical contacts obtained by the micrometallurgical method [21], which can operate in a wide temperature range (from −196 °C to +250 °C) without losing their properties. Hence, we conclude that these conditions for electrochemical synthesis are technologically advantageous and refer to them as the "functional electrodeposition range".

The coatings were consistently silver-white and fine-crystalline, without burns or spots, across the current density range of 3 to 15 mA cm^{-2}. They were dense, evenly distributed, and uniformly structured. At lower current densities (3 mA cm^{-2}) and room temperature, the Ag-Re deposit exhibited a uniform, closely packed, and finer crystalline structure (Figure 6a).

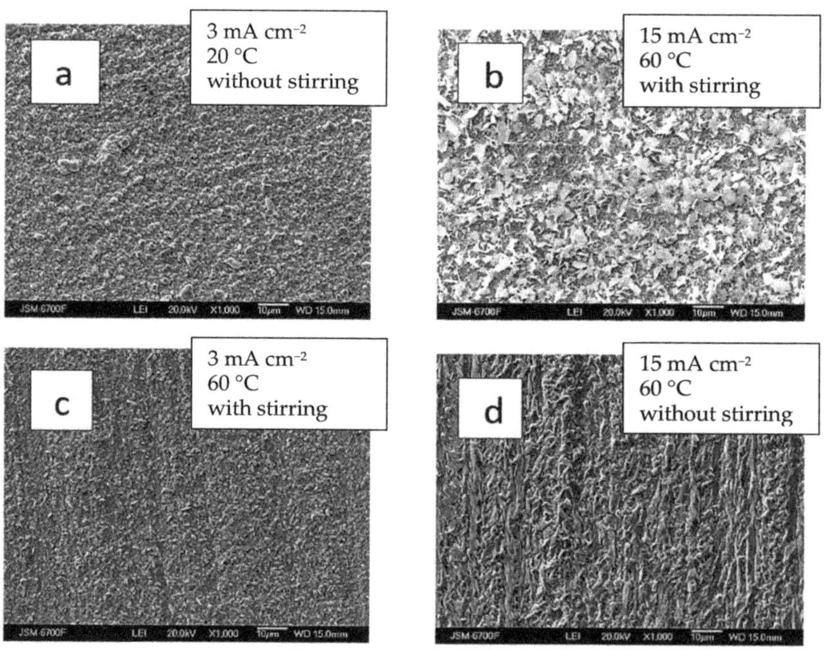

Figure 6. SEM images of samples obtained from bath #2 at different current densities, mA cm^{-2}: 3—(**a**,**c**); 15—(**b**,**d**); and temperature, °C: 20—(**a**); 60—(**b**–**d**); without (**a**,**d**) and with stirring (**b**,**d**). The thickness of the coatings is 2–2.5 μm.

The thin-layer Ag-Re coatings are uniformly silvery white, fine-crystalline, and matte. The coatings are well adhered to the copper substrate. Upon SEM analyzing the surface morphology, it was observed that most deposits closely replicate the copper substrate.

Chemical analysis revealed that despite the thickness of the coatings, the electrodeposits cover the substrate almost entirely. The deposition current density plays a role comparable to stirring in determining surface morphology. For example, at temperature 60 °C, the morphology of coatings obtained at 15 mA cm^{-2} with stirring resembled those formed at the same current density in a bath without stirring (Figure 6b,d).

The surface morphology of Ag-Re coatings obtained at different current densities and under intensifying factors such as stirring and elevated temperature (up to 60 °C) varies slightly with different applied deposition current densities (ranging from 3 to 15 mA cm^{-2}) (Figure 6b,c). Appearance of the coatings obtained at higher current density (15 mA cm^{-2}) are somewhat distinct from others.

3.2.2. Dependency of Deposit Composition and Morphology on Bath Temperature

Comparing the rhenium content in deposits obtained at a constant current density in an electrolyte at a $[Ag^+]:[ReO_4^-]$ ratio of 10:1, without forced convection but at varying temperatures from 20 to 60 °C, reveals a trend of decreasing rhenium content with rising temperature (Figure 7a).

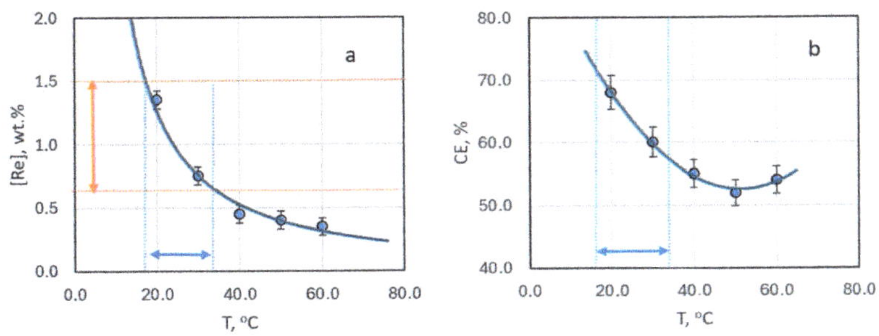

Figure 7. The mass fraction of rhenium in the deposit (**a**) and current efficiency (**b**) depending on the deposition temperature. The samples obtained from bath #2 at current density 3 mA cm^{-2} without stirring.

The reduction in rhenium content is most pronounced when the electrolysis temperature increases from 20 to 30 °C. In this case it drops almost threefold. Beyond this, the decrease in rhenium content becomes more gradual, with little change at temperatures between 40–60 °C (Figure 7a). Interestingly, the current efficiency demonstrates a dip, reaching a minimum at 50 °C, before increasing again (Figure 7b). SEM images of the coating surfaces obtained at 50 °C (Figure 6b) differ from those produced at other temperatures.

Figure 7a identifies the optimal temperature range (19–33 °C) for achieving coatings with a rhenium content in the "technologically advantageous" range of 0.7–1.5 wt.%.

3.2.3. Dependency of Deposit Composition and Morphology on Electrolyte Component Ratios

The silver and rhenium ion content ratio in the solution significantly influences the structure and morphology of Ag-Re deposits. Figure 8a,b illustrates that coatings from baths with a higher ratio of rhenium ions are more densely packed. In contrast to the white matte coating observed in baths with a predominance of silver, where scattered crystals of similar size are noticeable, the coatings from rhenium-rich baths appear more "metallic" and shiny. Nevertheless, both the deposit composition and the coating surface are more sensitive to factors like deposition current density and temperature.

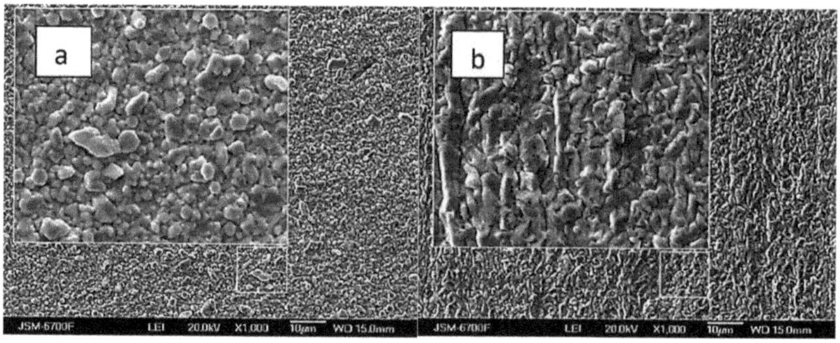

Figure 8. SEM images of samples obtained from baths with the $[Ag^+]$:[ReO_4^-] ratio: 10:1 (**a**), and 1:5 (**b**). Current density 10 mA cm^{-2}, temperature 20 °C.

It should be noted that the morphology of deposits obtained from Re-rich baths differs from Ag-rich baths. Analysis of the SEM-EDS mapping performed (Figure 9) demonstrates a uniform distribution of elements for samples deposited from solution with ratio $[Ag^+]$:$[ReO_4^-]$ = 10:1 (bath #2), and uneven for coatings deposited from the highest Re-rich solution with ratio $[Ag^+]$:$[ReO_4^-]$ = 1:10 (bath #10). This is clearly seen from the histograms of the distribution of elements shown in the same figure.

Deposits obtained from the Re-rich bath (bath #10) contain areas where dendritic silver is deposited, as well as disordered large (up to 85 μm) macrocrystals, which, as can be seen from the distribution of elements, consist mainly of rhenium and an oxygen containing phase. This indicates that Ag-Re deposits containing more than 10 wt.% Re may be a composite of silver and rhenium oxides. According to EDX analysis in the point of macrocrystal, the ratio of the atomic fractions of rhenium and oxygen did not allow us to unambiguously assume the stoichiometric composition of the oxide phase.

3.2.4. Dependency of Deposit Composition and Morphology on Surfactant Addition

Based on a voltammetric study, baths with higher ethanolamine additive content were chosen for further investigation. To evaluate the effect of ethanolamines additives on deposit composition and morphology, coatings were electrodeposited from baths with a silver to rhenium ion ratio of 2:1 (bath #5), with addition of 1.65 g L^{-1} monoethanolamine (bath #6) and 0.6 g L^{-1} triethanolamine (bath #7).

EDX analysis indicated that in bath #6 (with MEA) at 50 °C, the maximum rhenium content in deposits was 0.568 wt.%, whereas in bath #7 (with TEA) at the same temperature, rhenium content ranged between 0.17–0.485 wt.%. These values were lower than in baths without surfactants. There was not a clear correlation between rhenium content in the coating and deposition current density in these baths. Current efficiency remained high, almost constant up to a current density of 7.5 mA cm^{-2}, but then drops down to 66% at higher current densities (Figure 10). This pattern was observed with both MEA and TEA additives.

A slight decrease in rhenium content in the deposits from solutions with equal silver concentration but a fivefold higher concentration of perrhenate may be partially attributed to the increased pH levels in baths with ethanolamine additives. The pH was increased from 7.02 in bath #5 to 8.85 in bath #6 (with MEA) and 8.34 in bath #7 (with TEA), as shown in Table 1.

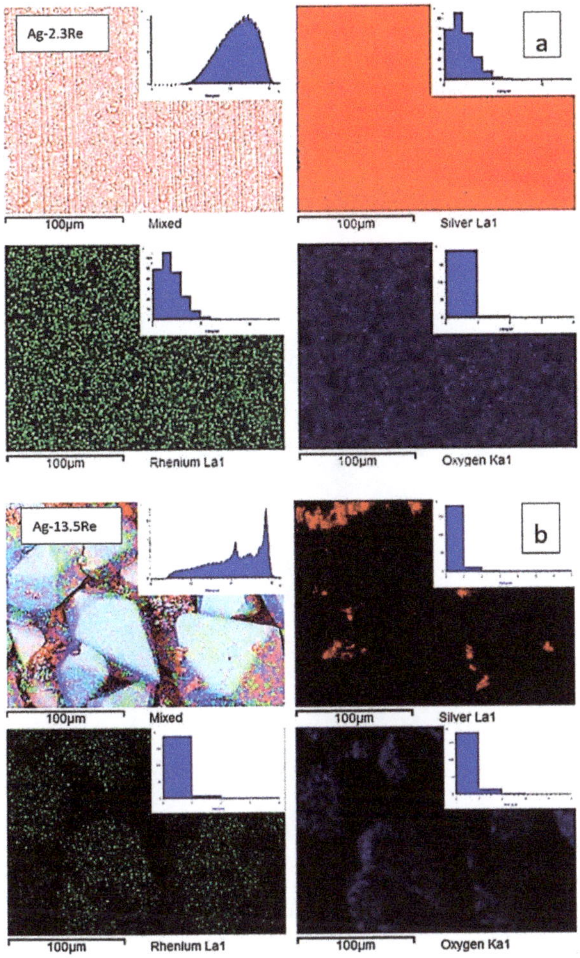

Figure 9. SEM images, corresponding elemental maps of Ag, Re, and O of sintered samples Ag-2.3Re (**a**) and Ag-13.5Re (**b**).

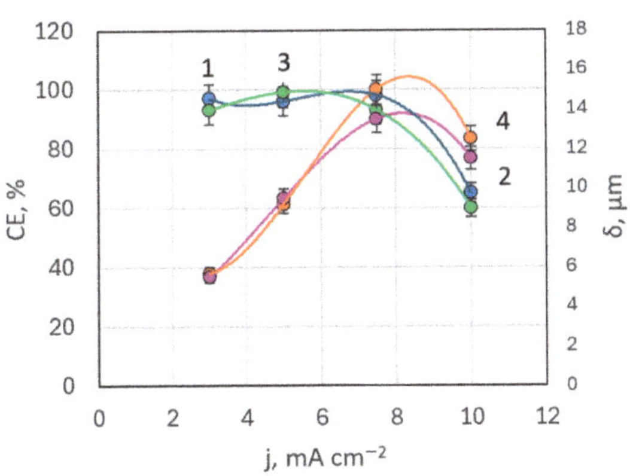

Figure 10. Dependences of current efficiency (curves 1, 3) and coating thickness (curves 2, 4) on deposition current density in the presence of TEA (curves 1, 2) and MEA (curves 3, 4).

In baths with ethanolamine additives, it was possible to produce coatings with thicknesses ranging from 5.0 to 13.7 microns. Deposits formed at lower current densities were silvery-white and of high quality, but dendritic growth was observed at current densities above 10 mA cm^{-2}.

Figure 11a–h demonstrates the impact of deposition current density on the morphology of Ag-Re coatings obtained from a silver dicyanoargentate–perrhenate electrolyte with triethanolamine and monoethanolamine as surfactants.

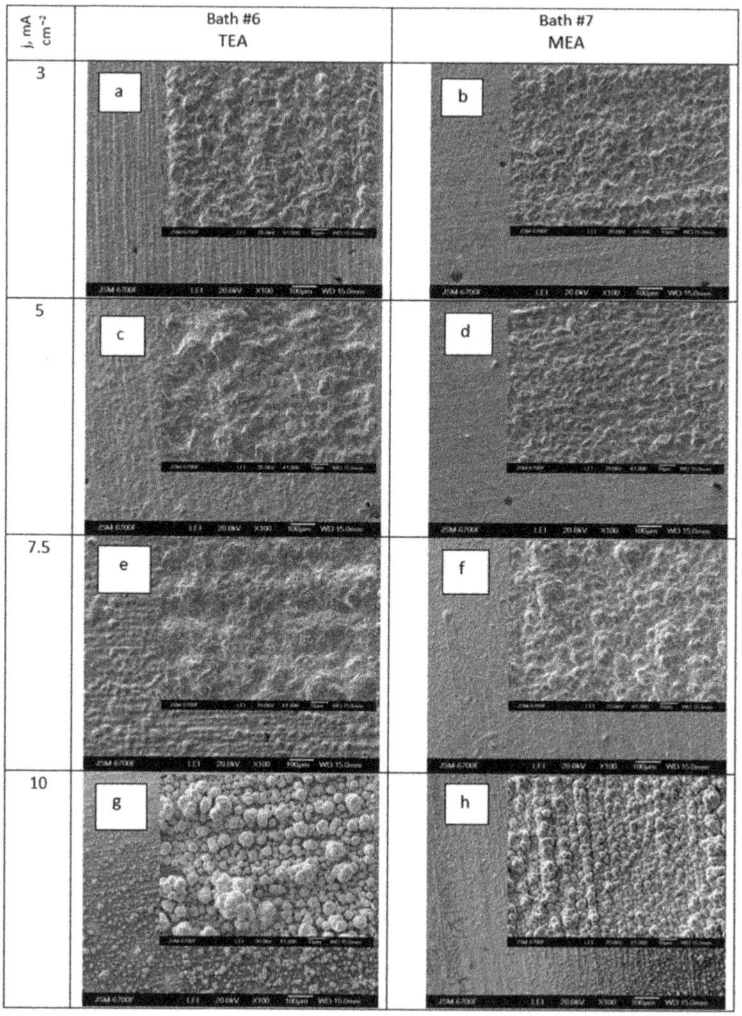

Figure 11. SEM images of samples obtained in dicyanoargentate–perrhenate electrolyte ($KAg(CN)_2$–0.11 M; $KReO_4$–0.055M) with additions of 0.06 mol L^{-1} triethanolamine (**a**,**c**,**e**,**g**) and 0.165 mol L^{-1} monoethanolamine (**b**,**d**,**f**,**h**) at current density, mA cm^{-2}: 3 (**a**,**b**); 5 (**c**,**d**); 7.5 (**e**,**f**); 10 (**g**,**h**). Temperature 50 °C.

Figure 11 shows that at low deposition current densities, the morphology of coatings obtained from baths with TEA and MEA additives differed insignificantly. Deposits from baths with MEA had a more densely packed structure. However, with an increase in deposition current density from 7.5 to 10 mA cm^{-2}, all deposits, regardless of the type of ethanolamine added, acquired a distinctly globular structure (Figure 11g,h). The size of the globules, under otherwise identical conditions, was larger in the case of baths with TEA. Deposits from baths with MEA additives exhibited a textured structure. There is a significant difference in the morphology of deposits obtained at the same temperature (50 °C) from baths without (Figure 6b) and with ethanolamine additives (Figure 11). Ag-Re coatings deposited at a current density of 10 mA cm^{-2} from bath #5 with the same component composition but without additives (Figure 8b) were characterized by a finer-grained structure.

Following the outcomes of voltammetric studies, baths with a higher concentration of ethanolamine additives (bath #6 and #7) were chosen for electrodeposition.

3.2.5. X-Ray Phase Analysis

To analyze the structure of coating obtained during the codeposition of silver with rhenium, Ag-Re samples with a higher (>1.5 wt.%) rhenium content were electrodeposited. Deposition was carried out from dicyanoargentate–perrhenate baths with a rhenium content up to ten times higher than silver (in a wider range of metal ion concentration ratios).

Structural analysis of phases and other structural parameters of electroplated silver-rhenium deposits and their comparisons with electrodeposited silver were performed using the XRD patterns that are shown in Figure 12. Diffraction patterns were obtained for coatings deposited at a current density of 30 mA cm^{-2} and a temperature of 18 °C from the baths #1, #2, #8, and #10, given in Table 1.

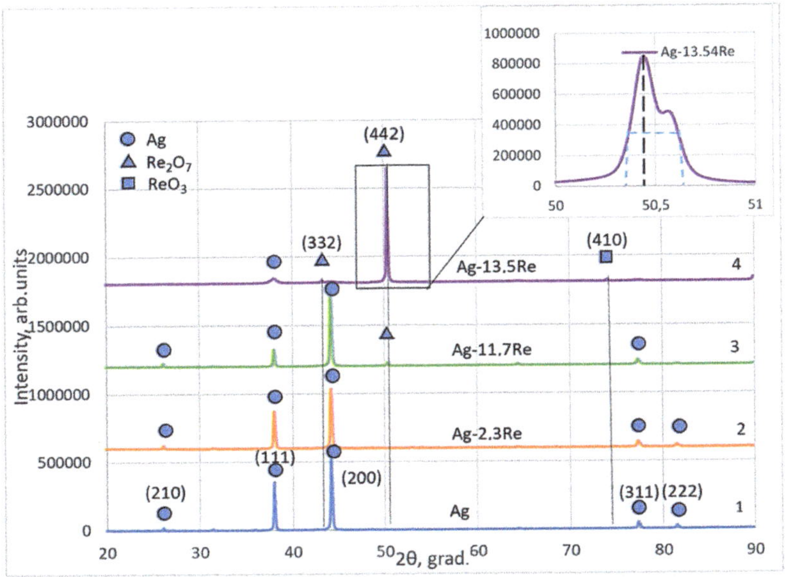

Figure 12. X-ray spectra electroplated silver (curve 1) and silver-rhenium coatings deposited from baths with different $[Ag^+]$:$[ReO_4^-]$ ratio: 10:1—2; 1:1.5—3; 1:10—4.

The obtained silver-rhenium coatings as well as silver are polycrystalline. The crystal structure of the electroplated Ag and Ag-Re alloys Ag-2.3Re, and Ag-11.7Re were investigated. The values of peaks at 38.12°, 44.28°, 77.47°, and 81.53° are attributed to silver crystal planes (210), (111), (200), (311), and (222) respectively (cf. Figure 12, curve 1–3). The Miller indexes in the diffraction pattern were assigned to each peak based on the data presented in JCPDS no 00–004-0783. The ratios of intensities (I_{hkl}) of diffraction peaks depend on plating conditions and are shown in Table 3.

Table 3. Diffraction peak intensities in the XRD patterns for electrodeposited silver and silver-rhenium coatings.

Sample	Intensities I_{hkl} of Diffraction Peaks (in Arbitrary Units)			
	I_{111}	I_{200}	I_{311}	I_{222}
Ag	55	100	3	0.6
Ag-2.3Re	62	100	2	0.7
Ag-11.7Re	19.5	100	4	0.9
Ag-13.5Re	100	~0	~0	~0

We previously determined [49] that silver deposits obtained from the BPC baths contain all possible textures like metallurgic silver exceptionally a texture (220). Two planes are predominant in electrodeposited silver, namely planes (111) and (200). Noticeably, the intensity of the peak revealed to the plane (200) is higher than to (111). This is typical for electrodeposited silver electroplated from the baths containing $Ag(CN)_2^-$. The intensity of other peaks corresponding to (311), (222), and (210) are negligibly low.

The obtained deposits have a strained lattice (not similar to metallurgic Ag [49]) with increased value of lattice parameter a_0 from 4.077 Å for BPC electrolyte up to 4.096 Å for Ag-Re alloys regardless of the predominating of high packing planes (200) and (111) (Table 4). Probably, the differences in a_0 values define the different properties of deposits.

Table 4. Lattice parameter (a_0) and interplanar spacing (d_{hkl}) for silver and silver-rhenium electroplated from various solutions.

		Bath #1	**Bath #2** $[Ag^+]:[ReO_4^-] = 10:1$	**Bath #9** $[Ag^+]:[ReO_4^-] = 1:1.5$	**Bath #10** $[Ag^+]:[ReO_4^-] = 1:10$
(111)	Θ, deg	19.07	19.06	19.06	-
	d_{hkl}, Å	2.3567	2.3579	2.3579	-
	a_0, Å	4.0770	4.0791	4.0791	-
(200)	Θ, deg	22.16	22.15	22.14	-
	d_{hkl}, Å	2.0413	2.0422	2.0431	-
	a_0, Å	4.0826	4.0844	4.0862	-
(311)	Θ, deg	38.72	38.74	38.74	-
	d_{hkl}, Å	1.2309	1.2318	1.2318	-
	a_0, Å	4.0824	4.0853	4.0853	-
(222)	Θ, deg	40.62	40.62	40.62	-
	d_{hkl}, Å	1.1827	1.1827	1.1824	-
	a_0, Å	4.0969	4.0969	4.0959	-
(442)	Θ, deg	-	-	-	25.22
	d_{hkl}, Å	-	-	-	1.8071
	a_0, Å	-	-	-	3.3202
(332)	Θ, deg	-	-	-	21.66
	d_{hkl}, Å	-	-	-	2.0861
	a_0, Å	-	-	-	2.2484

For Ag-11.7Re and especially Ag-13.5Re coatings, the values of peaks at 43.48°, 50.46°, and 73.68° were indexed. These may be attributed to Re_2O_7 crystal planes (332) and (442), respectively (JCPDS no 04–007-0368). For Ag-13.5Re, the obtained peak at 73.68° corresponds to the ReO_3 crystal plane (410) (JCPDS no 04–007-2352) (cf. Figure 12, curve 3, 4).

This is consistent with the results of the study [28], which established that increasing the concentration of perrhenate ions in the bath promotes the formation of rhenium oxides. However, considering the stepwise nature of the electroreduction process, the formation of an intermediate silver-rhenium(VI) oxide compound, such as Ag_xReO_3, cannot be ruled out. Additionally, since the XRD analysis of our samples was conducted several weeks after electrolysis, the possibility that the diffraction peaks correspond to rhenium bronze-type compounds, H_xReO_3, also cannot be excluded. In a study [50], structural transformations of ReO_3 oxide and diffraction patterns for typical polycrystalline powder sample ReO_3 recorded at different durations of time from the date of preparation were described. The additional peaks observed in such spectra (with 2theta values similar to ours) were attributed by the authors to the formation of H_xReO_3 compounds over time. Thus, this issue requires further in-depth study.

The calculation of crystallite size was performed using the Scherrer Equation (3). The average values of crystallite size for electrodeposited silver and silver-rhenium coatings are presented in Figure 13. As follows from the presented data, the obtained electrodeposits are nanocrystalline.

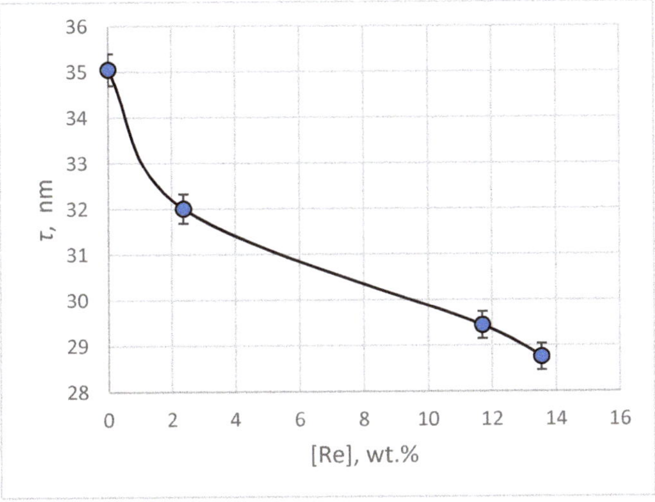

Figure 13. Average crystallite sizes of electrodeposited silver and silver-rhenium coatings depending on the rhenium content in the deposit.

The introduction of rhenium to silvering BPC electrolyte in an amount of 3–30 g L^{-1} leads to the grinding of the deposit and a decrease in the crystallites size from 35.05 to 28.75 nm at a current density of 3 mA cm^{-2}. The refinement of crystallites is probably due to adsorbed particles that prevent the grain growth.

3.2.6. Characterizing Coating Microhardness, Adhesion, and Porosity

The microhardness of obtained Ag-Re coatings depends on the current density and the content of rhenium ions in the electrolyte. Silver coatings deposited from BPC electrolyte without rhenium ions in the current density range of 3–15 mA cm^{-2}, have a hardness of 74.6 ± 9.7 kg mm^{-2}. The introduction of 0.011 M $KReO_4$ into the electrolyte (bath #2) leads to an increase in microhardness by 25% and higher, up to 92.6 ± 11.0 kg mm^{-2}. Electrodeposition from dicyanoargentate–perrhenate baths with a higher $[ReO_4^-]$:$[Ag^+]$ ratio ensured an increase in microhardness. With $[ReO_4^-]$:$[Ag^+]$ ratio of 1.5:1 (bath #8), the microhardness value was 107.3 ± 16.0 kg mm^{-2}. For the bath with $[ReO_4^-]$:$[Ag^+]$ ratio of 1.5:1 (bath #9, #10), coatings possessing a hardness of up to 116.1 ± 19.8 kg mm^{-2} were obtained (Figure 14). The H_V errors are not related to the measurement methodology but to the structural unevenness of the Ag-Re coating. The inclusion of rhenium in the coating provides greater stability of hardness over time.

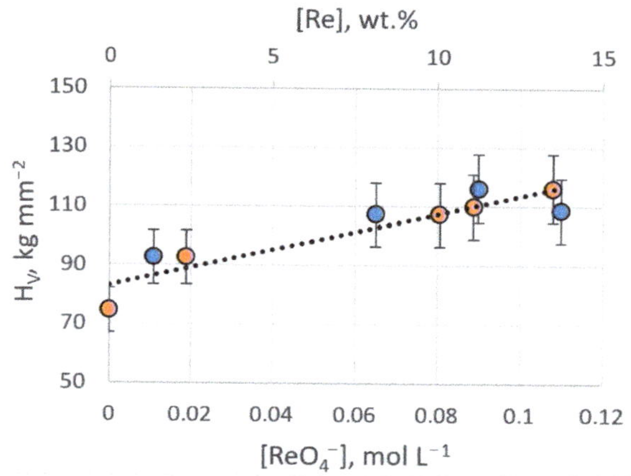

Figure 14. Dependence of the microhardness of Ag-Re coatings on the content of rhenium ions in the bath (blue dots) and the mass fraction of rhenium in the deposit (orange dots).

The obtained results are in good agreement with the crystallite size of the coatings determined from XRD data. The microhardness values are plotted in Figure 15 as a function of $\tau^{-1/2}$ (where τ is the crystallite size). A linear relationship, typical for Hall–Petch plots, is obtained, hence confirming that grain boundaries play a crucial role in the mechanical properties of these electrodeposited films. The decrease in crystallite size essentially creates a higher volume of grain boundaries and thus impedes the dislocation motion in these.

The adhesion strength of the deposits was determined using the marking method. A grid of 6 × 6 cuts was applied to the coating surface using a sharp blade. The cuts were made parallel and then at a 90° angle to form a grid. After cleaning the area with a brush, tape was applied and quickly removed. The adhesion was rated on a scale, where "0" indicates 100% adhesion and values from "1" (5% loss) to "5" (100% loss) represent decreasing adhesion.

The lattice cut results were similar across all samples, showing almost no visible signs of destruction or peeling at the cuts' edges and intersections, indicating high adhesion to the substrates. Silver-rhenium coatings from all studied baths showed high adhesion to the copper substrate, scoring between "0" to "1" points, even at higher current densities. Silver coatings from rhenium-free electrolytes (bath #1) displayed slightly lower adhesion, scoring "2" to "3" points, suggesting that rhenium inclusion enhances adhesion.

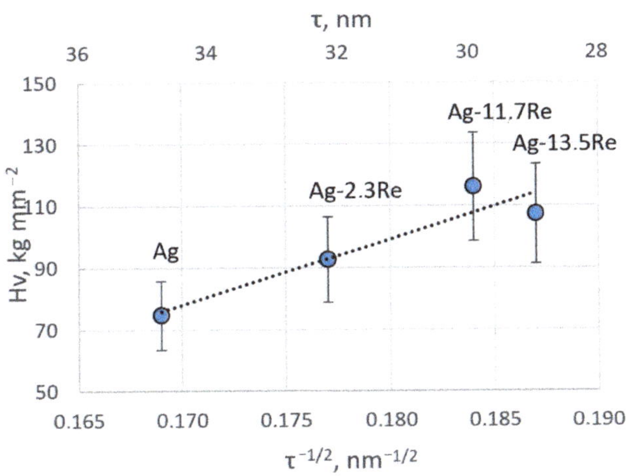

Figure 15. Hall–Petch plot for Ag-Re alloys showing the hardness as a function of the square root of the crystallite size determined from XRD measurements.

Porosity (symbatic with corrosion resistance) was assessed using filter paper moistened with a solution (g L^{-1}): $K_3[Fe(CN)_6]$—3, NaCl—10. Ag-Re coatings from bath #1 at current densities of 7.5–12.5 mA cm^{-2} and thicknesses of 6–8 μm were nearly non-porous. In bath #2, only at higher current densities were there an average of 3–5 pores per cm^2. Bath #3 coatings showed porosity under similar conditions. Coatings from baths with ethanolamine additives were almost non-porous, potentially influenced by the substitution of rhenium salt with potassium perrhenate.

High corrosion resistance was noted for coatings from all baths. After a year in an air atmosphere, no color or property changes were observed, indicating their corrosion resistance. Furthermore, the absence of pores in the coatings corroborates their corrosion-resistant properties.

4. Conclusions

A new dicyanoargentate–perrhenate electrolyte has been developed for the electrodeposition of Ag-Re alloys, based on a borate–phosphate–carbonate buffering silver-plating bath. Silver-white Ag-Re alloys with a thickness of 2–15 μm, containing from 0.15 to 13.5 wt.% Re, were deposited from baths with $[Ag^+]:[ReO_4^-]$ ratios ranging from 10:1 to 1:10. Increasing the current density from 3 to 15 mA cm^{-2} tended to decrease the rhenium content in the deposit and reduce current efficiency from 98% to 66%, irrespective of the temperature (20–60 °C) and stirring. The coatings from Re-rich baths were more metallic and shiny.

Voltammetric studies were conducted in baths with varying $[Ag^+]:[ReO_4^-]$ ratios of 10:1, 2:1, and 1:5. The potential of the limiting current on the polarization curves for all studied baths was −0.65 V. Electrodeposition occurred under mixed kinetics, and the main kinetic parameters of the silver and rhenium co-deposition process were established based on the temperature factor. The calculated activation energy from exchange currents and inverse temperature was 65 kJ mol^{-1}. The impact of ethanolamine additives (MEA and TEA) on the electrodeposition process was studied. TEA inhibited the electrode process, likely due to surface adsorption, while MEA activated the process and helped avoid silver anode passivation, possibly due to its complexing ability with silver (I) and the formation of a complex polyligand compound. The Re content in the alloy was lower in baths without surfactants.

X-ray analysis showed that the obtained Ag-Re coatings had a nanocrystalline structure with crystallite sizes ranging from 28.75 to 35.05 nm. The deposits obtained from Ag-rich baths had a strained lattice (dissimilar to metallurgic Ag) with an increased a_0 from 4.077 Å for electrodeposited Ag up to 4.096 Å for Ag-Re alloys, regardless of the predominance of high packing planes (200). Additional peaks were recorded in the diffraction pattern for Ag-13.5Re coatings deposited from a Re-rich bath. These peaks may be attributed to ReO_3 and Re_2O_7 crystal planes. To explain the peaks, taking into account the possibility of the formation of compounds such as Ag_xReO_3 or H_xReO_3, further study is required.

The microhardness of Ag-13.5Re deposits was more than one and a half times higher than that of pure silver (116.1 ± 19.8 vs. 74.6 ± 9.7 kg mm^{-2}). Ag-Re coatings demonstrated high adhesion to the copper substrate and were practically non-porous. A correlation between the main structural parameters of the deposit and its properties (linear relationship "microhardness–crystallite size" in Hall–Petch coordinates) was revealed.

The "functional electrodeposition range" (j = 2.5–6 mA cm^{-2}, t = 19–33 °C) of Ag-Re coatings with 0.7–1.5 wt.% Re as electrical contact materials operating in a temperature range from −196 °C to +250 °C has been established. Thus, the developed baths and electrolysis modes allow the electrochemical deposition method to synthesize functional Ag-Re coatings with a controlled, predetermined Re content, which are a promising material for use in the manufacture of electrical contacts operating in a wide temperature range.

Author Contributions: O.B. (Oksana Bersirova): Conceptualization, Methodology, Investigation, Data curation, Formal analysis, Writing—original draft, Writing—review and editing, Supervision, Funding acquisition. V.K.: Methodology, Writing—review and editing. S.K.: Investigation, Data curation. O.B. (Olena Bondar): Investigation. All authors have read and agreed to the published version of the manuscript.

Funding: This project has received funding through the MSCA4Ukraine project, which is funded by the European Union (Project No. 1233494). Views and opinions expressed are, however, those of the authors only and do not necessarily reflect those of the European Union. Neither the European Union nor the MSCA4Ukraine Consortium as a whole nor any individual member institutions of the MSCA4Ukraine Consortium can be held responsible for them.

Institutional Review Board Statement: Not applicable.

Informed Consent Statement: Not applicable.

Data Availability Statement: The original contributions presented in this study are included in the article. Further inquiries can be directed to the corresponding author.

Acknowledgments: We express our gratitude to our colleague, H. Cesiulis from Vilnius University, for his valuable contribution and suggestions in certain aspects of our research, as well as for the administrative assistance.

Conflicts of Interest: The authors declare no conflicts of interest.

References

1. Silver Based Materials. 2023. Available online: https://www.electrical-contacts-wiki.com/index.php?title=Silver_Based_Materials&mobileaction=toggle_view_desktop (accessed on 19 August 2024).
2. Cameron, D.S. Chemistry, Electrochemistry, and Electrochemical Application | Silver. In *Encyclopedia of Electrochemical Power Sources*; Garche, J., Ed.; Elsevier: Amsterdam, The Netherlands, 2009; pp. 876–882. [CrossRef]
3. Balakai, V.I.; Arzumanova, A.V.; Starunov, A.V.; Balakai, I.V. Properties of Electrolytic Silver-Based Alloy. *Inorg. Mater. Appl.* **2018**, *9*, 947–953. [CrossRef]
4. Księżarek, S.; Woch, M.; Klir, T.; Kołacz, D.; Karpiński, M.; Rudnicki, K. Environmentally friendly silver-based contact materials. *Ores Non-Ferr. Met.* **2012**, *57*, 549–555.

5. Kuzmar, I.; Lanin, V.; Pas, N.; Khmyl, A. Composite electroplated coatings based on silver for electronic products. *Technol. Electron. Ind.* **2006**, *6*, 58–60.
6. Bogush, N.V.; Khmyl, A.A.; Kushner, L.K.; Dezhkunov, N.V. Composition, structure and functional properties of silver-tungsten composition electrochemistry coatings formed with the help of ultrasound. *Dokl. BGUIR* **2021**, *19*, 23–31. [CrossRef]
7. Bogush, N.V.; Khmyl, A.A.; Kushner, L.K. Structure and physical-mechanical properties of silver-tungsten composite coatings obtained by electrochemical deposition. *Dokl. BGUIR* **2017**, *107*, 54–61.
8. Kołacz, D.; Księżarek, S.; Woch, M.; Karwan-Baczewska, J.; Staszewski, M.; Rudnicki, K.; Borkowski, P.; Sienicki, A. Nanocrystalline Ag-Re Composite as a Potential Material for Electric Contacts Fabrication. *Arch. Metall. Mater.* **2016**, *61*, 1847–1852. [CrossRef]
9. E.U. Turns. (General Dynamics, Ft. Worth, Texas) Industrial Development of the Silver-Rhenium Alloy Plating Process. In Proceedings of the 57th National Convention of the Arer. Electroplaters Soc, Montreal, QC, Canada, 21–25 June 1970.
10. Jones, T. Rhenium plating. *Met. Finish.* **2003**, *86–96*, 101. [CrossRef]
11. Jones, T. *Electrodeposition of the Lesser-Known Precious Metals: Osmium, Iridium, Rhodium, Rhenium, Ruthenium, Revised Edition*; Finishing Publications Ltd.: Warrington, UK, 2005; 214p.
12. Sadana, Y.; Wang, Z. Electrodeposition of Alloys. XX: Electrodeposition of Gold Rhenium Alloys from Acidic Citrate Solutions and Their Xray Structure. *Surf. Coat. Technol.* **1989**, *37*, 419–434. [CrossRef]
13. Nguyen, Y.N.; Yoon, J.; Shin, J.; Son, I. Electrodeposition of Pd-Ag alloy for electrical contacts. *Surf. Eng.* **2022**, *38*, 633–640. [CrossRef]
14. Povetkin, V.V.; Skifsky, S.V.; Koreshkova, E.V. Electrolyte for deposition of copper-rhenium alloy. Patent: RU 2 234 560 C1, 27 June 2004. Available online: https://patenton.ru/patent/RU2234560C1 (accessed on 20 August 2004).
15. Kołacz, D.; Księżarek, S.; Woch, M.; Karwan-Baczewska, J.; Staszewski, M.; Rudnicki, K.; Borkowski, P.; Sienicki, A. Nanocrystalline Ag-Re composite as a potential material for electric contacts fabrication. In Proceedings of the 27th International Conference on Electrical Contacts, Dresden, Germany, 22–26 June 2014; pp. 451–456. Available online: https://ieeexplore.ieee.org/document/6857211 (accessed on 17 July 2014).
16. Księżarek, S.; Woch, M.; Kołacz, D.; Kamińska, M.; Borkowski, P.; Walczuk, E. Progress in fabrication technology of silver-based contact materials with particular account of the Ag-Re and $AgSnO_2Bi_2O_3$ composites. In Proceedings of the 26th International Conference on Electrical Contacts and the 4th International Conference on Reliability of Electrical Products and Electrical Contacts, Beijing, China, 14–17 May 2012; pp. 186–193. Available online: https://core.ac.uk/download/pdf/53096993.pdf (accessed on 21 January 2016). [CrossRef]
17. Hauner, F.; Jeannot, D.; Pinard, J. AgFe and $AgFe_2O_3$ Contact Materials for Voltage Switcher. In Proceedings of the 19th International Conference on Electric Contact Phenomena, Nurnberg, Germany, 14–17 September 1998; pp. 317–323.
18. Kołacz, D.; Księżarek, S.; Borkowski, P.; Karwan-Baczewska, J.; Lis, M.; Kamińska, M.; Juszczyk, B.; Kulasa, J.; Kowalski, A.; Wierzbicki, Ł.; et al. The Influence of Mechanical Alloying and Plastic Consolidation on the Resistance to Arc Erosion of the Ag-Re Composite Contact Material. *Materials* **2021**, *14*, 3297. [CrossRef]
19. Borkowski, P.; Walczuk, E.; Frydman, K.; Wojcik-Grzybek, D. Switching properties of contacts made of silver-tungsten and silver-tungsten-rhenium composite. In Proceedings of the 59th IEEE Holm Conference on Electrical Contacts, Newport, RI, USA, 22–25 September 2013; pp. 197–206. [CrossRef]
20. Księżarek, S.; Kołacz, D.; Czepelak, M.; Śmieszek, Z.; Marszowski, K. Technological aspects of plastic working of the Ag-Re(5, 8 and 10) wt % composite contact material. *Metall. Metall. Eng. News* **2011**, *8*, 633–636.
21. Farmakovsky, B.V.; Somkova, E.A.; Sergeeva, O.S.; Yurkov, M.A.; Tochenyuk, D.A.; Bystrov, R.Y.; Semenov, A.S.; Peskov, T.V.; Gerashchenkov, D.A. Silver-Based Alloy for Nanostructured Coatings. Patent RU 2350673, 27 March 2009. Available online: https://patenton.ru/patent/RU2350673C1.pdf (accessed on 27 March 2009).
22. Bersirova, O.L.; Kochetova, S.A.; Bondar, O.O. Synthesis and characterization of electrolytic Ag-Re coatings. In Proceedings of the 2nd International Research and Practice Conference "Nanoobjects & Nanostructuring" (N&N-2022), Lviv, Ukraine, 23–25 September 2022; pp. 57–58. Available online: https://drive.google.com/file/d/1TROnAqz8nft88hJ9adnHajWc_JEpGWa9/view (accessed on 21 September 2022).
23. Shibleva, T.G.; Povetkin, V.V.; Ivanova, T.E. Electrolyte for Deposition of Silver-Rhenium Alloy. Patent RU 2459017C1, 20 August 2012. Available online: https://patenton.ru/patent/RU2459017C1 (accessed on 20 August 2012).
24. Ivanov, A.F.; Kruglikov, S.S.; Mikhailova, A.N.; Ivanova, G.D.; Goncharov, V.K. Electrolyte for Coating Silver-Rhenium Alloys. Copyright certificate RUS No. 627188. Application No. 2473785, 5 October 1978. Available online: https://patenton.ru/patent/SU627188A1 (accessed on 5 October 1978).

25. Ivanov, A.F.; Ginzberg, S.A. Electrolyte for the Deposition of Silver-Based Alloys. Patent SU 528356A1, 15 September 1976. Available online: https://patenton.ru/patent/SU528356A1 (accessed on 15 September 1976).
26. Khmyl, A.A.; Kushner, L.K.; Kuzmar, I. Peculiarities of the formation of functional coatings silver—Rhenium oxides. In *Creation of New and Improvement of Existing Technologies and Equipment for Applying Galvanic and Their Replacement Coatings, Proceedings of the Materials of the 5th Republican Scientific and Technical Seminar, Minsk, Belarus, 22–23 December 2015*; Zharsky, I.M., Dormeshkin, O.B., Chernik, A.A., Eds.; Belarusian State Technological University BSTU: Minsk, Belarus, 2015; pp. 4–7.
27. Turns, E.W.; Head, J.W.; Hoffman, H.C.; Porter, A.C. Process for Lubricating a Bearing Surface. U.S. Patent 3,342,708, 19 September 1967.
28. Cao, H.; Chai, D.; Wu, L.; Zheng, G. Communication—A Mechanistic Study on Electrodeposition of Rhenium from Acidic Solution of Ammonium Perrhenate. *J. Electrochem. Soc.* **2017**, *164*, D825. [CrossRef]
29. Bersirova, O.; Byk, S.; Kublanovsky, V. Electrodeposition of silver. In *Monograph*; Medinform: Kyiv, Ukraine, 2013; 168p. Available online: http://www.ionc.kiev.ua/deposite/Ag_electrodeposition.pdf (accessed on 16 June 2017).
30. Bersirova, O.; Kublanovsky, V.; Emelianov, V.; Anufriev, L. Electrodeposition of functional silver coatings from borate-phosphate-carbonate baths. *Chemija* **2003**, *14*, 16–21.
31. Bersirova, O.; Kublanovsky, V.; Cesiulis, H. Electrochemical Formation of Functional Silver Coatings: Nanostructural Peculiarities. *ECS Trans.* **2013**, *50*, 155–163. [CrossRef]
32. Hamdi, R.; Rached, A.; Massoudi, I.; Al-Zuraie, R.; Al-Hamadah, K.; Al-Otaibi, A.; Flemban, T.; Alonizan, N.; Ghrib, T. Electrodeposition Study of Silver: Nucleation Process and Theoretical Analysis. *J. Electron. Metarial* **2021**, *50*, 5507–5513. [CrossRef]
33. Burkat, G.K.; Safronova, I.V. Electrodeposition of silver from dicyanoargenate electrolyte in the presence of nanocarbon additives. *Mod. Probl. Sci. Educ.* **2014**, *2*, 1–8. Available online: https://s.science-education.ru/pdf/2014/2/513.pdf (accessed on 8 April 2025).
34. Bersirova, O.L.; Kublanovskii, V.S. Crystalline Roughness as a Morphological Characteristic of the Surface of Electroplated Silver Coatings. *Russ. J. Appl. Chem.* **2009**, *82*, 1944–1948. [CrossRef]
35. Bersirova, O.; Kublanovsky, V.; Bersirov, S. Modeling the Composition of the Pre-Cathode Layer in Dicyanoargentate Buffer Electrolyte without Excess Ligand. In Proceedings of the 12th International Conference Nanomaterials: Application & Properties 2022 NAP-2022, Krakow, Poland, 11–16 September 2022. [CrossRef]
36. Bersirova, O.; Kublanovsky, V. Electrochemical Behavior of Silver in Dicyanoargentate Electrolytes. In Proceedings of the 7th International Conference Nanomaterials: Application & Properties 2017 NAP-2017, Zatoka, Ukraine, 10–15 September 2017. Available online: https://ieeexplore.ieee.org/stamp/stamp.jsp?arnumber=8190157 (accessed on 14 December 2017).
37. Jasni, A.B.; Yoshihara, S. Electrodeposition of Silver in Cyanide-Free Solution Containing Pyrimidine Derivative as a Complexing Agent. *J. Electrochem. Soc.* **2023**, *170*, 092504. [CrossRef]
38. Bersirova, O.L.; Kublanovsky, V.S. Electrolytic Nickel-rhenium alloys: Synthesis, structure and corrosion properties. *Mater. Sci.* **2019**, *54*, 506–511. [CrossRef]
39. Kuznetsov, V.V.; Gamburg, Y.; Zhulikov, V.V.; Krutskikh, V.M.; Filatova, E.A.; Trigub, A.L.; Belyakova, O.A. Electrodeposited NiMo, CoMo, ReNi, and electroless NiReP alloys as cathode materials for hydrogen evolution reaction. *Electrochim. Acta.* **2020**, *354*, 136610. [CrossRef]
40. Cao, H.; Hu, L.; Zhang, H.; Hou, G.; Tang, Y.; Zheng, G. The significant effect of supporting electrolytes on the galvanic deposition of metallic rhenium. *Int. J. Electrochem. Sci.* **2020**, *15*, 6769–6777. [CrossRef]
41. Zhulikov, V.V.; Gamburg, Y.D. Eelectrodeposition of rhenium and its alloys. *Russ. J. Electrochem.* **2016**, *52*, 847–857. [CrossRef]
42. Naor, A.; Eliaz, N.; Gileadi, E. Electrodeposition of Alloys of Rhenium with Iron-Group Metals from Aqueous Solutions. *J. Electrochem. Soc.* **2010**, *157*, D422. [CrossRef]
43. Rhenium. Term Paper. File Archive of Students Studfile. Available online: https://studfile.net/preview/2474240/page:3/ (accessed on 25 March 2015).
44. DePew, J.R.; Larson, T.L. Electrodeposition of rhenium for wear resistance. A preliminary report. *Plating* **1970**, 475–478.
45. *ISO 2409*; Paints and varnishes—Cross-cut test. International Organization for Standardization: Geneva, Switzerland, 2020.
46. Brantsevich, V.K. Influence of electrolyte composition on the structure and properties of composite electrochemical coatings based on silver. In Proceedings of the Materials of the 48th Scientific Conference of Graduate Students, Undergraduates and Students, Minsk, Belarus, 7–11 May 2012; pp. 120–121.
47. Antropov, L.I. *Theoretical Electrochemistry*; Mir Publisher: Moscow, Russia, 1972.
48. Grall, M. *Electrochemical Study of Chemical Properties in Ethanolamine and Its Mixtures with Water*; CEA Fontenay-aux-Roses: Paris, France, 1964. Available online: https://inis.iaea.org/search/search.aspx?orig_q=RN:36065597 (accessed on 12 January 2025).

49. Cesiulis, H.; Bersirova, O.; Valiuniene, A.; Prosycevas, I.; Baltrunas, G. Structure and Morphology of Silver Electrodeposits. *Mater. Sci.* **2004**, *10*, 142–146.
50. Majid, C.A.; Hussain, M.A. Structural transformations in polycrystalline rhenium trioxide. *Phw. Chem. Solids* **1995**, *56*, 255–259. [CrossRef]

MDPI AG
Grosspeteranlage 5
4052 Basel
Switzerland
Tel.: +41 61 683 77 34

Materials Editorial Office
E-mail: materials@mdpi.com
www.mdpi.com/journal/materials